Handbook of Recycling Technology

Handbook of Recycling Technology

Edited by Clark Hubert

SYRAWOOD
PUBLISHING HOUSE

New York

Published by Syrawood Publishing House,
750 Third Avenue, 9th Floor,
New York, NY 10017, USA
www.syrawoodpublishinghouse.com

Handbook of Recycling Technology
Edited by Clark Hubert

International Standard Book Number: 978-1-68286-751-8 (Hardback)

Cataloging-in-Publication Data

Handbook of recycling technology / edited by Clark Hubert.
    p. cm.
Includes bibliographical references and index.
ISBN 978-1-68286-751-8
1. Recycling (Waste, etc.). 2. Recycling industry--Technological innovations.
3. Refuse and refuse disposal--Technological innovations. I. Hubert, Clark.
TD794.5 .H36 2019
363.7282--dc23

TABLE OF CONTENTS

PREFACE

The main aim of this book is to educate learners and enhance their research focus by presenting diverse topics covering this vast field. This is an advanced book which compiles significant studies by distinguished experts in the area of analysis. This book addresses successive solutions to the challenges arising in the area of application, along with it; the book provides scope for future developments.

Traditional waste disposal has tremendous negative impact on the environment. Recycling is a viable alternative because it helps to reduce greenhouse gas emissions and prevent wastage. It involves the conversion of waste materials into new materials and objects. The materials that can be recycled include paper, glass, metal, textiles, electronics and plastic. Biodegradable waste such as garden waste and food can also be recycled. Recyclable materials are brought to a collection center and then sorted, cleaned and reprocessed into new materials. Recycling plays a crucial role in managing industrial waste. This book is compiled in such a manner, that it will provide in-depth knowledge about the practices of recycling. This book is a valuable compilation of topics, ranging from the basic to the most complex advancements in recycling technology. It aims to serve as a resource guide for students and experts alike and contribute to the growth of the discipline.

It was a great honour to edit this book, though there were challenges, as it involved a lot of communication and networking between me and the editorial team. However, the end result was this all-inclusive book covering diverse themes in the field.

Finally, it is important to acknowledge the efforts of the contributors for their excellent chapters, through which a wide variety of issues have been addressed. I would also like to thank my colleagues for their valuable feedback during the making of this book.

Editor

Mechanical and Thermal Characterization of Melt-Filtered, Blended and Reprocessed Post-Consumer WEEE Thermoplastics

Erik Stenvall * and Antal Boldizar

Academic Editor: William Bullock

Materials and Manufacturing Technology, Chalmers University of Technology, SE-412 96 Gothenburg, Sweden; antal.boldizar@chalmers.se

*   Correspondence: erik.stenvall@borealisgroup.com

Abstract: A melt-blended and melt-filtered real post-consumer and recyclable waste electrical and electronic equipment plastics blend (WEEEBR) was studied, where the WEEEBR contained mainly acrylonitrile-butadiene-styrene copolymer (~40 wt %), high impact polystyrene (~40 wt %) and polypropylene (~10 wt %). The main aim was to better understand the influence of different reprocessing conditions on the mechanical and thermal properties of WEEEBR and to compare these properties with the corresponding properties of model material blends of samples from single screw extrusion, twin screw extrusion and injection molding. For all the reprocessing alternatives studied, WEEEBR was found to be processable and an acceptable surface character could be obtained within narrow processing condition windows. It was found in particular that the reprocessing conditions influenced the elongation at break of WEEEBR, and to a lesser extent also the width of the polypropylene melting temperature region. The highest yield stress and elongation at break of WEEEBR was obtained after twin-screw extrusion at low barrel temperatures (180–200 °C) and a low screw rotation rate (60 rpm). Injection molding produced brittle materials with low impact strength, possibly due to molecular orientation effects.

Keywords: WEEE; blend; extrusion; reprocessing

---

1. Introduction

Legislation provides strong incentives for increasing the recycling of plastics from waste electrical and electronic equipment (WEEE) [1]. Previous studies have indicated that WEEE consists of about 20–33 weight % (wt %) plastics [2,3]. A higher plastics content has been found in small WEEE, around 35 wt % [4] and as much as 60 wt % in waste from cell phones [5]. The plastics are mainly re-meltable thermoplastics of many different types, but acrylonitrile-butadiene-styrene copolymer (ABS), high impact polystyrene (HIPS) and polypropylene (PP) have been found to constitute the main part of the WEEE plastics fraction [4,6–9].

Although there are many different waste management options for WEEE, landfilling is probably still the most frequently employed option globally [10]. In the European Union, the reuse and recycling rates are relatively high, the problem being that the collection rate in many countries is low and imposes significant uncertainties in the WEEE management [11]. The main alternatives for waste management of plastics, apart from landfilling, have been presented as energy recovery, feedstock recycling and mechanical recycling [12]. Feedstock recycling has not been commercialized to any significant extent, mainly due to the high costs of the processes and the low quality of the feedstock produced compared to that of normal grade commercial feedstock [13]. It has previously been shown from a life cycle assessment perspective that there are significant benefits in mechanical recycling of waste plastics rather than using the waste for energy recovery, with regard to both resource utilization and greenhouse gas emissions [14]. Considerable efforts have been devoted to the mechanical recycling of plastics by separation of the different plastic types normally occurring in the mixed WEEE stream [15]. The separation of different plastic types has been associated with large investment costs, and many of the separation technologies used have a limited efficiency, often producing substantial residue fractions or partially contaminated materials. For example, density-based separation techniques have limited efficiency in separating plastics containing different amounts of additives, since the additives may have a large influence on the density of the plastic [16]. This means that it is also necessary to consider blending mixed WEEE plastics when performing mechanical recycling, especially when the waste volumes or investment possibilities are limited. One often mentioned drawback of blending mixed WEEE plastics is the variation in composition. To some extent, the variation in composition cannot be avoided, but on the other hand, waste management sampling studies have shown the main components in such a blend to roughly be found in similar amounts in several different studies [6,8,9]. Most likely, additives and compatibilisers are also necessary to enhance and even out the properties of mixed WEEE plastic blends. In particular, antioxidants should be important considering recycling of such blends, to avoid premature degradation of the plastics during the reprocessing and use-phase and to not exclude the possibility of repeated recycling.

It has however been found that a WEEE plastics stream that can be considered to be recyclable may also contain 1–2 wt % non-thermoplastic contamination that cannot be melted in a conventional plastics processing machine and thus significantly lowers the quality of the recycled material [9]. Different types of melt-filtration have previously been used to remove this non-thermoplastic contamination in a continuous recycling process [17].

The present study considers the possibility of recycling a WEEE plastics blend (free of brominated flame retardants) by continuous melt-filtration followed by various reprocessing alternatives. This approach builds on previous studies of the WEEE plastics composition [9] and on the mechanical and thermal properties of model material blends [18]. The two main aims of this work were to study the mechanical performance of a melt-filtered WEEE plastics blend and to compare it with model material blends of similar compositions and also to study the influence of the processing conditions during recycling on the mechanical and thermal properties of the product.

Previous studies have indicated that ABS and HIPS are miscible at least in some proportions, since the mechanical properties followed the rule of mixtures and the blend had a common glass transition (T_g) well predicted by the Fox equation [19–21]. Although most mechanical properties of a HIPS/ABS blend seem to follow the rule of mixtures, blending the two plastics may have a detrimental effect on the impact strength [22]. It has also been reported that some reprocessing procedures may result in a partial separation of the ABS from the HIPS component. In particular, ABS has been seen to form a skin around a more HIPS rich core in injection molding [23]. Although HIPS and ABS have been found to be compatible under some conditions, blends of polyolefins and polystyrenes are normally considered immiscible with a low interfacial adhesion [24,25]. The incompatibility of these blends is also reflected in the mechanical properties, such as impact, tensile and flexural strength. Lower mechanical properties than would be expected from the rule of mixtures have, for instance, been reported for blends of ABS and PP [22]. A contamination content of 6 wt % PP in recycled ABS has been reported to lead to a significantly lower impact strength and yield stress, and the elongation at break was reduced to almost one fourth of that of the uncontaminated recycled ABS [16]. Poor injection molding processability was also reported for ABS blends contaminated with PP [22]. In general, the commonly used mechanical characterization methods for blends with different combinations of HIPS, ABS and PP are tensile testing and impact testing [19,22]. Differential scanning calorimetry (DSC) and dynamic mechanical thermal analysis (DMTA) have also been used to study the miscibility of polystyrene and polyolefin blends [19,20,22].

2. Experimental Section

2.1. Materials

The material used was a melt-blended and melt-filtered WEEE plastics blend of recycled material (WEEEBR) from post-consumer waste. This material should be compliant with the European directive on the "Restriction of the use of certain Hazardous Substances in electrical and electronic equipment" (RoHS) [26]. This collected post-consumer waste plastics blend was obtained from Stena Technoworld in Halmstad (5 July 2011), and it has previously been analyzed with respect to its composition [9]. Prior to being reprocessed in our laboratory, the material had undergone a dust- and surface-cleaning, melt-blending, melt-filtration and hot die granulation at Next Generation Recycling Machinen in Feldkirchen, Austria. The recycling equipment used was an S:GRAN 85 in conjunction with an Ettlinger rotating drum melt-filter and hot die granulator. The extruder was run at a screw rotation rate of 145 rpm, a throughput of 280 kg/h and a temperature profile of 210-230-190-190-210-230 °C. It was found that the continuous Ettlinger melt-filter filtered out about 1 wt % of mainly non-thermoplastic contamination in order to form the WEEEBR material. WEEEBR was compared to two model material blends; one ternary blend of virgin ABS, HIPS and PP (TBV, Ternary Blend of Virgin plastics) and one ternary blend of collected and separated fractions of ABS, HIPS and PP (TBR, Ternary Blend of Recycled plastics). A more detailed description of the compositions of the TBV and TBR can be found in a previous work [18].

2.2. Processing Equipment

WEEEBR was melt processed by single screw extrusion (SSE), twin-screw extrusion (TSE) or injection molding (IM). The SSE used was a Collin type 132 single screw extruder with a 50 × 1.5 mm slit die. It had a conventional polyolefin processing screw with a diameter (D) of 25 mm, a length of 25D and compression ratio of 1.93. The screw rotation rate was kept at 60 rpm and the cylinder temperature profile was straight at 180, 200 or 220 °C. The extrudate was oriented and flattened by a 3 + 2 roll puller, type Brabender 843316003.

The TSE was a co-rotating Werner & Pfleiderer ZSK 30 M9/2 (1984) with five heating zones along the barrel and one heating zone for the die. The length of the screws was 966 mm and the barrel bore diameter 31 mm. The configuration of the screws had been optimized for mixing and compounding, as shown schematically in Figure 1. In this case, screw rotation rates of 60, 90 and 120 rpm were used and the temperature profiles are shown in Table 1. A humped barrel temperature profile was applied for all twin-screw extrusion runs to avoid over-heating and resin degradation [27]. The roll puller used for SSE was also used for TSE. In both cases, the roll puller was set to achieve a draw down ratio of approximately four, which had previously been found to yield the highest ductility of TBV [18].

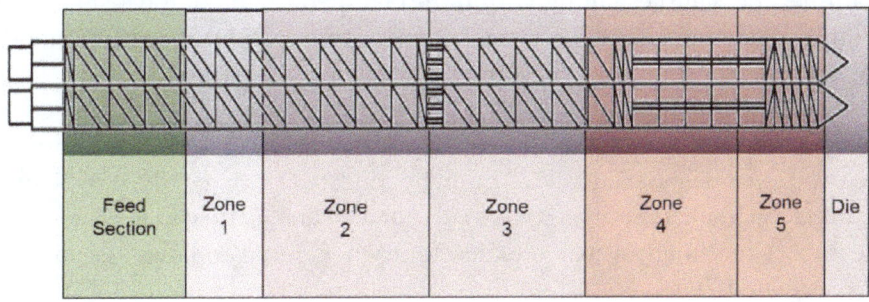

Figure 1. The twin-screw extruder (TSE) screw configuration and zone distribution along the screw. The feed section and Zones 1–2 consist mainly of fast transportation elements and end with one pressure element, then one mixing and four fast transportation elements follow in Zone 3. Zones 4–5 consists of two pressure elements, five mixing elements (kneading blocks), one reversed flow pressure element, and five pressure elements.

Table 1. Twin-screw extrusion (TSE) temperature profiles used.

Twin Screw Extrusion	Temperature Profile ("Hopper to Die")
T1	140-170-170-180-180-170 °C
T1.5	150-180-180-190-190-180 °C
T2	160-190-190-200-200-190 °C
T2.5	170-200-200-210-210-200 °C
T3	180-210-210-220-220-210 °C

The IM was performed with an Arburg Allrounder 221M 250-55 (1996) with a maximum clamping force of 250 kN. The dimensions of the screw were 25 mm in diameter (D) and 20 L/D (length/D). From hopper to nozzle, the temperature profile used was 140-180-180-190-180 °C (LT, Low Temperature) or 160-220-220-230-220 °C (HT, High Temperature). The mold was oil-tempered to 60 °C. The sample shapes produced were according to ISO (International Organization for Standardization) 3167-B (single cavity) for impact testing and ISO 527-2-5B (twin cavity) for tensile testing. The processing conditions used to produce the samples are listed in Table 2.

Table 2. Injection Molding (IM) setup, ISO 3167-B for impact testing and ISO 527-2-5B for tensile testing.

Cavity Standard: Parameter/Dimensions	ISO 3167-B	ISO 527-2-5B
Thickness	4 mm	1 mm
Mass (including sample, runners, sprue)	14 g	3.9 g
Total Cycle Time	**30 s**	**11 s**
Injection Time	0.7 s	0.2 s
Holding Pressure Time	8 s	2 s
Total Cooling Time	24 s	5 s
Holding Pressure	700 bar	800 bar

2.3. Material Characterization

The modulus of elasticity, yield strength and elongation at yield and break were determined. For some brittle materials, the yield strength was not obtainable and the stress at break was instead measured. The tensile properties were measured with a Zwick/Z 2.5 tensile tester equipped with pneumatic grips and a 500 N load cell. The test bars were produced from extruded strips by die punching and had a thickness of 0.5–0.8 mm. Tensile tests were performed at 22 ± 2 °C and 40% ± 10% relative humidity (RH). The cross-head speed during tension was set to obtain a strain rate of 10% elongation per min and a pre-load of 1 N was used. Seven tensile test specimens were evaluated for each type of sample, and samples failing within the clamped region of the test bars were discarded from the evaluation. Average values for seven samples were calculated and the standard deviation was taken as a measure of statistical uncertainty unless otherwise stated.

The impact properties were evaluated by a Charpy Edgewise single notch test according to ISO 179/1eA. The impact tester was a CEAST 9850 with impact energy of 0.5 J. The samples were notched with a CEAST AN50 to a final notch depth of 2 mm by repeated cutting of 0.1 mm 20 times at a cutting speed of 16 m/min. The impact specimens were cut, notched and evaluated at Swerea IVF (in Mölndal, Sweden). Nine impact specimens were evaluated for each material and average values and standard deviation were calculated.

Thermal properties were studied by differential scanning calorimetry (DSC), with a Perkin Elmer DSC7, equipped with an intracooler. Samples approximately 10 mg in weight were punched out from thin extruded strips or from injection molded test bars and placed in aluminum crucibles. The transitions were studied by heating from 50 to 200 °C at a rate of 10 °C/min in a nitrogen environment. Two measurements per material type were used and the results were averaged.

3. Results and Discussion

3.1. Mechanical Properties

To better understand the relationship between material quality and reprocessing, WEEEBR has been reprocessed by SSE, TSE and IM, and for each melt-blending technique, different processing conditions have been used. A small part of the study concerned the influence of moisture on the properties of the WEEEBR. Although most thermoplastics found in WEEE are non-hygroscopic, ABS normally requires drying prior to processing due to the acrylonitrile content, for example, four hours at around 90 °C [27]. It has also been reported that talc-filled PP and plastics containing carbon black absorb moisture and require drying [28]. Since all these plastics have previously been found in the studied WEEE stream [29], it was important to study the moisture-sensitivity of WEEEBR. It was found that WEEEBR absorbed less than 0.2 wt % of water, after saturation by immersion in water. A comparison study of seven samples dried according to the recommendations for ABS and seven samples stored at 40% ± 10% RH, all processed in the same way by TSE, showed no significant differences in stiffness (E), elongation at break (ε_b) or yield stress (σ_y). This applies to the short-term properties, and it is not clear whether the long-term properties might be affected by drying.

In Figure 2, stress at break (σ_b) is shown instead of σ_y, since WEEEBR did not yield before breaking when SSE was used, except for one sample processed at 180 °C. As can be seen in Figure 2, processing at 200 °C or higher resulted in very low ε_b and σ_b, probably due to excessive degradation of WEEEBR already around 200 °C, as was previously reported for TBV [18].

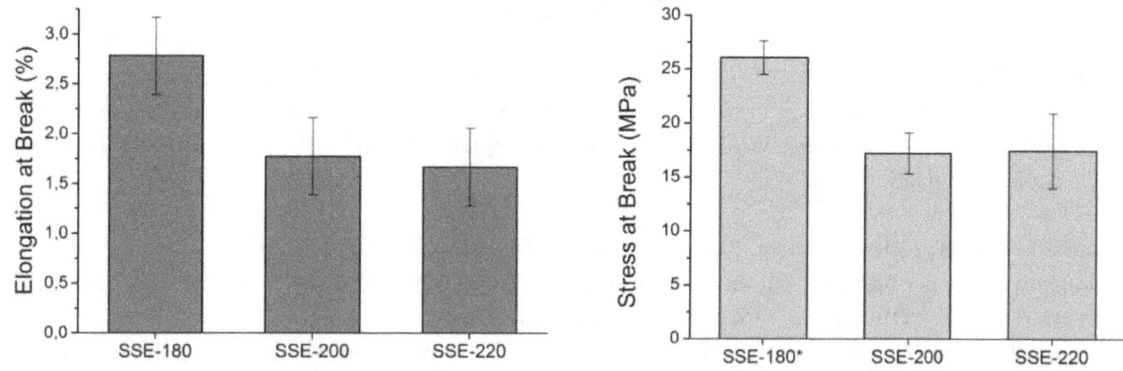

Figure 2. The influence of SSE barrel temperature (straight profile) on ε_b and σ_b for WEEEBR processed at 60 rpm. * Only samples breaking before yield have been included. Error bars show ± one standard deviation.

In the TSE reprocessing of WEEEBR, the screw rotation rate and barrel temperature were varied independently in order to study their influence on the mechanical properties of the collected and reprocessed blend. It is, however, well known that these parameters are not independent, as the melt-blending of WEEEBR is considered to be a complex interplay between screw design, feed rate, barrel temperature and shear rate. Figure 3 shows that processing at 60 rpm and avoiding the highest used temperature profile T3 were beneficial in increasing the ductility. A low screw rotation rate was previously found to result in a higher strength and ε_b of both TBV and TBR [18]. Processing at 60 rpm and T2 may be considered to be a favorable processing condition with TSE for WEEEBR, since it yielded the highest values of both ε_b and σ_y. Higher screw rotation rates are expected to result in considerable degradation due to the high shear rates which might, at least locally, increase the temperature load on the material.

Table 3 summarizes the main results from the TSE runs of WEEEBR, TBV and TBR under different processing conditions. Both E and σ_y tend to be slightly higher in the recycled materials than in the virgin materials, as has also been observed previously for HIPS, PP and compatibilised blends of these two [30]. No significant differences were found in the elongation at yield (ε_y), except perhaps for TBV processed at 60 rpm and T1 that was the most ductile material, and exhibited a slightly higher ε_y than the other materials. An important characteristic of the materials was whether they exhibited a yield point. All TBV and TBR materials showed a yielding behavior, but for WEEEBR, it was mainly the materials processed at low screw rotation rates (60 rpm) or low barrel temperatures (T1 or T2) that showed a consistent yielding.

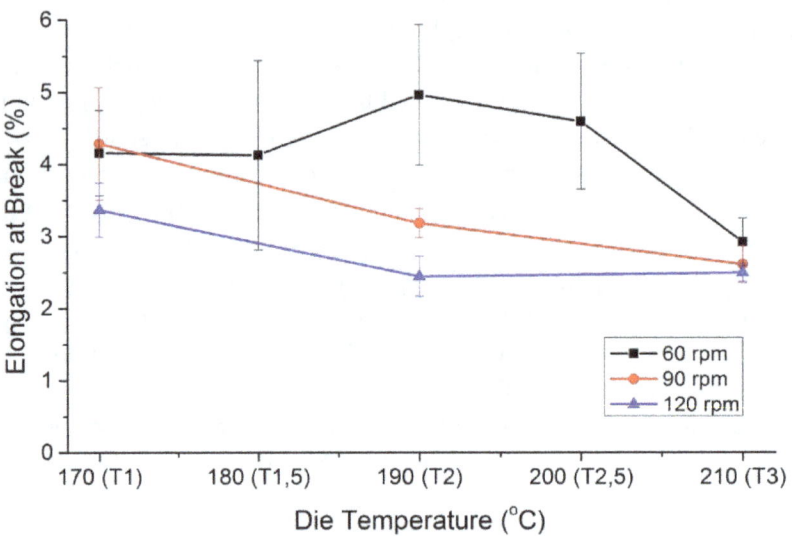

Figure 3. Elongation at break (ε_b) with respect to TSE temperature of WEEEBR at 60, 90 and 120 rpm screw rotation rate. The temperatures are referring to the die zone temperature. The complete temperature profiles are given in parenthesis and are explained in Table 1. The error bars show $\pm$ one standard deviation.

Table 3. Tensile properties for twin-screw extrusion (TSE) samples of waste electrical and electronic equipment plastics blend (WEEEBR), a ternary blend of virgin plastics (TBV) and a ternary blend of recycled plastics (TBR). The standard deviation were based on 7 samples except for; * average and standard deviation based on 6 samples; ** too few samples to calculate standard deviation, average based on remaining samples.

Samples	Tensile Modulus (GPa)	Yield Stress (MPa)	Stress at Break (MPa)	Elongation at Break (%)	Elongation at Yield (%)	Number of Samples Yielding
WEEEBR TSE 60-T1	1.20 ± 0.02	30.1 ± 0.4	-	4.2 ± 0.6	3.1 ± 0.2	7/7
WEEEBR TSE 60-T2	1.24 ± 0.01	31.6 ± 0.7	-	5.0 ± 1	3.2 ± 0.1	7/7
WEEEBR TSE 60-T3	1.23 ± 0.01	30.5 **	28.8 ± 0.5 *	2.9 ± 0.3	3.1 **	1/7
WEEEBR TSE 90-T1	1.24 ± 0.02	31.5 ± 0.5 *	31.4 **	4.3 ± 0.8	3.2 ± 0.1 *	6/7
WEEEBR TSE 120-T1	1.25 ± 0.01	30.4 **	29.9 **	3.4 ± 0.4	3.0 **	4/7
WEEEBR TSE 120-T3	1.29 ± 0.02	-	28.1 ± 1.1	2.5 ± 0.1	-	0/7
TBV TSE 60-T1	1.05 ± 0.03	28.6 ± 1.2	-	22 ± 9	3.4 ± 0.2	7/7
TBV TSE 60-T3	1.16 ± 0.01	30.7 ± 0.5	-	5.4 ± 0.6	3.2 ± 0.1	7/7
TBV TSE 120-T3	1.19 ± 0.03	30.3 ± 0.4	-	5.4 ± 0.9	3.0 ± 0.0	7/7
TBR TSE 60-T1	1.24 ± 0.04	33.1 ± 1.1	-	13 ± 6	3.2 ± 0.1	7/7
TBR TSE 60-T3	1.32 ± 0.06	34.7 ± 0.4	-	8.7 ± 5	3.1 ± 0.1	7/7

The IM of WEEEBR was very sensitive to the processing conditions, the level of holding pressure and the injection speed for instance, had an essential influence on the material quality. Only within a narrow processing conditions window could the material be injection molded. The first impression of some industrial partners was that the surface appearance could be considered acceptable for many applications.

The tensile properties of WEEEBR, TBR and TBV processed by SSE, TSE and IM under similar processing conditions are shown in Figure 4 and 5. The same screw rotation rate and draw down ratio was used in both SSE and TSE. Figure 4 indicates that E and σ_y for the recycled materials TBR and WEEEBR were higher than those of TBV for all processing methods, as is also shown for TSE in Table 3. Recycled plastics are often found to exhibit poorer mechanical properties than virgin ones, but the change in properties is not unidirectional [31], while TBR and WEEEBR were found to exhibit poorer ductility (Figure 5), they also exhibited higher E and σ_y than TBV (Figure 4). Figure 5 shows that IM at LT resulted in very brittle materials that did not yield and had very low ε_b-values. This was possibly due to the relatively fast mold cooling, producing a layered structure and different molecular orientations in the layers, as described by Pisciotti [32]. A comparison of the blends in Figure 5 shows that TBV had the highest ε_b-values and that WEEEBR had significantly lower ε_b than both model material blends. The higher ε_b-values of the blends produced by TSE than by SSE were probably due to better mixing in TSE, as the kneading blocks in the TSE can be expected to lead to a more efficient distributive mixing [33].

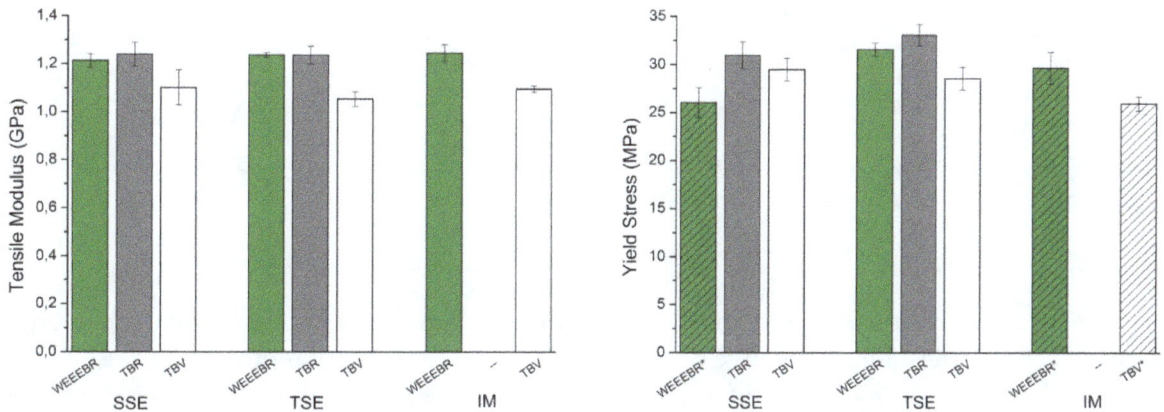

Figure 4. Comparison of E and σ_y when reprocessing WEEEBR, TBR and TBV by SSE, TSE and IM. The processing conditions were similar, about 180 °C barrel temperature and 60 rpm. * σ_b is used instead of σ_y for samples not yielding. Error bars show $\pm$ one standard deviation.

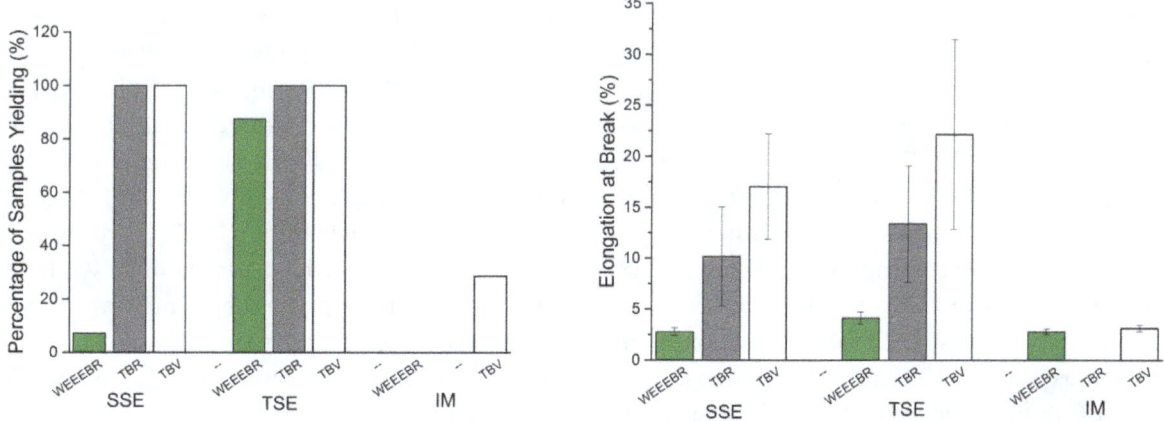

Figure 5. Percentage of samples yielding and ε_b when reprocessing WEEEBR, TBR and TBV by SSE, TSE and IM. The processing conditions were similar, about 180 °C barrel temperature and 60 rpm. Error bars show $\pm$ one standard deviation.

The impact strength shown in Figure 6 could indicate that injection molded WEEEBR and TBV had a limited compatibility between the different polymeric phases. Drying seems to have no significant influence on the impact strength, as was also concluded earlier when drying prior to tensile testing.

A slight increase in impact strength was however seen for WEEEBR injection molded at a temperature about 40 °C higher (HT) than LT.

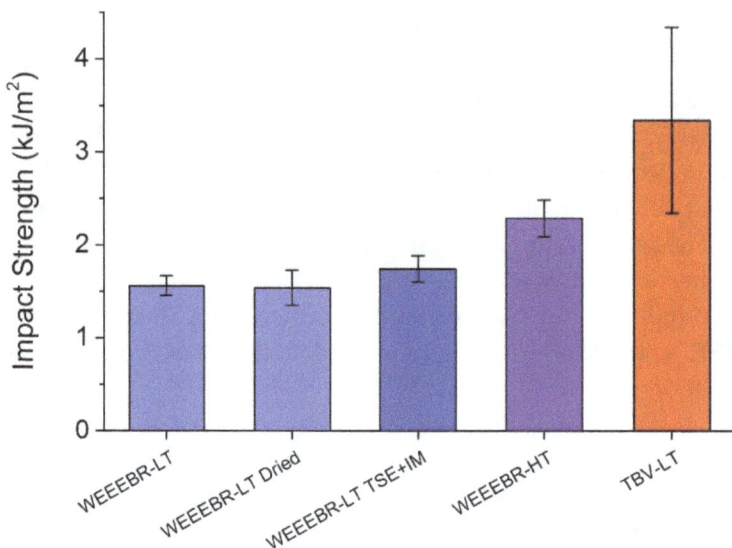

Figure 6. Charpy single notch impact strength of injection molded WEEEBR and TBV. Error bars show ± one standard deviation based on nine samples.

3.2. Thermal Properties

The influence of processing conditions on the thermal properties was more subtle than on the mechanical properties. Representative DSC curves for different ways of reprocessing WEEEBR are shown in Figure 7. An indication of a glass transition can be seen within the temperature range 100–115 °C. The individual glass transitions of the main components HIPS and ABS were, however, not distinguishable. The thermal analysis was focused on the melting of polyethylene (PE) and PP shown in Table 4. It has previously been reported that WEEEBR contains 1–2 wt % PE and around 10 wt % PP [9]. Since TBV is a ternary virgin blend, the PP fraction was free of any PE contamination, but both TBR and WEEEBR had PE in the blends. Both PP and PE are semi-crystalline thermoplastics and the amount of crystallinity can be expected to be of importance for WEEEBR because of the reinforcing effect of the crystals. Table 4 indicates that the onset temperature of the PP melting differs slightly depending on the reprocessing conditions of the WEEEBR. It was seen that WEEEBR processed by TSE and IM at HT exhibited the higher onset temperatures of the PP melting and a narrower peak distribution than, for instance, WEEEBR granules (before being reprocessed) and WEEEBR processed by IM at LT. The approximately 5 °C higher onset temperature of PP in WEEEBR processed by TSE and IM at HT indicated thicker crystals and a narrower crystal lamellar thickness distribution [34,35]. This may be related to a slower cooling rate at the crystallization temperature [36], which corresponds well with previous studies of a thin oriented skin region formed with IM at high processing temperatures [37]. The enthalpy of fusion of the PP fraction is indicated in Table 4 to be approximately 10 J/g for all the materials except TBR. Although this does not fully explain the lower enthalpy of fusion of TBR, a contributing factor may be the PE contamination of the separated PP fraction, resulting in a slightly smaller mass fraction PP than expected in TBR. From the enthalpy of fusion, a rough estimate of the degree of crystallinity of PP can be calculated according to Equation (1) [38,39]

$$X_c = \frac{\Delta H_{PP}(T_m)}{\Delta H_{PP}^0(T_m)(1 - m_a)} \tag{1}$$

where X_c is the weight fraction crystallinity, $\Delta H_{PP}(T_m)$ is the measured enthalpy of fusion for the PP melting in Table 4, $\Delta H_{PP}^0(T_m)$ is the theoretical enthalpy of fusion for the PP melting of a 100%

crystalline polymer and m_a is the mass fraction of non-polypropylene material. The value of $\Delta H_{PP}^0 (T_m)$ has previously been reported to be 207 J/g for PP [40]. From Table 4, the weight fraction crystallinity was calculated to approximately 40 wt % for WEEEBR and TBV, while it was slightly lower for TBR. The calculated degree of crystallinity may be considered reasonable for PP, although the results calculated from Table 4 must be considered to be merely indicative due to the approximated figures for both the measured enthalpy of fusion and the weight fraction of PP in the blends.

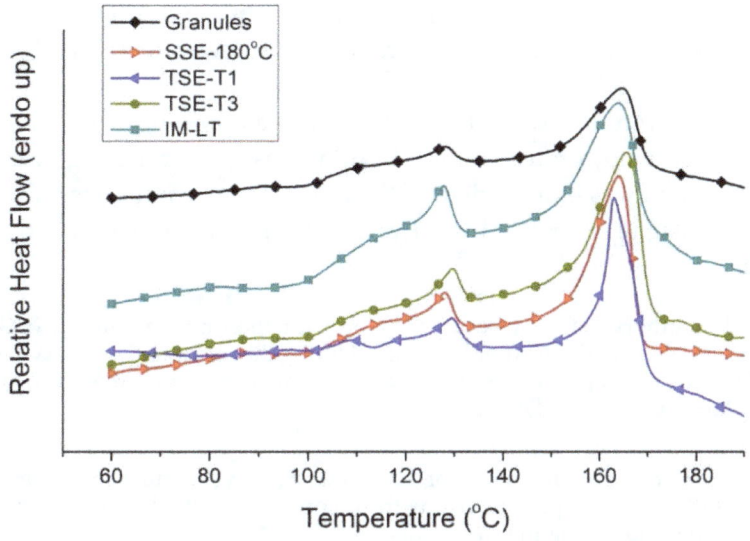

Figure 7. Some DSC curves of WEEEBR reprocessed in different ways.

Table 4. Thermal characterization of the melting of polyethylene (PE) and polypropylene (PP) by differential scanning calorimetry (DSC).

Samples	PE Melting Peak (°C)	PP Melting Onset (°C)	PP Melting Peak (°C)	PP Melting Point (°C)	PP Enthalpy of Fusion (J/g)
WEEEBR Granules	128	153	164	169	8
WEEEBR SSE-180	128	155	164	168	10
WEEEBR SSE-220	129	155	165	169	9
WEEEBR TSE-T1	129	158	163	170	9
WEEEBR TSE-T3	130	156	166	170	9
WEEEBR TSE + IM-LT	128	154	164	169	10
WEEEBR IM-LT	128	153	164	170	8
WEEEBR IM-HT	130	158	165	171	9
TBR TSE-T1	129	152	163	169	4
TBV TSE-T1	-	151	161	168	7
TBV TSE-T3	-	152	162	167	7

4. Conclusions

WEEEBR was found to have a potential to be processed by SSE, TSE or IM. Under favorable processing conditions, the surface character of WEEEBR was considered acceptable for many applications. It was also seen that the processing parameters used in SSE, TSE and IM had a significant influence on the mechanical properties, especially the properties related to ductility. The ε_b-value of WEEEBR was observed to be particularly sensitive to high processing temperatures or high screw rotation rates.

A comparison of SSE, TSE and IM showed that there were relatively different characteristics of the reprocessed materials. The reprocessing of WEEEBR by TSE was favorable for both σ_y and ε_b. Both TBV and TBR had higher ε_b values when processed by TSE, but σ_y was not significantly different from that of SSE. Almost all the materials produced by TSE at low temperatures exhibited a yield point, which is an important material character for many applications. Materials produced by IM were more brittle than those produced by extrusion, probably because of their different states of molecular orientation. The brittleness was also seen in the impact strength, which indicated limited compatibility in the blends. Interestingly, higher impact strength was obtained for WEEEBR injection molded at a high temperature (HT), probably because it had a thinner skin region and more uniform orientation distribution in the cross-section.

WEEEBR was found to exhibit a low degree of water absorption. Drying the material prior to processing did not influence the properties significantly, as was seen for TSE and IM and has also been shown previously for TBV produced by SSE [18].

It can be concluded from the DSC studies that PE was present in all the real post-consumer recycled materials. The enthalpy of fusion of WEEEBR indicated that the degree of crystallinity of the PP (approximately 40 wt %) was independent of the reprocessing conditions. Although the degree of the crystallinity was similar, reprocessing by TSE or IM at HT seemed to narrow the PP crystal melting temperature region.

Acknowledgments: The authors thank Stena Technoworld for the material and Next Generation Recycling Machinen (NGR) for their help in melt-blending and melt-filtering the material. Swerea IVF in Mölndal is thanked for the measurement of the impact properties. The authors are also very grateful for the financial support from the Chalmers Area of Advance Material Science.

Author Contributions: Erik Stenvall; main author, has taken part in planning the experiments, experimental work, evaluating the results and main part of writing the manuscript. Antal Boldizar; supervisor, has taken part in planning the experiments and evaluating the results.

References

1.   European Parliament; Council of the European Union. Directive_2012/19/EU Directive 2012/19/EU of the European Parliament and of the Council of 4 July 2012 on Waste Electrical and Electronic Equipment (WEEE). *Off. J. Eur. Union* **2012**, *L 197*, 38–71.

2.   Flaris, V.; Singh, G.; Rao, A.R. Recycling Electronic Waste. *Plast. Eng.* **2009**, *65*, 10–15.

3.   EPA. *Recycling and Disposal of Electronic Waste*; The Swedish Environmental Protection Agency: Bromma, Sweden, 2011.

4.   Dimitrakakis, E.; Janz, A.; Bilitewski, B.; Gidarakos, E. Small WEEE: Determining recyclables and hazardous substances in plastics. *J. Hazard. Mater.* **2009**, *161*, 913–919. [CrossRef] [PubMed]

5.   UNEP. *Converting Waste Plastics into a Resource*; United Nations Environmental Programme: Nairobi, Kenya, 2009.

6.   Wäger, P.; Böni, H.; Buser, A.; Morf, L.; Schluep, M.; Streicher, M. Recycling of Plastics from Waste Electrical and Electronic Equipment (WEEE)—Tentative Results of a Swiss Study. In Proceedings of the R'09 World Congress, Davos, Switzerland, 14–16 September 2009.

7.   Freegard, K.; Tan, G.; Morton, R. *Develop a Process to Separate Brominated Flame Retardants from WEEE Polymers*; The Waste & Resources Action Programme: Banbury, UK, 2006.

8.   Martinho, G.; Pires, A.; Saraiva, L.; Ribeiro, R. Composition of plastics from waste electrical and electronic equipment (WEEE) by direct sampling. *Waste Manag.* **2012**, *32*, 1213–1217. [CrossRef] [PubMed]

9.   Stenvall, E.; Tostar, S.; Boldizar, A.; Foreman, M.R.S.; Möller, K. An analysis of the composition and metal contamination of plastics from waste electrical and electronic equipment (WEEE). *Waste Manag.* **2013**, *33*, 915–922. [CrossRef] [PubMed]

10.   Robinson, B.H. E-waste: An assessment of global production and environmental impacts. *Sci. Total Environ.* **2009**, *408*, 183–191. [CrossRef] [PubMed]

11.   Tojo, N.; Fischer, C. *Europe as a Recycling Society European Recycling Policies in Relation to the Actual Recycling Achieved*; European Topic Centre on Sustainable Consumption and Production (ETC/SCP): New York, NY, USA, 2011.

12. Dodbiba, G.; Fujita, T. Progress in Separating Plastic Materials for Recycling. *Phys. Sep. Sci. Eng.* **2004**, *13*, 165–182. [CrossRef]

13. IPTS. *End-of-Waste Criteria for Waste Plastic for Conversion*; JRC European Comission, IPTS: Seville, Spain, 2013.

14. Dodbiba, G.; Takahashi, K.; Sadaki, J.; Fujita, T. The recycling of plastic wastes from discarded TV sets: Comparing energy recovery with mechanical recycling in the context of life cycle assessment. *J. Clean. Prod.* **2008**, *16*, 458–470. [CrossRef]

15. Freegard, K.; Tan, G.; Frisch, S. *WEEE Plastics Separation Technologies*; DEFRA Waste Research Team: Bramhall, UK, 2007.

16. Tall, S. Recycling of Mixed Plastic Waste—Is Separation Worthwhile? Ph.D. Thesis, Royal Institute of Technology, Stockholm, Sweden, 2000.

17. Dvorak, R.; Evans, R.; Kosior, E. *Commercial Scale Mixed Plastics Recycling*; WRAP: Banbury, UK, 2009.

18. Stenvall, E.; Tostar, S.; Boldizar, A.; Foreman, M.R.S.J. The Influence of Extrusion Conditions on Mechanical and Thermal Properties of Virgin and Recycled PP, HIPS, ABS and Their Ternary Blends. *Int. Polym. Process.* **2013**, *28*, 541–549. [CrossRef]

19. Brennan, L.B.; Isaac, D.H.; Arnold, J.C. Recycling of acrylonitrile-butadiene-styrene and high-impact polystyrene from waste computer equipment. *J. Appl. Polym. Sci.* **2002**, *86*, 572–578. [CrossRef]

20. Tarantili, P.A.; Mitsakaki, A.N.; Petoussi, M.A. Processing and properties of engineering plastics recycled from waste electrical and electronic equipment (WEEE). *Polym. Degrad. Stab.* **2010**, *95*, 405–410. [CrossRef]

21. De Souza, A.C.; Ereio, A.V. *ABS/HIPS Blends Obtained from Waste Electrical and Electronic Equipment (WEEE)*; PlasticsEurope: Nuremberg, Germany, 2013.

22. Lazzaro, E.; Sbarski, I.; Bishop, J. *Recycling of Engineering Thermoplastics Used in Consumer Electrical and Electronic Equipment*; IRIS, Swinburne University of Technology: Victoria, Australia, 2007; pp. 717–724.

23. Lindsey, C.R.; Barlow, J.W.; Paul, D.R. Blends from reprocessed coextruded products. *J. Appl. Polym. Sci.* **1981**, *26*, 9–16. [CrossRef]

24. Li, J.; Li, H.; Wu, C.; Ke, Y.; Wang, D.; Li, Q.; Zhang, L.; Hu, Y. Morphologies, crystallinity and dynamic mechanical characterizations of polypropylene/polystyrene blends compatibilized with PP-g-PS copolymer: Effect of the side chain length. *Eur. Polym. J.* **2009**, *45*, 2619–2628. [CrossRef]

25. Wycisk, R.; Trochimczuk, W.M.; Matys, J. Polyethylene-polystyrene blends. *Eur. Polym. J.* **1990**, *26*, 535–539. [CrossRef]

26. European Parliament; Council of the European Union. Directive_2011/65/EU Directive 2011/65/EC of the European Parliament and of the Council of 8 June 2011 on the Restriction of the Use of Certain Hazardous Substances in Electrical and Electronic Equipment. *Off. J. Eur. Union* **2011**, *32*, 88–110.

27. Giles, H.F., Jr.; Wagner, J.R., Jr.; Mount, E.M., III. *Extrusion—The Definitive Processing Guide and Handbook*; William Andrew Publishing: New York, NY, USA, 2005.

28. Calhoun, A.; Golmanavich, J. *Plastics Technician's Toolbox, Volumes 1–6*; Society of Plastics Engineers: Bethel, CT, USA, 2002.

29. Stenvall, E. Electronic Waste Plastics Characterisation and Recycling by Melt-Processing. Licentiate Thesis, Chalmers University of Technology, Gothenburg, Sweden, 2013.

30. Santana, R.M.C.; Manrich, S. Studies on morphology and mechanical properties of PP/HIPS blends from postconsumer plastic waste. *J. Appl. Polym. Sci.* **2003**, *87*, 747–751. [CrossRef]

31. Vilaplana, F.; Karlsson, S. Quality Concepts for the Improved Use of Recycled Polymeric Materials: A Review. *Macromol. Mater. Eng.* **2008**, *293*, 274–297. [CrossRef]

32. Pisciotti, F. *On Defect Generation and the Appearance of Injection-Moulded Polymers*; Chalmers Univeristy of Technology: Gothenburg, Sweden, 2004.

33. Lawal, A.; Kalyon, D.M. Mechanisms of mixing in single and co-rotating twin screw extruders. *Polym. Eng. Sci.* **1995**, *35*, 1325–1338. [CrossRef]

34. Gedde, U.W. *Polymer Physics*; Chapman & Hall: London, UK, 1999.

35. Chang, H.; Zhang, Y.; Ren, S.; Dang, X.; Zhang, L.; Li, H.; Hu, Y. Study on the sequence length distribution of polypropylene by the successive self-nucleation and annealing (SSA) calorimetric technique. *Polym. Chem.* **2012**, *3*, 2909–2919. [CrossRef]

36. Maier, C.; Calafut, T. *Polypropylene: The Definitive User's Guide and Databook*; Elsevier Science: Amsterdam, Netherlands, 2008.

37. Wu, P.C.; Huang, C.F.; Gogos, C.G. Simulation of the mold-filling process. *Polym. Eng. Sci.* **1974**, *14*, 223–230. [CrossRef]

38. Patel, A.; Bajpai, R.; Keller, J.M. On the crystallinity of PVA/palm leaf biocomposite using DSC and XRD techniques. *Microsyst. Technol.* **2014**, *20*, 41–49. [CrossRef]

39. Barone, J.R. Polyethylene/keratin fiber composites with varying polyethylene crystallinity. *Compos. Part A Appl. Sci. Manuf.* **2005**, *36*, 1518–1524. [CrossRef]

40. Akinci, A. Mechanical and structural properties of polypropylene composites filled with graphite flakes. *Arch. Mater. Sci. Eng.* **2009**, *35*, 91–94.

The Influence of Compatibilizer Addition and Gamma Irradiation on Mechanical and Rheological Properties of a Recycled WEEE Plastics Blend

Sandra Tostar [1,*], Erik Stenvall [2], Mark R. St J. Foreman [1] and Antal Boldizar [2]

[1]   Chemistry and Chemical Technology, Chalmers University of Technology, Gothenburg SE-412 96, Sweden; foreman@chalmers.se

[2]   Materials and Manufacturing Technology, Chalmers University of Technology, Gothenburg SE-412 96, Sweden; erik.stenvall@borealisgroup.com (E.S.); antal.boldizar@chalmers.se (A.B.)

*   Correspondence: sandra.tostar@chalmers.se

Academic Editor: William Bullock

Abstract: Waste electrical and electronic equipment (WEEE) is growing rapidly, and the plastics within WEEE have an important role in fulfilling the recovery and recycling targets defined in the European WEEE Directive. This study considers recycling of WEEE plastics by making a blend of the different plastics instead of separating them. The mechanical and thermal properties can be enhanced by adding a compatibilizer. It was found that one compatibilizer, a styrene-b(ethylene-co-butylene)-b-styrene (SEBS) copolymer named Kraton® G1652 E, had a large impact on the ductility of the recycled WEEE plastics blend. By adding 2.5 weight % (wt%) of this copolymer, the elongation at break increased by more than five times compared with the non-compatibilized samples, with only a small decrease in stiffness and strength. The storage modulus (G') decreased slightly with increasing compatibilizer amounts while the impact strength increased with increasing amounts of compatibilizer, from 2.1 kJ/m^2 (reference material) to 3.6 kJ/m^2 (5 weight % (wt%) compatibilizer). It was found that Kraton® FG1901 E (styrene-b(ethylene-co-butylene)-b-styrene (SEBS) grafted with maleic anhydride (MAH)), Royaltuf® 372P20 (styrene acrylonitrile (SAN) modified with ethylene-propylene-diene elastomers (EPDM)) and Fusabond® P353 (polypropylene (PP) with a high degree of grafted MAH) were ineffective as compatibilizers to the blend. Gamma irradiation (50 kGy) did not improve the mechanical properties however: the impact strength of the gamma-irradiated samples was lower than that of the non-irradiated samples.

Keywords: WEEE; recycling; compatibilizer; SEBS; plastics blend

1. Introduction

Electronic waste is one of the fastest growing waste streams today, making it important to find new recycling strategies for the different materials included, for environmental, legal and economic reasons. The reported annual amount of globally-generated WEEE in 2014 was 42 million tons [1,2], and considering that the plastic fraction of WEEE was 10 wt%–30 wt% [3–5], that is 4.2 to 12.6 million tons to reuse or recycle. The WEEE plastics contain up to 15 different types which makes it difficult and costly to separate the plastics from each other, which is how plastics material recycling mainly is done today. Plastic fillers and overlapping plastics densities makes the separation even more troublesome. While near infrared spectroscopy (NIR) might be able to distinguish between polymers it is often hindered by the pigments present in "black plastics" [6], while the electrostatic separation methods will be impeded by the thin layer of moisture which exists on all surfaces when the humidity is high [7].

Additionally, washing and de-dusting of the plastics are common. Previous composition studies on WEEE plastics made by the authors have revealed that it consists mainly of high impact polystyrene (HIPS, 42 wt%), acrylonitrile-butadiene-styrene copolymer (ABS, 38 wt%), and polypropylene (PP, 10 wt%) [8], but there are also other plastics, such as polyethylene (PE), polyurethane (PUR), and polycarbonate (PC), depending on the waste stream, reported by others [8–12]. The conventional way of recycling plastics is by means of different sorting and separation steps followed by melt-blending of the different types [9]. Such a recycling method may leave large rest fractions that are excluded from recycling. In this study, another approach is taken. We have investigated the use of a compatibilizer to enhance the mechanical properties of a blend material of all of the different thermoplastics within WEEE, with the exclusion of thermosets and other contaminants that could impact the processing equipment or recycled blend negatively [13]. This process would simplify recycling by reducing many of the sorting, separation and washing steps. A composition control of the blended material may though appear. The influence of gamma irradiation on the plastic properties was also investigated since it is known that it can, firstly, cause free radicals and with that either create crosslinks or chain breakage of the polymer chains [14], and, secondly, it can be used to graft one polymer onto another [15].

2. Experimental Section

2.1. Materials

The material used was a melt-blended and melt-filtered WEEE plastics blend of recycled material (WEEEBR) determined by combustion analysis to be C, 83.46; H, 8.77; N, 2.20; O, 1.27; Cl, 0.13; Br, <0.10% from post-consumer flame-retardant free waste. This collected post-consumer waste was obtained from Stena Technoworld in Halmstad (5 July 2011) and has previously been analyzed with respect to its composition [8]. The material should be compliant with the European Directive on the Restriction of the Use of Certain Hazardous Substances in Electrical and Electronic Equipment (RoHS) [16]. Before the reprocessing in our laboratory, the material had undergone dust and surface cleaning, melt blending, melt-filtration, and hot die granulation at Next Generation Recycling Maschinen in Feldkirchen, Austria. The recycling equipment used for melt-blending, filtration, and granulation was an S:GRAN 85 in conjunction with an Ettlinger rotation drum melt-filter and hot die granulator. The extruder screw rotation rate was 145 rpm with a throughput of 280 kg/h, and it filtered out about 1 wt% of mainly non-thermoplastic contamination in order to form the WEEEBR material. The pre-made compatibilizers studied are shown in Table 1. To make the cost of the blended material comparable to virgin material, low amounts of the compatibilizers were added.

Table 1. The four different compatibilizers (wt%) blended with WEEEBR for tensile testing.

Compatibilizers Used	Description	Amounts of Compatibilizer
Kraton® G1652 E	SEBS (30 wt% styrene)	0.83, 1.25, 2.5, 5, 10, 20 wt%
Kraton® FG1901 E	SEBS-g-MAH	2.5, 5, 10 wt%
Royaltuf® 372P20	SAN modified with EPDM	5, 10, 20 wt%
Fusabond® P353	PP-g-MAH	0.83, 2.5, 5, 10 wt%

Kraton Performance Polymers is the producer of Kraton® G1652 E and FG1901 E, both based on a linear triblock copolymer of styrene-b(ethylene-co-butylene)-b-styrene (SEBS), with a styrene content of 30 wt%. In addition, FG1901 E was grafted with 1.4 wt%–2 wt% maleic anhydride (MAH). Addivant provided Royaltuf® 372P20, which is styrene acrylonitrile (SAN) modified with ethylene-propylene-diene elastomers (EPDM). Fusabond® P353 is based on polypropylene (PP) with a high degree of grafted MAH and produced by DuPont™.

2.2. Processing Equipment

The WEEEBR was blended with the different compatibilizers in a melt process with a co-rotating twin screw extruder, Werner and Pfleiderer ZSK 30 M9/2 (1984), with five heating zones along the barrel and one heating zone for the die. The screw length was 966 mm with a barrel bore diameter of 31 mm. The screw configuration was optimized for mixing and compounding. The temperature profile was set to 160-190-190-200-200-190 °C to avoid over-heating and resin degradation and the screw rotation rate was 60 rpm [17]. The extruded material was oriented and flattened with a 3 + 2 roll puller, Brabender 843316003, to a thickness of 0.4–0.9 mm. Test specimens, in the shape of dog bones, were produced by die punching to a shape according to ISO 527-5A. For the impact test, new batches with compatibilized materials were prepared in another co-rotating twin screw extruder, Coperion ZSK 26K, 10.6, manufactured by Coperion, Germany. The residence time in the extruder was 50 s, the screw speed 180 rpm, and the throughput 6 kg·h^{-1}. The blends are presented in Table 2. It is noteworthy that none of the properties of the material produced using the second machine were inconsistent with the properties of the product made using the first machine.

Table 2. WEEEBR blends studied for impact testing with the compatibilizer Kraton® G1652 E (0–10 wt%).

Material	Amount of Kraton® G1652 E (wt %)
WEEEBR reference	0
WEEEBR	2.5
WEEEBR	5
WEEEBR	10
WEEEBR (gamma-irradiated granulate)	0
WEEEBR (gamma-irradiated granulate)	2.5
WEEEBR (gamma-irradiated granulate)	5
WEEEBR (gamma-irradiated dog bones)	0

The gamma irradiation used in the impact test was performed in a Gammacell 220 (Atomic Energy of Canada, now trading as Norion) with a dose rate of 8 kGy·h^{-1} on average, and the dose rate was measured with the ferrous-cupric sulphate dosimeter test (4 February 2014). The temperature within the chamber was approximately 50 °C. The WEEEBR granulate and dog bones were irradiated with a dose of 50 kGy.

2.3. Material Characterization

The mechanical properties of modulus of elasticity, yield strength and elongation at break were measured with a Zwick/Z 2.5 tensile tester equipped with pneumatic grips and a 500 N load cell. The measurements were performed at 22 ± 2 °C and 40 % ± 10 % relative humidity. Seven specimens were evaluated for each type of sample and the average values were calculated with ± one standard deviation.

The Charpy impact properties were evaluated according to ISO 179/1eA by an Edgewise single notch test and compared with un-notched samples. The samples were notched with an Instron CEAST AN50, 0.1 mm each time to a final notch depth of 2 mm. The impact test equipment was an Instron CEAST 9050, with an impact energy of 0.5, 1.0, and 4.0 J. The impact specimens were prepared and evaluated at Swerea IVF. The results reported were based on ten test specimens evaluated for each material and given as average values and ± one standard deviation.

The thermal storage modulus and loss tangent were determined by Dynamic Mechanical Thermal Analysis (DMTA) using Rheometrics Solids Analyzer RSA II together with TA Orchestrator RSA2 software. Two curves were produced for each material. Unless they matched, a third measurement was made. The first of the matching curves was used. The melt flow rate was determined according to

ISO 1133 and measured on a Modular Melt Flow 7024, produced by Ceast, Italy. One measurement was performed per material.

3. Results and Discussion

Previous tests performed on WEEEBR have shown low elongations at break (εb) and low impact strengths [13], indicating poor compatibility between the phases, suggesting it is a very brittle material. To increase the ductility, compatibilization with compatibilizers was considered.

3.1. Tensile Properties

As expected, it can be seen in Figure 1 that the most effective compatibilizer was a styrene-ethylene/butadiene-styrene triblock copolymer (SEBS G1652 E). This substance caused a significant increase in the ductility of WEEEBR. Even at as low as 2.5 wt% compatibilizer, the εb-values increased over five times. This is considerably higher than has previously been reported for PS/PP blends compatibilized by SEBS (up to 25 wt%) [18]. At least two explanations exist for this improving effect of the SEBS. Firstly, the SEBS is a block copolymer with sections that are similar and, thus, compatible, to the two types of polymers (polystyrene and polypropylene) in the blend. It could act as a surface-active material, reducing the surface tension between the two liquid phases during processing [19]. This reduction in interfacial tension should result in a reduction of the domain size can be expected to improve the properties of the mixture [20]. An alternative explanation offered by La Mantia is that the SEBS is not a true compatibilizer, instead it reduces the brittleness of at least one of the phases instead of accumulating at the interfaces [21].

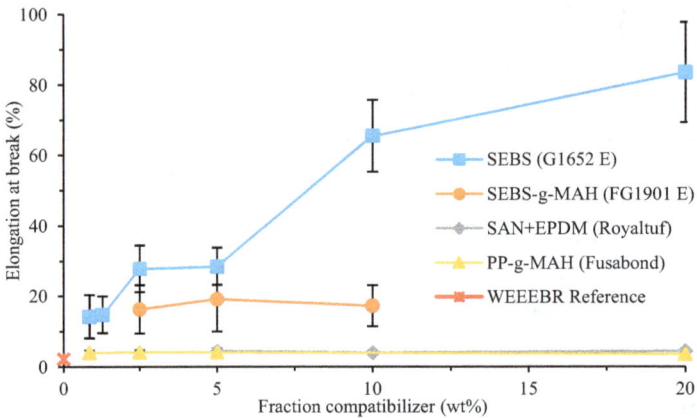

Figure 1. Elongation at break measured for WEEEBR with the compatibilizers used and non-compatibilized WEEEBR (as reference). The error bars represent the standard deviation.

However, while SEBS grafted with maleic anhydride (MAH) (FG1901 E) increased the εb-value, this increase was smaller than for the unmodified SEBS. Unlike the unmodified SEBS, increasing the quantities of the MAH-grafted SEBS did not cause further improvement in the plastic.

While it could be thought that a MAH-grafted PP should be a good compatibilizer for the WEEE blend, it turned out to be ineffective. It has previously been reported by Parameswaranpillai *et al.* that a PP-g-MAH compatibilizer for a PP/PS was effective for stabilizing the morphology and improving the impact strength. However, the other mechanical properties such as tensile strength, flexural strength, and elongation at break gave irregular results compared with the reference blend [22]. A study by Yilu *et al.* reported a novel approach to graft PP with dual monomers (PP-*g*-(MAH-*co*-St)), overcoming the inability of MAH to homopolymerize, which causes little MAH to be grafted onto PP and severe degradation of the PP backbone [23,24]. The sluggish reaction of MAH with the electron-poor radicals, formed by the reaction of MAH with a radical, can be overcome by adding styrene (St) as a second monomer onto PP. This not only increases the grafting proportion of MAH onto

PP (gMAH) but also reduces the PP chain scission degradation [25]. The styrene works as a medium to bridge the gap between the PP macroradicals and the MAH. The co-polymer St can first graft onto the PP backbone and then react with the MAH [26].

Focusing on the best SEBS-containing compatibilizer, different amounts (1.25 wt%–20 wt%) were blended with the WEEEBR, and the transition from a brittle (non-compatibilized) to a ductile material is shown in Figure 2. The strain at break increased significantly for all added amounts, but from an engineering point of view an addition of about 5 wt% compatibilizer, giving an increase in strain at break of about 25 %, can be considered enough to replace virgin plastics used in electronic equipment with WEEEBR.

As seen in Figures 3 and 4 both the stiffness (σ_b) and the yield stress (σ_y) decreased with increasing amounts of compatibilizer. The stiffness was lower compared with the reference sample for all the compatibilized samples and appeared to have a linear dependency on the compatibilizer content. On the other hand, low compatibilizer levels (<2.5 wt%) resulted in a slight decrease in yield stress but had only a very small influence on the σ_y-values between 2.5 wt% and 20 wt%.

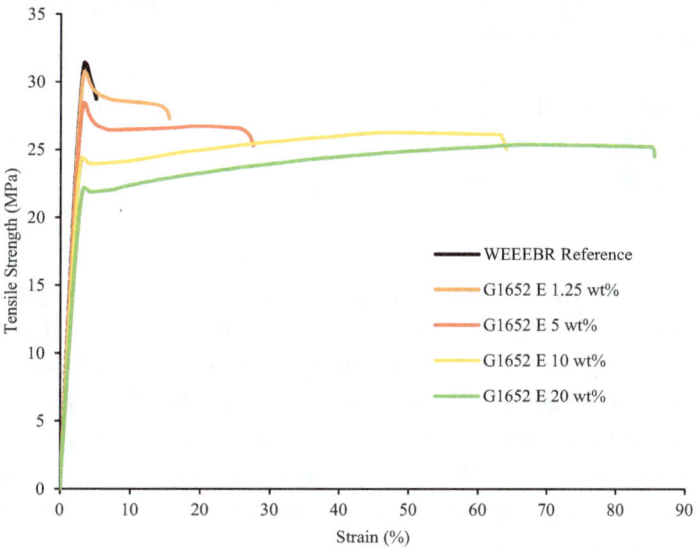

Figure 2. Tensile strength *versus* strain for non-compatibilized WEEEBR (reference) and compatibilized WEEEBR samples studied.

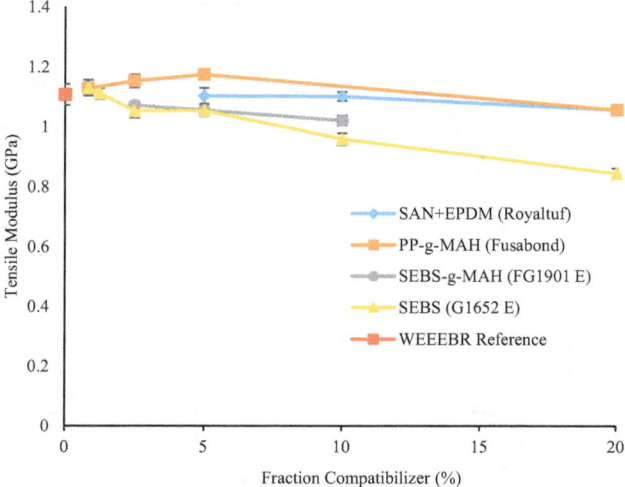

Figure 3. Tensile modulus of WEEEBR with the compatibilizers used and non-compatibilized WEEEBR. The error bars represent the standard deviation.

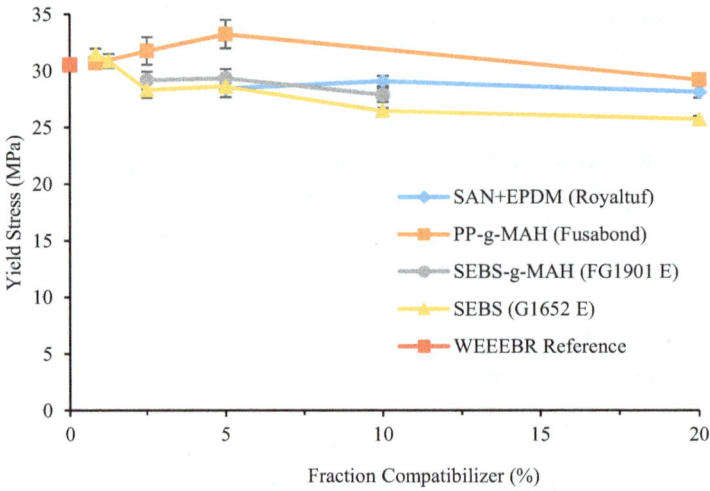

Figure 4. Yield stress of WEEEBR with the compatibilizers used and non-compatibilized WEEEBR. The error bars represent the standard deviation.

It is interesting to note that while an increase in toughness is normally associated with a decrease in stiffness and strength [18,27], in our case the addition of the SEBS only caused a moderate decrease in these important mechanical properties.

3.2. DMTA

The DMTA results of the storage modulus (G') and loss tangent (tan δ) studied in compression mode for WEEEBR mixed with different amounts of the compatibilizer Kraton® G1652 E can be seen in Figures 5 and 6. The storage moduli were around 1 GPa (room temperature, 22 °C), in compliance with previous results by the authors [13], and fairly stable up to 85 °C. As expected from Figure 3, the storage modulus at temperatures below 85 °C was somewhat lower with increasing amounts of compatibilizer. It is interesting to note, however, that the compatibilizer does not seem to have any negative influence on the thermal stability, since no real difference in the softening (>100 °C) of the materials could be seen between the reference and the compatibilized samples.

The T_g can be found as the tan δ peak of WEEEBR and the compatibilizer in Figure 5 at about 110 °C.

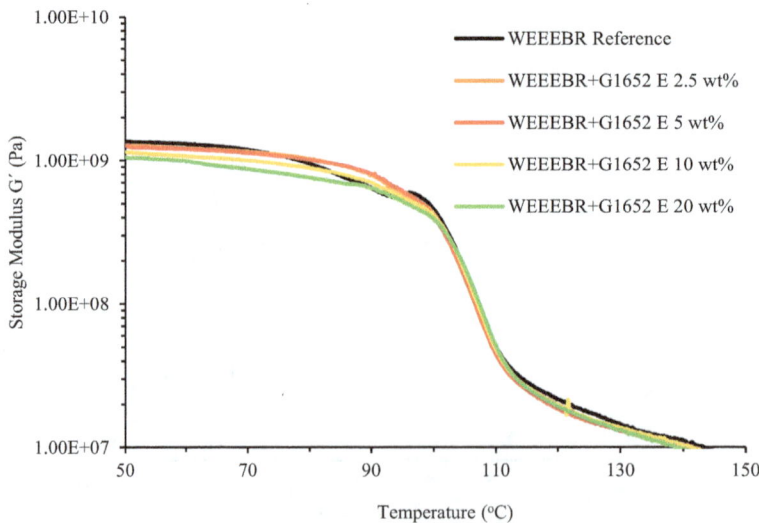

Figure 5. Storage modulus as a function of temperature for WEEEBR with different amounts of compatibilizer.

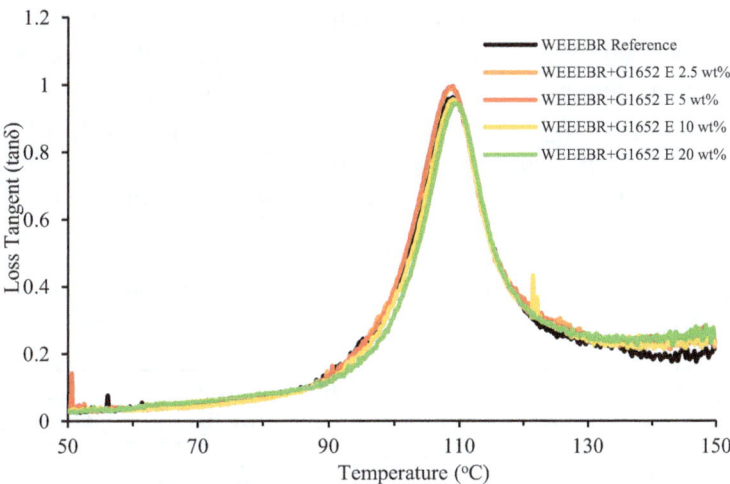

Figure 6. The loss tangent as a function of temperature for WEEEBR reference and WEEEBR with different amounts of compatibilizer.

3.3. Impact Strength

The impact strength increased with increasing amounts of compatibilizer for both notched and un-notched samples; see Table 3. For the notched samples, the impact strength increased by 28.6% for the 2.5 wt% compatibilized sample, 81.0% for the 5 wt% compatibilized sample, and an astonishing 138.1% for the 10 wt% compatibilized sample. The standard deviation is based on ten samples. Un-notched samples gave a higher standard deviation than notched samples, as could be expected.

Table 3. Impact testing of WEEEBR blended with the compatibilizer Kraton® G1652 E (0–10 wt%) and gamma-irradiated samples (50 kGy).

Material	Amount of Kraton® G1652 E (wt%)	Notched Samples		Un-Notched Samples	
		kJ/m^2	Standard Deviation	kJ/m^2	Standard Deviation
WEEEBR reference	0	2.1	0.3	7.9	1.0
WEEEBR	2.5	2.7	0.1	13.2	1.5
WEEEBR	5	3.8	0.3	16.8	2.0
WEEEBR	10	5.0	0.1	24.5	4.1
WEEEBR (50 kGy)	0	1.3	0.1	6.7	0.9
WEEEBR (50 kGy)	2.5	2.3	0.1	10.7	1.1
WEEEBR (50 kGy)	5	2.6	0.0	12.5	1.9
WEEEBR (50 kGy) *	0	2.1	0.5	8.0	1.1

* Gamma-irradiated dog bones.

The impact strength of gamma-irradiated samples decreased while the compatibilizer showed similar behaviour to that within the non-irradiated samples.

No change was seen in the mechanical properties of gamma-irradiated samples (same value as WEEEBR reference). This effect could be due to the fact that hydroperoxides require thermal activation to form hydroxyl and alkoxy radicals [28]. The gamma irradiation of polyvinyl chloride (PVC) results in the formation of hydroperoxides, which later form these radicals [28]. The hydroxyl radicals will abstract hydrogens at random sites, and the alkoxy radicals can undergo a beta-scission reaction forming a carbonyl group and a new but smaller macromolecular radical. This thermal activation of the hydroperoxides leading to chain scission has been specifically reported for polypropylene [29]. Thus, it is the granules that need to be gamma-irradiated, and then mechanically and thermally processed for a reaction resulting in a change of mechanical properties to occur.

When moslding samples of WEEEBR were prepared for the impact test, it was noted that the injection molding pressure increased with increasing amounts of compatibilizer and also that the increase was smaller for the gamma-irradiated samples; see Table 4. These observations indicated a viscosity decrease in the gamma-irradiated blends. A decrease in the mechanical strength of one component in a composite material can cause the mechanical properties of the material taken as a whole to decrease, as polypropylene is known to be more sensitive to radiation than polystyrene [30] and to undergo mainly chain scission, rather than crosslinking [31]. We reason that the loss of strength in the polypropylene domains can explain the effects of irradiation followed by thermal processing of the WEEEBR.

Table 4. Injection molding pressures used for molding the WEEEBR blends with the compatibilizer Kraton® G1652 E (0–10 wt%).

Material	Amount of Kraton® G1652 E (%)	Injection Molding Pressure (bar)	Increase in Injection Molding Pressure (%)
WEEEEBR reference	0	435	-
WEEEBR	2.5	508	17
WEEEBR	5	544	25
WEEEBR	10	544	25
WEEEBR (50 kGy)	0	435	0
WEEEBR (50 kGy)	2.5	471	8
WEEEBR (50 kGy)	5	508	17

3.4. Melt Flow Rate

The MFR decreased with increasing amounts of compatibilizer for the WEEEBR; see Table 5. Thus, the material became more viscous, indicating better compatibility between the phases, which can be explained by better phase adhesion [32] and possibly co-continuous structures [33]. It is notable in Table 5 that the major decrease in MFR is seen between 0 and 2.5 wt% SEBS and further increases of the compatibilizer amount only resulted in a slight decrease of MFR. This indicated that 2.5 wt% SEBS, or less, was enough to significantly change the flow characteristics of the polymer blend and should have a positive effect on the phase adhesion in the WEEEBR material. Gamma irradiation was instead found to increase the MFR, which can be explained by a degradation of the polymer chains subjected to gamma irradiation expectedly resulting in chain scission and decreased molecular weight. The increase in MFR with gamma irradiation was seen for all the different compatibilization amounts compared with the non-irradiated samples; hence, the viscosity decreased and the material flowed more easily.

Table 5. Melt Flow Rate of WEEEBR blended with the compatibilizer Kraton® G1652 E (0–10 wt%) and gamma-irradiated samples (50 kGy).

Material	Amount of Kraton® G1652 E (wt%)	Melt Flow Rate (g/10min) @220 °C, 5 kg	
		Non-Irradiated	50 kGy Gamma-Irradiated
WEEEBR reference	0	29	45
WEEEBR	2.5	21.7	38.2
WEEEBR	5	21.1	34
WEEEBR	10	19.6	-

- Means not measured.

4. Conclusions

The addition of SEBS greatly increased the ductility of WEEEBR, even at low levels (2.5 wt%). On the other hand, maleic anhydride-grafted polypropylene failed to improve the properties of the mixture of plastics. The addition of SEBS to the mixture caused only a small decrease in the storage modulus, while a considerable increase in the impact strength was seen, from 2.1 kJ/m^2 to 3.8 kJ/m^2, with 5 wt% compatibilizer. At temperatures below 85 °C, the storage modulus was reduced as the amount of compatibilizer increased. It was interesting to note, however, that the compatibilizer did not seem to influence the glass temperature of the plastics blend, since no significant difference in the softening of the materials was seen between the reference and the compatibilized samples. The MFR decreased with increasing amounts of compatibilizer for the WEEEBR and, thus, the material became more viscous, indicating better compatibility between the phases, which can be explained by better phase adhesion and possibly co-continuous structures. As the addition of SEBS only caused a moderate increase in the viscosity of the molten plastic it will not prevent the use of either extrusion or injection molding as means of forming objects from the recycled plastic. It was found that gamma irradiation (50 kGy) before thermal processing expectedly caused a moderate reduction in the mechanical strength of the plastic due to chemical degradation of the plastic.

Acknowledgments: The authors want to thank the Chalmers Area of Advance Material Science for the financial funding.

Author Contributions: Sandra Tostar; main author, has taken part in planning the experiments, experimental work, evaluating the results and main part of writing the manuscript. Erik Stenvall; has taking part in planning the experiments, experimental work, evaluating the results and partly writing the manuscript. Mark R. St J. Foreman; supervisor, has taken part in planning the experiments and evaluating the results. Antal Boldizar, supervisor, has taken part in planning the experiments and evaluating the results.

References and Notes

1. El-Kretsen. *El-Kretsen Verksamheten 2014*; El-Kretsen: Stockholm, Sweden, 2014.
2. Menikpura, S.N.M.; Santo, A.; Hotta, Y. Assessing the climate co-benefits from Waste Electrical and Electronic Equipment (WEEE) recycling in Japan. *J. Clean. Prod.* **2014**, *74*, 183–190. [CrossRef]
3. Toxics Link. *Improving Plastic Management in Delhi—A Report on WEEE Plastic Recycling*; Toxics Link: Delhi, India, 2011.
4. Hirayama, D.; Saron, C. Characterisation of recycled acrylonitrile-butadiene-styrene and high-impact polystyrene from waste computer equipment in Brazil. *Waste Manag. Res.* **2015**, *33*, 543–549. [CrossRef] [PubMed]
5. Wang, R.; Xu, Z. Recycling of non-metallic fractions from waste electrical and electronic equipment (WEEE): A review. *Waste Manag.* **2014**, *34*, 1455–1469. [CrossRef] [PubMed]
6. Waste & Resources Action Programme. *Separation of Mixed WEEE Plastics Final Report (WRAP Project MDD018 and MDD023)*; WRAP: Scotland, UK, 2009.
7. Dascalescu, L.; Samuila, A.; Iuga, A.; Morar, R.; Csorvasy, I. Influence of material superficial moisture on insulation-metal electroseparation. In Proceedings of the Conference Record of the 1992 IEEE Industry Applications Society Annual Meeting, Huston, TX, USA, 4–9 October 1992; Volume 2, pp. 1472–1478.
8. Stenvall, E.; Tostar, S.; Boldizar, A.; Foreman, M.R.S.J.; Moller, K. An analysis of the composition and metal contamination of plastics from waste electrical and electronic equipment (WEEE). *Waste Manag.* **2013**, *33*, 915–922. [CrossRef] [PubMed]
9. Ramesh, V.; Biswal, M.; Mohanty, S.; Nayak, S.K. Recycling of engineering plastics from waste electrical and electronic equipments: Influence of virgin polycarbonate and impact modifier on the final performance of blends. *Waste Manag. Res. ISWA* **2014**, *32*, 379–388.
10. Wäger, P.A.; Böni, H.; Buser, A.; Morf, L.; Schluep, M.; Streicher, M. *Recycling of Plastics from Waste Electrical and Electronic Equipment (WEEE)—Tentative Results of a Swiss study*; R'09 World Congress: Davos, Switzerland, 2009.
11. Freegard, K.; Tan, G.; Morton, R. *Develop a Process to Separate Brominated Flame Retardants from WEEE Polymers*; Waste Resources Action Programme: Banbury, UK, 2006.
12. Schlummer, M.; Gruber, L.; Maeurer, A.; Woiz, G.; van Eldik, R. Characterisation of polymer fractions from waste electrical and electronic equipment (WEEE) and implications for waste management. *Chemosphere* **2007**, *67*, 1866–1876. [CrossRef] [PubMed]
13. Stenvall, E. *Functional Properties and Morphology of Recycled Post-Consumer WEEE Thermoplastic Blend*; Chalmers University of Technology: Gothenburg, Sweden, 2015.
14. Clegg, D.W.; Collyer, A.A. *Irradiation Effects on Polymers*; Springer: Berlin, Germany, 1991.

15. Wang, F.; Zheng, K.; Tong, B. Nonisothermal Crystallization Behavior of Polypropylene Grafted Silane and Styrene Using gamma-Ray Irradiation. *J. Macromol. Sci. Part B Phys.* **2011**, *50*, 942–951. [CrossRef]

16. Directive 2011/65/EU. Available online: http://eur-lex.europa.eu/legal-content/EN/TXT/?uri=CELEX: 32011L0065 (accessed on 8 June 2011).

17. Giles, H.F.J.; Wagner, J.R.J.; Mount, E.M.I. *Extrusion—The Definitive Processing Guide and Handbook*; William Andrew Publishing: New York, NY, USA, 2005.

18. Sani Amril, S.; Hassan, A.; Mokhtar, M.; Syed Mustafa Syed, J. Effect of SEBS on the mechanical properties and miscibility of polystyrene rich polystyrene/polypropylene blends. *Prog. Rubber Plast. Recycl. Technol.* **2005**, *21*, 261–276.

19. Kallel, T.; Massardier-Nageotte, V.; Jaziri, M.; Gerard, J.F.; Elleuch, B. Compatibilization of PE/PS and PE/PP blends. I. Effect of processing conditions and formulation. *J. Appl. Polym. Sci.* **2003**, *90*, 2475–2484.

20. Deanin, R.D.; Manion, M.A. *Handbook of Polyolefins*, 2nd ed.; CRC Press: New York, NY, USA, 2000.

21. La Mantia, F.P. Recycling of heterogeneous plastics wastes. II—The role of modifier agents. *Polym. Degrad. Stab.* **1993**, *42*, 213–218. [CrossRef]

22. Parameswaranpillai, J.; Joseph, G.; Jose, S.; Hameed, N. Phase morphology, thermomechanical, and crystallization behavior of uncompatibilized and PP-g-MAH compatibilized polypropylene/polystyrene blends. *J. Appl. Polym. Sci.* **2015**, *132*. [CrossRef]

23. Bettini, S.H.P.; de Mello, L.C.; Munoz, P.A.R.; Ruvolo-Filho, A. Grafting of maleic anhydride onto polypropylene, in the presence and absence of styrene, for compatibilization of poly(ethylene terephthalate)/(ethylene-propylene) blends. *J. Appl. Polym. Sci.* **2013**, *127*, 1001–1009. [CrossRef]

24. Zheng, Y.Y.; Zhao, S.F.; Cheng, L.; Li, B.M. Synthesis and reaction kinetics model of suspension phase grafting polypropylene with dual monomers. *Polym. Bull.* **2010**, *64*, 771–782. [CrossRef]

25. Li, Y.; Xie, X.M.; Guo, B.H. Study on styrene-assisted melt free-radical grafting of maleic anhydride onto polypropylene. *Polymer* **2001**, *42*, 3419–3425. [CrossRef]

26. Zhang, Y.L.; Guo, Z.F.; Zhang, L.M.; Pan, L.S.; Tian, Z.; Pang, S.J.; Xu, N.; Lin, Q. Mechanochemistry: A novel approach to graft polypropylene with dual monomers (PP-g-(MAH-co-St)). *Polym. Bull.* **2015**, *72*, 1949–1960.

27. Metals, A.S.O. *Characterization and Failure Analysis of Plastics*; ASM International: Novelty, OH, USA, 2003.

28. Colombani, J.; Labed, V.; Joussot-Dubien, C.; Perichaud, A.; Raffi, J.; Kister, J.; Rossi, C. High doses gamma radiolysis of PVC: Mechanisms of degradation. *Nucl. Instrum. Methods Phys. Res. Sect. B Beam Interact. Mater. At.* **2007**, *265*, 238–244. [CrossRef]

29. Scott, G. Environmental stability of polymers. In *Polymers and the Environment*; The Royal Society of Chemistry: London, UK, 1999; pp. 38–67.

30. Choppin, G.; Liljenzin, J.-O.; Rydberg, J. *Radiochemistry and Nuclear Chemistry*, 3rd ed.; Butterwort-Heinemann: Oxford, UK, 2002.

31. Swallow, A.J. *Radiation Chemistry of Organic Compounds*; Pergamon Press: Oxford, UK, 1960; p. 380.

32. Santana, R.M.C.; Manrich, S. Studies on morphology and mechanical properties of PP/HIPS blends from postconsumer plastic waste. *J. Appl. Polym. Sci.* **2003**, *87*, 747–751. [CrossRef]

33. Tall, S.; Karlsson, S.; Albertsson, A.C. Improvements in the properties of mechanically recycled thermoplastics. *Polym. Polym. Compos.* **1998**, *6*, 261–267.

End-of-Life Strategies for Used Mobile Phones Using Material Flow Modeling

Kuniko Mishima [1],*, Michele Rosano [2], Nozomu Mishima [3] and Hidekazu Nishimura [1]

[1] Graduate School of System Design and Management, Keio University, Yokohama 223-8526, Japan; h.nishimura@sdm.keio.ac.jp

[2] Sustainable Engineering Group, Curtin University, Perth, WA 6845, Australia; M.Rosano@curtin.edu.au

[3] Graduate School of Engineering and Resource Science, Akita University, Akita 010-8502, Japan; nmishima@gipc.akita-u.ac.jp

* Correspondence: mishima-ssdb2172@jcom.zaq.ne.jp

Academic Editor: William Bullock

Abstract: In order to secure valuable materials and to establish better circular economy practice, new legislation to promote recycling of small-sized e-waste including used mobile phones started in April 2013, in Japan. In order to consider appropriate methods to reduce material usage in mobile phone production, an examination of appropriate strategies in handling used mobile phone products is warranted. This paper investigates an analysis of material flow model for used mobile phones. Then, by analyzing the model, it tries to find suitable strategies to reduce the material consumption associated with mobile phone production and consumption. Although material recycling is an important strategy in Japan, other waste management options exist. This research indicates which factors are keys in reducing material consumption and CO_2 emission, and establishing resource efficient production. The study concludes that "domestic product reuse" and "official recycling networks" are equally good in reducing the consumption of virgin materials associated with mobile phone production. However, in doing so, it is necessary to establish a system in which consumers can properly return their used mobile phones for recycling. Such an end-of-life waste management system can reduce both waste and resource consumption and the environmental impacts associated with increasing mobile phone production. Further research investigating the value of increasing the product reuse rate and the collection return rate for mobile phones is also warranted.

Keywords: used mobile phone; end-of-life option; material recycling; product reuse; material consumption; CO_2 emission; sensitivity analysis

1. Introduction

Japan is not a resource-rich country, with most natural resources needing to be imported from other countries. However, manufactured waste in Japan is increasingly seen as being an important source of resource material given that end-of-life appliances typically stay in the domestic market many years after their "useful life" has ended. Such end-of-life (EOL) products have become a source of material for urban mining [1,2]. For example, 16% of the world reserve of gold exists in Japan [3] as urban mining material. The importance and effectiveness of utilizing precious metals, critical metals and common metals in urban mines has been well noted in previous research [4]. From April 2013, legislation to promote the recycling of small-sized e-waste started in Japan in order to recover currently unutilized but valuable resources [5] in advance of the enforcement of the legislation. This legislation only focuses on the material recycling associated with consumer behavior with used mobile phones [6], where typically a certain volume of small-sized e-waste products/materials like mobile phones at end-of-life are re-utilized as second-hand products with the same functionality as the original products. This second-hand use needs to be taken into account in this smart phone era with increasing pressure on resource efficiency and scarcity. Previous research by Mishima (2015) [7] has shown the value and importance of motivating consumers to put their used products in recycling bins to provide second-hand use for these end-of-life products. However this study has also highlighted the need for new systems to be put in place to assist with increasing the recyclability of used mobile phones in Japan.

In order to establish a consumer-friendly recycling system, more understanding of Japanese consumer behavior for mobile phones is needed These investigations required a better understanding of the "movement" of these end-of-life products through a modeling and analysis of the material flow. Previous research [8–10] and a public report [11] have investigated the material flow of some used electronic equipment such as Personal Computers and mobile phones, in Japan and in other countries. Although these studies have been helpful in clarifying some of the current problems in consumer recycling systems, situations can be different in nowadays "smartphone era." A new investigation based on the current situation in Japan is necessary. One of the studies [12] focused on environmental impacts of recycling considering material compositions of mobile phones. Another study [13] shows factors to affect statistics regarding material recycling of WEEE combining material flow analysis and structural analysis. These previous studies have mainly focused on material recovery from used products. However, material recycling is not the only end-of-life waste management option. Such used products can become stocks of potential second-hand products and their associated reusable components. As a result, further comprehensive studies are necessary to investigate other recycling strategies for sustainable end-of-life waste management options for e-waste products like mobile phones. There is also a study [14] that discusses appropriate end-of-life management strategies based on scenario analysis. This study is also driven by a similar motivation and takes a different approach. A simple sensitivity analysis is chosen to analyze the illustrated model.

The mobile phone is one of the most important e-waste products in terms of its recycling potential given the amount of critical metals they contain [5]. If all Japanese used mobile phones were collected and recycled, 2%–3% of (Japanese) annual consumption of gold, silver and palladium could be reduced [5]. This amount is not very large, but should be considered in the context of potential economic and political changes in resource-exporting countries. It is well-known that there are behavioral challenges with the collection of used mobile phones via recycling bins. Mishima (2015) [7] noted the potential for financial incentives (reward money) in encouraging the placement of used mobile phones in recycling bins. However, this same research also suggested that a financial incentive for encouraging the placement of the used mobile phone in the recycling bin was unlikely to encourage second hand product use [6]. Mishima (2015) suggested that the total residual value of recoverable materials in a used mobile phone was only around 100 JPY [15]. In order to build a sustainable recycling system for mobile phones, it is considered necessary to recover larger residual values from used products. Since resource recovery and efficiency are increasingly becoming an economic imperative for industrialized countries [16], producers are increasingly having to consider how to reduce the usage of virgin materials and the recovery of waste materials in their production activities. The objective of this study is to investigate the current material flow of used mobile phones to provide an enhanced understanding of the potential EOL waste management options for mobile phones in Japan.

2. Material Flow Modeling of Used Mobile Phone

2.1. Collection of Used Mobile Phone

This paper investigates the material flow of used mobile phones currently in Japan. The collection of used mobile phones in Japan has been previously investigated by the MRN (mobile phone recycle network) [17]. This organization is a collaboration between private enterprises and collects used phones for material recycling. Figure 1. illustrates the material flow model of used mobile phones in Japan in 2010, focusing on the primary organizations collecting the phones. Except for illegal dumping and hibernation, the three main collectors of used mobile phones are the MRN, other similar collectors and local government. Only approximately 20% of end-of-life mobile phones in Japan are collected by the MRN. The remaining 80% are either "hibernated" or collected by other organizations. Once the used phones are collected by other organizations, it is difficult to account for their material flows. For the purposes of this research, the MRN process is assumed to provide an indicative value for the enhanced recycling of mobile phones in Japan, assuming that recovery rates are able to be significantly increased. It is then considered possible to extrapolate the potential for improving end-of-life strategies of used mobile phones in Japan in both reducing the mobile phones sent to landfill and in improving both their domestic recycling and export potential.

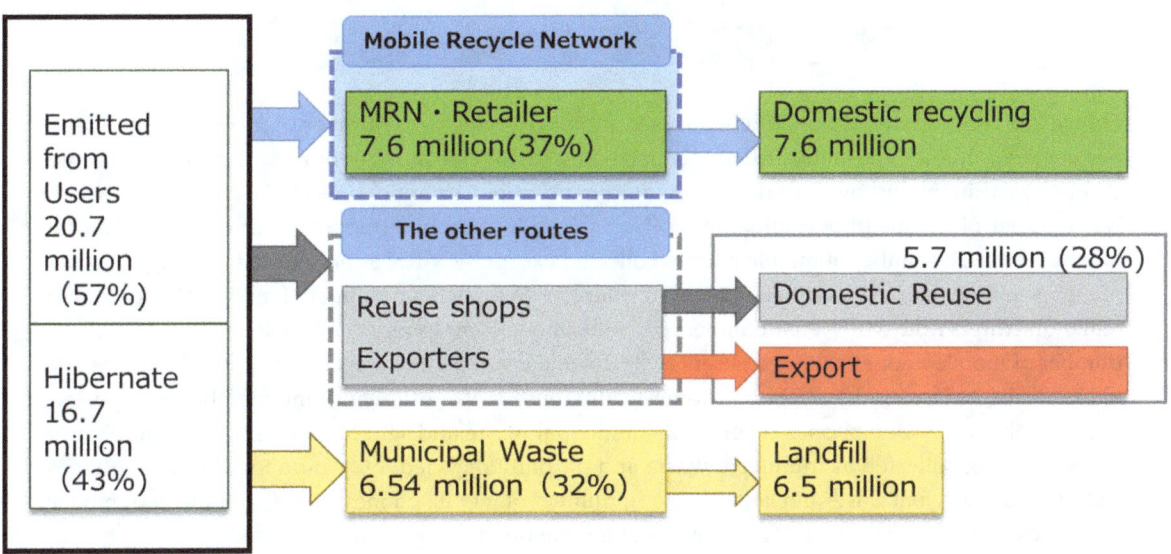

Figure 1. Overall flow of the used mobile phones in Japan in 2010 [18].

As is shown in Figure 1, 28% of mobile phones will be reused in domestic or international markets, 37% will be recycled through the MRN and 32% end up in landfill. As discussed previously, mobile phones contain critical precious metals, and some rare earth materials. These materials are completely wasted if they are landfilled as municipal waste. Currently, therefore, the MRN has a key role to play in the recycling of mobile phones in Japan, as it is the only direct route to recycling.

2.2. Detailed Material Flow Model

The MRN data set, however, is not able to confirm how long the mobile phones are in use or the final destination of the used mobile phones after the first collection by MRN and others. It should also be noted that some consumers may try to place their products in recycling bins sometime after having stopped using the mobile once they have salvaged the phones necessary data, in which case hibernation may not be the final destination. The actual material flows of mobile phones can, as a result, be quite complex. Figure 2 is a material flow model of used mobile phones that includes all possible routes after use. Some routes might not be currently feasible, but are considered in order to define the full potential for recycling routes available for mobile phones. The number of each route is defined as the following.

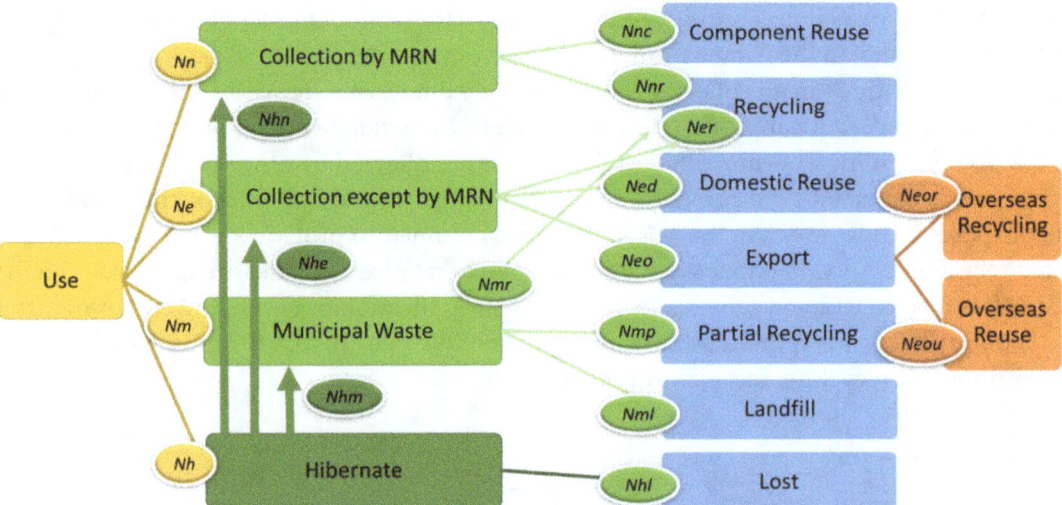

Figure 2. Precise material flow model of used mobile phones. U_y: number of mobile phones used per year; N_n: number of mobile phones collected by MRN; N_e: number of mobile phones collected except by MRN (including undertake); N_m: number of mobile phones collected as municipal waste; N_h: number of mobile phones hibernated; N_{hn}: number of mobile phones collected by MRN after hibernation; N_{he}: number of mobile phones collected except by MRN after hibernation; N_{hm}: number of mobile phones collected as municipal waste after hibernation; N_{nc}: number of mobile phones sent to remanufacturers from MRN; N_{nr}: number of mobile phones sent to recyclers(smelters) from MRN; N_{er}: number of mobile phones collected except by MRN and domestically recycled; N_{ed}: number of mobile phones collected except by MRN and domestically reused; N_{eo}: number of mobile phones exported; N_{mr}: number of mobile phones collected as municipal waste and sent to recyclers; N_{mp}: number of mobile phones collected as municipal waste and partially recycled; N_{ml}: number of mobile phones collected as municipal waste and landfilled; N_{hl}: number of mobile phones lost; N_{eor}: number of mobile phones exported and recycled; N_{eou}: number of mobile phones exported and reused.

2.3. Quantification of the Flow

The reuse of mobile phones extends product life and reduces material consumption in both domestic and international mobile phone production. The potential downside to these recycling efforts include the fact that harmless and efficient recycling processes have not been well established in many markets for used mobile phones and the outflow of used electronic products from Japan does result in a need to replace the equivalent amount of materials for new mobile phone production. To quantify Figure 2, surveys [9,17–19] of end-of-life situations of Japanese mobile phones were referred. Assumptions 1 to 7, made in relation to consumer recycling behavior [20], are also included.

1. The consumer will return their used mobile phones to MRN, when they purchase new phones. Thus, N_{hn} is negligible.
2. Once the used mobile phones have been hibernated, independent collectors do not collect (purchase) these products. Thus, N_{he} is negligible.
3. The amount of flow from non-MRN agencies to recyclers is considered negligible, since non-reusable products will not be undertaken or purchased. Thus, N_{er} is negligible.
4. Some products will be sent to recyclers from municipal waste. The total of the three major options is 20.2 million units. The difference between this number and the total number of used mobiles is the number of independently recycled mobiles.
5. "Lost" can be assumed that the used products are dumped unconsciously. Thus, N_{hl} should be added to the numbers of municipal waste.

6. The difference between the number collected by MRN and the number sent to recyclers is equivalent to the number sent from municipal waste to recyclers (N_{mr}).

7. Used mobile phones collected except by MRN will be domestically reused or exported.

An additional survey [21] of the policies of local government in the handling of small-sized e-waste products highlighted that among 25 Local governments, only 11 of them recover limited materials from municipal waste and put products/materials into the recycling process. When a recycling process of small-sized e-waste depends on hand-pick from waste and recovers only iron and copper, it is assumed that very limited numbers of mobile phones are recycled. Finally, Table 1 reflects all the investigated and assumed data regarding the material flow model of used mobile phones in Japan. The numbers of the phones available in the previous surveys are shown in the table with reference numbers. The other numbers were decided by the above-mentioned assumptions and calculations.

Table 1. Estimated quantitative data of Mobile phone market in Japan [17–19].

Variables	Meaning of the Variables	Amount (Million Units)
N_n	number of mobile phones collected by MRN	6.97 [17]
N_e	number of mobile phones collected except by MRN (including undertake)	5.72 [18]
N_m	number of mobile phones collected as municipal waste	7.21 (calculated)
Nl	Number of mobile phone landfilled	6.54 [18]
N_{hn}	number of mobile phones collected by MRN after hibernation	0 (assumed)
N_{he}	number of mobile phones collected except by MRN after hibernation	0 (assumed)
N_{hm}	number of mobile phones collected as municipal waste after hibernation	0.13 [19]
N_{nc}	number of mobile phones sent to remanufacturers from MRN	0.02 (calculated)
N_{nr}	number of mobile phones sent to recyclers(smelters)	7.60 [18]
N_{er}	number of mobile phones collected except by MRN and domestically recycled	0 (assumed)
N_{ed}	number of mobile phones collected except by MRN and domestically reused	0.14 [18]
N_{eo}	number of mobile phones exported	5.58 (calculated)
N_{mr}	number of mobile phones collected as municipal waste and sent to recyclers	0.65 (calculated)
N_{mp}	number of mobile phones collected as municipal waste and partially recycled	0.15 (calculated)
N_{ml}	number of mobile phones collected as municipal waste and landfilled	6.54 [18]
N_{eor}	number of mobile phones exported and recycled	5.58 in total (calculated)
N_{eou}	number of mobile phones exported and reused	

In the next section, we evaluate the different end-of-life options for mobile phone recycling in terms of reducing the material consumption and environmental impacts associated with their production.

3. Quantitative Analysis of the Material Flow Model

3.1. Method of the Analysis

The purpose of this analysis is to clarify which end-of-life option for used mobile phones results in the best way to reduce the material consumption or environmental impacts associated with mobile phone production depending on the different material flows of the recycling options considered. A coefficient of partial differentiation is used to show how sensitive the final output (amount of material consumption) is to the change of parameters in the material flow model. In clarifying what are the most efficient recycling strategies in detail, including from "increasing the MRN collection rate" to "increasing the domestic reuse rate". Based on the material flow model shown in Figure 2, the amount of material consumption involved in new mobile phone production can be estimated. Whilst the actual material composition of the mobile phones is not considered, differing values are given to the recycling rates assuming a certain percentage in weight of the product will be recycled.

3.2. Quantification of Matetrial Usage

If we consider the annual production of mobile phones increases 5% per year [22] and the average use life is about three years [22], the number of mobile phones in the market after one generation of product life can be expressed by Equation (1). Equation (2) represents the total material consumption in mobile phone production. Obviously, total mobile production can be increased by reducing the proportion of mobiles recycled. It is also assumed that not every phone can be reused and not all of the mobile phone material by weight can be recycled. A "recycling rate" (R_r) has been defined. It is also assumed that not every mobile phone component can be reused [23]. A reuse rate for these components is defined R_c, with the balance of component materials put into the material recycling process. Used products sent to municipal waste landfill are also assumed to include some limited recycling potential for the recovery of iron and copper. This landfill recycling process rate is defined as R_{pr}.

As the majority of mobile phones used in the Japanese market are imported, the domestic reuse and domestic material recycling rates have an effect on reducing mobile imports. This analysis considers this effect is equivalent to a direct reduction of the total number of mobiles produced. As a result, the annual number of production is expressed by Equation (3).

$$U_{y+3} = U_y \cdot PI^3 \tag{1}$$

$$W_{y+3} = N_{y+3} \cdot M \tag{2}$$

$$N_{y+3} = (U_{y+3} - N_{ed} - N_{nc} \cdot R_c - (N_{nr} + N_{er} + N_{mr}) \cdot R_r - N_{mp} \cdot R_{pr}) \tag{3}$$

U_y: number of mobile phones put into the market in year y [18]
PI: production increase rate [22]
W_y: weight of material used for mobile phone production in year y
N_y: number of annual production of year y
M: average weight of a mobile phone
R_r: material recycling rate
R_c: component reusable rate
R_{pr}: partial recycling rate in municipal waste

There are also some end-of-life options for consumers that need to be considered. To express the different consumers' choices available, three independent parameters, the "collection rate by MRN; R_n," the "collection rate except by MRN; R_e," and the "municipal waste rate; R_m" were defined. Each rate is calculated by dividing the corresponding numbers by the annual output. These consumers' choices are, theoretically, available after the used mobile phones are hibernated. The rate of mobile phones collected by MRN after hibernation, the rate collected by non-MRN agencies after hibernation and the rate dumped as municipal waste to landfill are defined as R_{hn}, R_{he} and R_{hm}, respectively.

The practical material flow data show most of the used mobile phones collected by MRN are sent to recyclers directly. However, some can be used for component reuse. The weight ratio of recyclable materials and recoverable components are defined as R_{nr} and R_{nc}, respectively.

It is also assumed that a certain proportion of collected mobile phones by non-MRN agencies are domestically reused with the remainder exported. We define the ratio of domestically reused mobile phones by non-MRN agencies as R_{ed}. It is expected that some mobiles are sent to recyclers, as not all products will actually be reused. Thus, the ratio of the used products sent to recyclers from non-MRN agencies (independent collectors) is R_{er}.

After collection as municipal waste, some used mobile phones are recovered from the waste and sent to recyclers. Some are put into low-tech recycling processes to recover Iron and Copper. The ratio of those sent to recyclers is defined as R_{mr}, and the ratio sent for iron/copper recovery is defined as R_{mp}. Using these parameters, and Equations (1) and (2), Equation (3) can be re-written as Equation (4).

The ratios expressing the first end-of-life options such as R_n, is calculated by dividing the corresponding the number N_n by the total number U_y. Those expressing the second end-of-life option such as R_{hn} is expressing the relevant ratio with the previous option. Using these ratios, the numbers in Equation (3) can be expressed as Equations (4) to (9).

$$N_{ed} = (R_e + R_{he}) \cdot R_{ed} \cdot U_y \tag{4}$$

$$N_{nc} = (R_n + R_{hn}) \cdot R_{nc} \cdot U_y \tag{5}$$

$$N_{nr} = (R_n + R_{hn}) \cdot R_{nr} \cdot U_y \tag{6}$$

$$N_{er} = (R_e + R_{he}) \cdot R_{er} \cdot U_y \tag{7}$$

$$N_{mr} = (R_m + R_{hm}) \cdot R_{mr} \cdot U_y \tag{8}$$

$$N_{mp} = (R_m + R_{hm}) \cdot R_{mp} \cdot U_y \tag{9}$$

By rewriting Equation (3) using Equations (1), (2), (4) to (9), Equation (10) can be obtained. The list of the defined ratios is shown below. Equation (10) represents how each rate of recycling affects the total amount of material usage in mobile production.

$$W_{y+3}/(M \cdot U_y) = PI^3 - (R_e + R_{he}) \cdot R_{ed} - (R_n + R_{hn}) \cdot R_{nc} \cdot R_c$$
$$-(R_n + R_{hn}) \cdot R_{nr} \cdot R_r - (R_e + R_{he}) \cdot R_{er} R_r - (R_m + R_{hm}) \cdot R_{mr} \cdot R_r - (R_m + R_{hm}) \cdot R_{mp} \cdot R_{pr} \tag{10}$$

R_n: collection rate by MRN

R_e: collection rate except by MRN

R_m: municipal waste rate

R_{nr}: sent rate for material recycling from MRN

R_{nc}: sent rate for component reuse from MRN

R_{hn}: collection rate by MRN after hibernation

R_{he}: collection rate except by MRN after hibernation

R_{hm}: collection rate as municipal waste after hibernation

R_{mr}: rate of used products recovered from municipal waste and sent to recycler

R_{mp}: rate of used products put into a low-tech recycling process

3.3. Calculation of Matetrial Usage Using Sensitivity Analysis

Each usage rate was defined in Table 2 as the number divided by the total number of mobile phones in each recycling market. The current values of the flow rates are calculated in Table 2. The other ratios are negligible. Based on these values, a sensitivity analysis is carried out to determine the importance of each recycling parameter in reducing material usage in total mobile phone production. By calculating the partial differential of each parameter in Equation (10), it is possible to determine the value of each parameter in improving the end-of-life options for used mobile phones. In these equations, parameters are set to the values shown in Table 2. R_r, R_c and R_{pr} are set by previous surveys [9,17,24] and the other ratios were calculated. This analysis helps to suggest the best routes to resource efficiency in examining EOL options for mobile phones in Japan. In order to clarify the properness of the end-of-life options for consumers, two new ratios, R_{comp} and R_{reuse}, were introduced. These two ratios express the overall ratios of domestic reuse and component reuse, respectively, and are expressed by the multiple of R_n and R_{nc}, and of R_e and R_{ed}, respectively.

Table 2. Current value of the flow rates.

Flow Rates	Current Value
R_n	0.186
R_e	0.153
R_m	0.193
R_{nr}	0.997
R_{nc}	0.003
R_{ed}	0.025
R_{eor}	0.975 (in total)
R_{eou}	
R_{mr}	0.089
R_{mp}	0.020
R_{ml}	0.891
R_{hm}	0.008
R_r	0.7 [17]
R_c	0.3 [24]
R_{pr}	0.1 [9]

3.4. Calculation of CO$_2$ Emission

As was mentioned in the beginning, not only material consumption but also CO$_2$ emission is an important aspect to determine end-of-life strategies. This study also investigated the CO$_2$ emissions associated with the material flow of each recycling strategy. Table 3 shows the variables defined in the previous study by Sugiyama *et al.* (2015) [25] representing the CO$_2$ emissions associated with each activity in the material flow model. Here, the environmental impacts of other activities such as the refurbishment for reuse and disassembly for component reuse, are assumed to be negligible. Using these variables, the illustrated material flow model shown in Figure 2 and Table 1, is expressed in Equation (11) in terms of their environmental impact. It is assumed that transportation distances are negligible for domestic reuse and component reuse.

Table 3. Defined variables for environmental impact.

Variable	Corresponding Activity
E_{total}	Total
E_p	Production
E_u	Use
E_{td}	Domestic transportation
E_{to}	Overseas transportation
E_{rd}	Domestic recycling
E_{ro}	Overseas recycling
E_l	Landfill

$$
\begin{aligned}
E_{total} = {} & E_p \cdot (U_{y+3} - (R_e + R_{he}) \cdot R_{ed} \cdot U_y - (R_n + R_{hn}) \cdot R_{nc} \cdot R_c \cdot U_y - \\
& ((R_n + R_{hn}) \cdot R_{nr} + (R_e + R_{he}) \cdot R_{er} + (R_m + R_{hm}) \cdot R_{mr})) \cdot R_r \cdot U_y - \\
& (R_m + R_{hm}) \cdot R_{mp} \cdot R_{pr}) + E_u \cdot U_{y+3} + E_{td} \cdot ((R_n + R_{hn}) \cdot R_{nr} + (R_e + \\
& R_{he}) \cdot R_{er} + (R_m + R_{hm}) \cdot R_{mr}) \cdot U_y + E_{to} \cdot (R_e + R_{he}) \cdot (1 - R_{er} - R_{ed}) \cdot \\
& U_y + E_{rd} \cdot ((R_n + R_{hn}) \cdot R_{nr} + (R_e + R_{he}) \cdot R_{er} + (R_m + R_{hm}) \cdot R_{mr}) \cdot U_y \\
& + E_{ro} \cdot (R_e + R_{he}) \cdot (1 - R_{er} - R_{ed}) \cdot U_y + E_l \cdot (R_m + R_{hm}) \cdot R_{ml} \cdot U_y
\end{aligned}
\tag{11}
$$

4. Results and Discussion

4.1. Calculation of Material Reduction

Partial differential equations were calculated for seven parameters (R_n, R_e, R_m, R_{he}, R_{comp}, R_{reuse}, and R_r) in Equation (10). In order to calculate the sensitivity of material consumption by changing "collection rate by MRN" (R_n), partial differential by R_n is applied to Equation (10). Then, the effect of the "collection rate by MRN" (R_n) (Equation (12)) is applied. Equations for the other six parameters are shown as Equations (13) to (18). Equation (13) is the partial differential of Equation (10) by R_e, and so on. Figure 3 is the result of the calculation of seven equations using the current rates shown in Table 3. The bar graph shows the relative effectiveness of increasing each parameter in reducing overall material consumption in mobile phone usage. The effects in reducing material consumptions are shown as positive values.

$$\frac{\partial(W_{y+3}/(U_y \cdot M))}{\partial R_n} = -R_{nr} \cdot R_r - R_{nc} \cdot R_c \tag{12}$$

$$\frac{\partial(W_{y+3}/(U_y \cdot M))}{\partial R_e} = -R_{ed} \tag{13}$$

$$\frac{\partial(W_{y+3}/(U_y \cdot M))}{\partial R_m} = -R_{mr} \cdot R_r - R_{mp} \cdot R_{pr} \tag{14}$$

$$\frac{\partial(W_{y+3}/(U_y \cdot M))}{\partial R_{hn}} = -R_{nr} \cdot R_r - R_{nc} \cdot R_c \tag{15}$$

$$\frac{\partial(W_{y+3}/(U_y \cdot M))}{\partial R_{comp}} = -R_c \tag{16}$$

$$\frac{\partial(W_{y+3}/(U_y \cdot M))}{\partial R_{reuse}} = -1 \tag{17}$$

$$\frac{\partial(W_{y+3}/(U_y \cdot M))}{\partial R_r} = -R_n \cdot R_{nr} - R_m \cdot R_{mr} \tag{18}$$

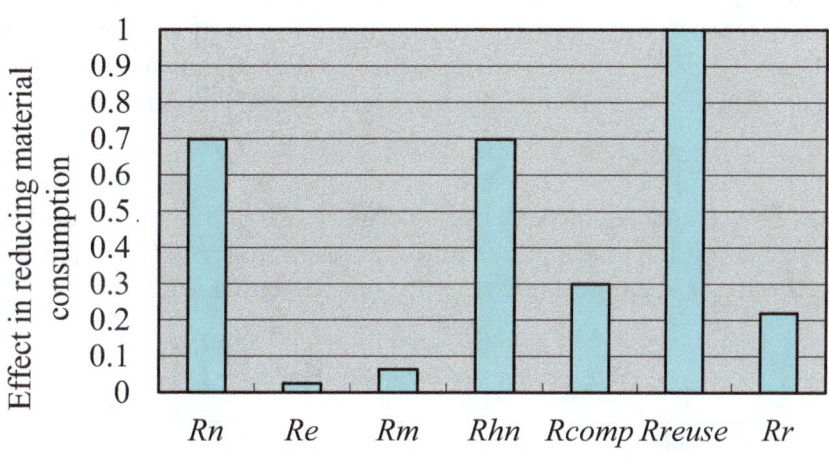

Figure 3. Effects of end-of-life options on material reduction.

4.2. Discussion on Material Reduction

Based on the above analysis of e-waste flow from mobile phone recycling, it is suggested that promoting domestic reuse is the best strategy to reduce the overall material consumption associated with mobile phone usage in Japan, as domestic reuse has a direct effect on reducing the material consumption associated with new production. Material recycling via the official mobile phone recycling network is the second best strategy to reduce mobile phone related resource consumption. As is shown in Equation (14), the effect of increasing collection rate by MRN after hibernation (R_{hn}) is the same as that of collection rate by MRN (R_n). Thus, for material reduction, it is equally good to promote collection of hibernated phones through MRN. In this calculation, the material recycling rate was set at 0.7 from the previous survey [17]. If the material recycling rate is higher than this, then material recycling by the MRN can be a stronger strategy and closer in effectiveness to the domestic reuse option. If the recycling rate is much lower than this value, then product reuse remains the best overall strategy. However, if there is no assurance that the products will be domestically reused, then component reuse could be a better strategy. Based on current end-of-life practices for used mobile phones, strategies to reduce the material consumption associated with the production of mobile phones are suggested below. Whilst the first three strategies are the strongest, 4, 5, 6 and 7 are still potentially viable dependent on individual market, social and technical conditions.

1.   Collection of used mobile phones for secondhand use in domestic market.
2.   Collection of used mobile phones by the MRN.
3.   Collection of hibernated phones by MRN.
4.   Collection of used mobile phones for component reuse.
5.   Technological development to increase material recycling rate.
6.   Collection of used mobile phones for municipal waste landfill.
7.   Collection of used mobile phones by non-MRN agencies for any purpose.

Although the technological development to increase the material recycling rate is important, such efforts may not be very effective under current collection rates. Efforts to establish an enhanced system to encourage consumers to put their products in recycling bins is considered very important. Such efforts should be considered first. The results also show that even if the collected mobile phones are dumped in landfill, this option is still better than a mobile phone "lost" in hibernation, since some limited material recycling can be applied to municipal waste. However, this research highlights that proper recycling strategies, such as those involving a network like the MRN, provide much better outcomes in terms of resource efficiency. Developing programs to encourage consumers to increase their efforts to place potentially recyclable manufactured e-waste products like mobile phones in the recycling system, is obviously a clear winner in terms of sustainability outcomes.

4.3. Calculation of CO_2 Reduction

Values for the environmental impact associated with each of the activities are noted in Table 3 and can be estimated according to a previous study by Takeshima *et al.* (2006) [26] and an inventory database by the Ministry of Environment (2015) [27], as noted as Table 4.

Table 4. Estimated environmental impact.

Variable	Environmental Impact (kg-CO_2/unit)
E_p	43
E_u	1
E_{td}	negligible
E_{to}	0.5
E_{rd}	1.8
E_{ro}	3.0
E_l	2.7

Using the values shown in Table 4, it is possible to calculate the environmental impacts associated with each end-of-life option. Again, by applying partial differential equations to Equation (11) and the aforementioned seven recycling options, Equations (19) to (25) can be obtained. Using the current value of each ratio indicated in Table 2, the values of the equations are calculated. The results presented in Figure 4 show the effects of increasing the seven parameters on reducing CO_2 emissions. Again, R_{comp} is the multiple of R_n and R_{nc}, and R_{reuse} is the multiple of R_e and R_{ed}.

$$\frac{\partial(E_{total}/U_y)}{\partial R_n} = -E_p(R_{nr} \cdot R_r + R_{nc} \cdot R_c) + (E_{td} + E_{rd}) \cdot R_{nr} \tag{19}$$

$$\frac{\partial(E_{total}/U_y)}{\partial R_e} = -E_p \cdot R_{ed} + E_{td} \cdot (R_{ed} + R_{er}) + E_{to}(1 - R_{er} - R_{ed}) \\ + E_{rd} \cdot R_{er} + E_{ro}(1 - R_{er} - R_{ed}) \tag{20}$$

$$\frac{\partial(E_{total}/U_y)}{\partial R_m} = -E_p(R_{mr} \cdot R_r + R_{mp} \cdot R_{pr}) + (E_{td} + E_{rd}) \cdot R_{mr} + E_l \cdot R_{ml} \tag{21}$$

$$\frac{\partial(E_{total}/U_y)}{\partial R_{hn}} = -E_p(R_{nr} \cdot R_r + R_{nc} \cdot R_c) + (E_{td} + E_{rd}) \cdot R_{nr} \tag{22}$$

$$\frac{\partial(E_{total}/U_y)}{\partial R_{comp}} = -E_p \cdot R_c \tag{23}$$

$$\frac{\partial(E_l^{total} U_y)}{\partial R_{reuse}} = -E_p - E_{to} - E_{ro} \tag{24}$$

$$\frac{\partial E_{total}}{\partial R_r} = -E_p((R_n + R_{hn}) \cdot R_{nr} + (R_e + R_{he}) \cdot R_{er} + (R_m + R_{hm}) \cdot R_{mr}) \tag{25}$$

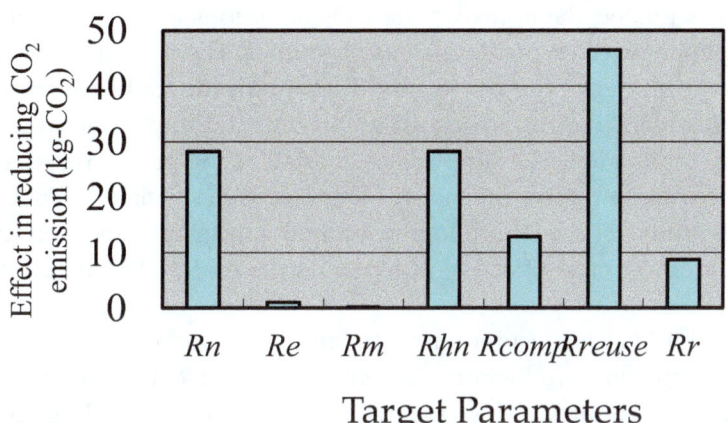

Figure 4. Effects of end-of-life options on CO_2 reduction.

4.4. Discussion on CO_2 Reduction

As shown in Figure 4, the general tendency of each parameter to reduce CO_2 emissions are very similar to the results presented in Figure 3. A good strategy to reduce material consumption is also a good to strategy to reduce the CO_2 emissions associated with reducing the material consumption associated with mobile phone usage. The best strategies are still to increase the domestic reuse rate and enhance material recycling. However, if the purchased product will be reused overseas, the product reuse strategy is not a very good option in terms of improving material consumption efficiency or reducing CO_2 emissions. In terms of environmental impact, increasing the material recycling rate is better than increasing the collection rate except when the mobiles are collected by the MRN. Although domestic reuse is a very good end-of-life option, transboundary movements of mobile phones can be a problem. Considering the risk of outflow, an appropriate strategy is to promote material recycling through public recycling networks.

5. Conclusions

Based on a survey of the current end-of-life usage of mobile phones in Japan, a material flow model was used to examine the best strategies for reducing the material consumption associated with mobile phone usage. Since some values were unknown, they were estimated on the basis of qualitative considerations of consumer behavior noted in previous research. Parameters to indicate the rates of different material flows were defined. An equation to calculate the material consumption associated with the production of mobile phones was developed. Since a large proportion of mobile phones are imported into Japan, recycled materials are not directly used for mobile phone production in Japan. However, the overall effect is to both reduce the material consumption associated with mobile phone usage and to extend the life of mobile phones being used. While this research uses a simple assumption that recycled materials contribute to reduce material consumption, this nexus is considered a fundamental parameter in resource efficiency.

In order to determine the relative value of seven different recycling options, "collection rate by the MRN", "collection rate of non-MRN agencies", "municipal waste recycling rate", "collection rate by the MRN after hibernation", "domestic reuse rate", "component reuse rate" and "recycling rate", a sensitivity analysis was carried out.

The results suggest that domestic product reuse and a proper recycling system such as the MRN network are the two most effective strategies in terms of reducing the material consumption associated with mobile phone usage in Japan. It also suggested that efforts to increase the recycling rates of mobile phones given current material flows will require increasing the collection rate of used mobile phones from consumers.

This study also estimated the approximate CO_2 emissions associated with the material flow lifecycle of a mobile phone based on previous research studies. These results were similar to the results determined for resource efficiency improvements. The significance of the domestic reuse option was also made clearer when considering associated environmental impact options. Hibernation as an end-of-life option is to be avoided. The significance of establishing proper recycled material collection systems to enhance resource efficiency in end-of-life production systems for mobile phones has been clearly shown in this study. The methodology used in this research has been helpful in indicating which end-of-life strategy is most effective in reducing the material consumption associated with mobile phone usage.

Future research should also consider other strategies, such as prolonging product life, reducing material intensity and reducing product obsolescence timeframes. In addition, it would be useful to apply this method to other countries, since the method is based on modeling and can be carried out even if precise material flow data are lacking. A public organization should be involved in determining end-of-life strategies.

Author Contributions: Kuniko Mishima and Nozomu Mishima formalized the equations of the material flow and carried out the sensitivity analysis of the equation. Michele Rosano reviewed the paper structure, checked English expression and analyzed the calculation results. Nozomu Mishima provided the basic idea to apply sensitivity analysis to the material flow. Hidekazu Nishimura modified the material flow equation to be analyzed and the partial derivatives. Kuniko Mishima wrote the paper.

References

1.   Nanjyo, M. Urban Mine, New Resource for the Year 2000 and Beyond. Bull. Res. *Inst. Miner. Dress. Metall.* **1998**, *43*, 239–251.

2.   Shiratori, T.; Nakamura, T. Concept of "Artificial Deposit" 2—Transition of the metal potential of spent electric and electronic appliances. *J. MMIJ* **2007**, *4*, 171–178. [CrossRef]

3.   Halada, K.; Ijima, K.; Shimada, M.; Katagiri, N. A possibility of urban mining in Japan. *J. Jpn. Inst. Metals* **2009**, *73*, 151–160. [CrossRef]

4. Yamane, E.; Minamoto, R.; Numata, T.; Nakajima, K.; Murakami, S.; Daigo, I.; Hashimoto, S.; Okumura, H.; Ishihara, K. Novel Evaluation Method of Elemental Recyclability from Urban Mine—Concept of Urban Ore TMR. *Mater. Trans.* **2009**, *50*, 1536–1540.

5. Report of Social Experiment on Promoting Collection of Used Mobile Phones as of Fiscal Year 2009. Available online: http://www.meti.go.jp/meti_lib/report/2010fy01/0020863.pdf (accessed on 20 January 2016). (In Japanese)

6. Murakami, S.; Ohsugi, H.; Murakami-Suzuki, R.; Mukaida, A.; Tsujimura, H. Average lifespan of mobile phones and in-use and hibernating stocks in Japan. *J. Life Cycle Assess. Jpn.* **2009**, *5*, 139–145. (In Japanese) [CrossRef]

7. Mishima, K.; Nishimura, H. Requirement Analysis to Promote Small-sized e-waste Collection from Consumers. *Waste Manag. Res.* **2015**. in printing. [CrossRef] [PubMed]

8. Yoshida, A.; Tasaki, T.; Terazono, A. Material flow analysis of used personal computers in Japan. *Waste Manag.* **2009**, *29*, 1602–1614. [CrossRef] [PubMed]

9. Murakami, S. Flow Scenarios and Resource Potentials for End-of-life Electric Appliances. *Mater. Cycles Waste Manag. Res.* **2009**, *20*, 237–244. [CrossRef]

10. Ministry of Environment, Ministry of Economy, Trade and Industry. Report of Committee for Recycling Systems and Re-Utilization of Valuable Metals in Small-Sized e-Waste. Available online: http://www.env.go.jp/council/former2013/03haiki/yoshi03-24.html (accessed on 20 January 2016). (In Japanese)

11. Case Study on Critical Metals in Mobile Phones Final Report, OECD. 2011. Available online: http://www.oecd.org/env/waste/Case%20Study%20on%20Critical%20Metals%20in%20Mobile%20Phones.pdf (accessed on 20 January 2016).

12. Soo, V.K.; Doolan, M. Recycling Mobile Phone Impact on Life Cycle Assessment. *Procedia CIRP* **2014**, *15*, 263–271. [CrossRef]

13. Duygan, M.; Meylan, G. Strategic management of WEEE in Switzerland—Combining material flow analysis with structural analysis. *Resour. Conserv. Recycl.* **2015**, *103*, 98–109. [CrossRef]

14. Neira, J.; Favret, L.; Fuji, M.; Miller, R.; Mahdavi, S.; Blass, V.D. End-of-Life Management of Cell Phones in the United States, Master's of Environmental Science and Management for the Donald Bren School of Environmental Science and Management. Available online: http://www.bren.ucsb.edu/research/documents/cellphonethesis.pdf (accessed on 20 January 2016).

15. Takahashi, K.; Nakamura, J.; Otabe, K.; Tsuruoka, M.; Matsuno, Y.; Adachi, Y. Resource Recovery from Mobile Phone and the Economic and Environmental Impact. *J. Jpn. Inst. Metals* **2009**, *73*, 747–751. [CrossRef]

16. G7 Summit Declaration. Available online: http://www.bundesregierung.de/Content/DE/_Anlagen/G8_G20/2015-06-08-g7-abschluss-eng.pdf?__blob=publicationFile&v=5 (accessed on 26 December 2015).

17. Web page of Mobile Recycle Network. Available online: http://www.mobile-recycle.net/ (accessed on 20 January 2016).

18. Web page of Ministry of Economy, Trade and Industry, Material Flow of Used Products. Available online: http://www.meti.go.jp/committee/summary/0003198/pdf/report01_02_02.pdf (accessed on 29 January 2016). (In Japanese)

19. Web page of Polis Department of Tokyo. Available online: http://www.keishicho.metro.tokyo.jp/toukei/bunsyo/toukei23/pdf/kt23d132.pdf (accessed on 20 January 2016).

20. Telecommunications Carriers Association, Recycling Statistics of Mobile Phones. Available online: http://www.tca.or.jp/press_release/pdf/150623.pdf (accessed 29 January 2016). (In Japanese).

21. Ministry of Economy, Trade and Industry, Case Studies of Small-Sized e-Waste Recycling in Local Governments. Available online: http://www.meti.go.jp/policy/recycle/main/admin_info/committee/o/27/hairi27_01-02.pdf (accessed on 29 January 2016). (In Japanese).

22. Internal Affairs and Communications. Available online: http://www.soumu.go.jp/main_content/000299758.pdf (accessed 26 December 2015). (In Japanese)

23. Repro Electric Co. Available online: http://www.reproele.jp/e/ (accessed 5 January 2016).

24. Yamaguchi, H.; Itsubo, N.; Tahara, K.; Inaba, A. Evaluation of CO_2 Emissions of Cellurar Phone Manufacturing. 2005. Available online: https://www.jstage.jst.go.jp/article/ilcaj/2005/0/2005_0_9/_pdf (accessed on 20 January 2016).

25. Sugiyama, K.; Honma, O.; Mishima, N. Quantitative Analysis of Material Flow of Used Mobile Phones in Japan. In Proceedings of the 13th GCSM, Binh-Duong, Vietnam, 16–18 September 2015.

26. Takeshima, A.; Fujinami, T.; Yamada, K.; Nakamura, R.; Yamaguchi, H.; Itsubo, T. LCA including the disposal and recycling stages of a mobile telephone. In Proceedings of the 2nd Conferene of the Institute of Life Cycle Assessment Japan, Tsukuba, Japan, 14–15 November 2006.

27. LCA Inventory Database by Ministry of Environment. Available online: http://www.env.go.jp/earth/ondanka/supply_chain/com04/ref02.pdf (accessed on 5 January 2016).

4

Do Eco-Fees Encourage Design for the Environment? The Relationship between Environmental Handling Fees and Recycling Rates for Printed Paper and Packaging

Calvin Lakhan

Department of Geography, Wilfrid Laurier University, Waterloo, Ontario, ON L6S2X5, Canada; lakh2440@mylaurier.ca

Academic Editor: Michele Rosano

Abstract: This study undertook a critical examination of Ontario's extended producer responsibility scheme for the residential "Blue Box" recycling program, specifically examining the relationship between packaging fee rates and material-specific recycling rates. Using data collected for each of the 23 materials found in the residential recycling program over the past decade, a regression model was developed to gauge what relationship (if any) packaging recycling rates have with fee rates, costs of material management and revenue from the sale of recyclable material. The modeling in this study indicates that packaging fee rates have no effect on packaging recycling rates. Recycling rates were positively correlated with material revenue and negatively correlated with material management costs. There is no evidence that suggests that Ontario's fee model used to allocate environmental handling fees to individual materials encourages waste diversion or design for the environment. The disconnect in the results and the intended function of packaging fee rates calls into question the appropriateness of Ontario's fee rate methodology.

Keywords: recycling; extended producer responsibility; fee model; diversion

1. Introduction

Extended producer responsibility (EPR) is becoming a favored public policy approach to manage post-consumption waste in most developed economies. Generally speaking, EPR shifts the financial (and sometimes physical) responsibility for the end-of-life (EOL) management of used packaging from consumers to the producer of the original packaging [1]. A producer, in this context, is commonly defined as the brand owner of a packaged product or the first person to import into a specific jurisdiction (typically the distributor or retailer who first receives the product in that jurisdiction). Although EPR policy formulation and programs for used packaging have existed for the better part of three decades, EPR thinking is still in its infancy: The ever-widening range of government initiatives, program implementation models and new enterprises forming in response to these changes highlight the relative immaturity of the field [2]. The primary reason for adopting EPR policies, to date, has been the relatively narrow issue of post-consumer waste management. While grounded within the broader sustainability framework, most program initiatives have focused on the collection and diversion of designated wastes from disposal, with increasing attention being paid to waste reduction and product/packaging design [3].

Ontario's transition to an EPR scheme in the early 2000s marked a shift in the cost of managing EOL products from the local tax base to packaging producers [4]. To date, Ontario's partial EPR scheme remains the foundation for managing and financing the provincial Blue Box program in Ontario; the distribution of recycling system costs is performance-based, with the fees paid by packaging producers being in direct proportion to their recycling performance. All other things being equal, materials with high rates of waste diversion will pay a smaller percentage of recycling system costs when compared to materials with low recycling rates. This is designed to incentivize producers to increase the relative recycling performance of their materials by investing in technologies, end markets and infrastructure to ensure the recyclability of their packaging. The underlying intuition behind Ontario's fee model is that materials who have their costs partially subsidized will want to increase their recycling rates such that they continue to transfer costs onto poorer performing materials. Conversely, materials who subsidize the costs of other materials will want to increase their recycling rates such that they can minimize the impact of the fee. Thus far, the effectiveness of this approach has yet to be evaluated. Policy decisions have been made predicated under the assumption that the fee methodology employed in Ontario improves recovery of household recyclables by encouraging packaging producers to make recyclable packaging choices. This study seeks to test this hypothesis by evaluating how Ontario's incentive-based EPR fee model has influenced recycling rate performance for printed paper and packaging materials. The main objective of this research is to examine whether fee incentives/disincentives for packaging producers increase recycling rates for packaging materials.

The analysis in this study builds upon the existing research, shifting the research focus away from individual consumers and households to packaging producers. To date, no study has evaluated how incentivization affects recycling behavior of packaging producers and brand owners. Doing so provides unique insights into the effectiveness of performance-based fees, particularly as Ontario's model of EPR funding spreads to other jurisdictions (both within Canada and abroad).

2. Literature Review

Extended producer responsibility for packaging waste, and, by proxy, packaging fees, has received relatively little attention from the research community; while studies on EPR do exist [5–7]), they have tended to focus on deposit return schemes for beverage containers, waste electronics and hazardous waste. Most of the research on packaging eco-fees in Canada has generally been grey literature—work carried out by consulting firms or local governments. Most of the information that is currently available has reflected either local circumstance that can differ substantially from one area to the next and/or has reflected a particular focus or interest of the author.

As noted by Mayers and Butler [8], the primary public policy arguments for implementing EPR for packaging include:

1. To transfer the costs of managing packaging waste from the local tax base to the producer and user of the product.
2. To provide a direct economic incentive for the producer of the package to reduce packaging materials and design packaging for improved recyclability.
3. As an initial step towards the development of a circular materials economy—where waste materials serve as feedstock for new processes (as opposed to the current norm: a linear extraction/production/consumption/disposal economic system).
4. To make the producer and consumer of the packaging fully responsible for the environmental impacts of it production, use and EOL management.

Notably absent in most EPR practices, to date, has been the ability to design and implement a program based upon a broader product and packaging lifecycle assessment [9]. This will likely change in the future to include considerations of greenhouse gas (GHGs) emissions, water impacts, hazardous materials and the use of renewable materials and renewable energy.

Most OECD (Organization for Economic Cooperation and Development) countries have adopted one or more of these EPR approaches for EOL product management and packaging. Approximately 61% of the OECD population currently has EPR policies for packaging in place. The significant majority of the remainder not covered is the United States (27%) [10]. Initial packaging EPR program models were predominately based upon the creation of a single national packaging compliance scheme [11].

More recent EPR policy trends have focused on assigning the legal responsibility for EOL management of packaging waste to individual producers, and on allowing each producer, operating individually (as part of a group or as a member of a producer responsibility organization), to discharge their legal obligation [12]. In most cases, EPR programs for non-packaging products and wastes require the producer to pay 100% of the program costs. Existing EPR programs for used packaging assign partial or full financial responsibility to producers, but there is a clear trend in Europe and Canada of assigning the full program costs to producers. Table 1 below summarizes EPR trends in European countries.

Table 1. European Union Programs Summary.

Approach	Countries	Trends
Producers pay 100% of costs	15	Move towards competing compliance schemes
Producers pay shared costs	10	Move to increasing industry cost share + costs of disposal for packaging not recycled
Tradable credits schemes	2	Provides only indirect price support for municipal recycling; focus on transport packaging
Packaging taxes	2	Add carbon costs as well as recycling costs; new government revenue source

The European Recovery & Recycling Organization (ERRA) launched packaging recycling pilots across Europe to demonstrate effective and efficient approaches to package recycling [13]. ERRA supported European-wide packaging legislation to stimulate wider adoption of packaging recycling schemes and minimize trade distortions in the common market. In Canada, major brand owners and grocery retailers have promoted voluntary approaches to recycling through Corporations Supporting Recycling for more than a decade. A lack of financial resources and the proliferation of provincial packaging regulations led CSR (Corporate Social Responsibility) to advocate for EPR programs for used packaging. In the United States, voluntary industry efforts have tended to be single-material or packaging-specific approaches, and the overall recycling rate for used packaging has remained relatively flat over the last decade.

To date, no study has examined whether packaging fees encourage packaging producers to select the most recyclable material. While environmental handling fees as a whole have been demonstrably successful in having companies internalize the costs of EOL waste management, the link between fees and product design for the environment remains poorly understood.

3. Materials and Methods

3.1. Description of Study Site

Ontario remains at the forefront of recycling initiatives and legislation in Canada, recognized as one of only three provinces in Canada to implement an extended producer responsibility scheme (EPR) for household recyclables. Residential and commercial waste diversion programs exist for MHSW (Material Hazardous or Special Waste), WEEE (Waste Electrical and Electronics Equipment), automobile tires, and printed paper and packaging (Blue Box) materials. Each of these programs exists under the oversight of Waste Diversion Ontario (WDO), a non-Crown corporation created under Ontario's 2002 Waste Diversion Act [14]. WDO was established to develop, implement and manage waste diversion programs for stakeholders from both private and public sectors [14].

Under provincial regulation O. Reg. 274/04, all producers of printed paper and packaging are required to pay a fee to finance the EOL management of material generated in the province. Producers are financially obligated to contribute 50% of reported municipal costs for the operation and maintenance of the Blue Box program.

Conversely, under provincial regulation O. Reg. 101/94, every municipality with a population of 5000 or more residents are obligated to operate a Blue Box program accepting at least five mandatory materials, plus three optional materials. A total of 23 packaging types have been classified as being eligible for inclusion in the Blue Box.

Data for Ontario's residential recycling system was obtained from the Waste Diversion Ontario municipal data call. Each year, WDO requests that every municipality within Ontario report detailed recycling and cost information regarding the management of their waste diversion programs [14]. Municipalities are required to log into the Waste Diversion Ontario web site and fill out an electronic questionnaire that solicits information that includes information on the amount of material recovered, the types of material recovered and the operating and capital costs associated with the management and collection of recyclables. All data used in this study pertains to printed paper and packaging recyclables found in the residential recycling stream, *i.e.*, newsprint, cardboard, glass, aluminum, steel, composite packaging and plastics.

3.2. Description of Stewardship Ontario Fee Model

The information collected by WDO is used to calculate material-specific costs by Stewardship Ontario using a "Pay-In Model" (PIM) [15]. This model allocates municipal recycling costs to individual materials using a three-step process.

These include:

1. Determine Blue Box Program Costs
2. Allocate Costs to Individual Materials
3. Determine Fee Rates

Each year, representatives from Stewardship Ontario, the Association of Municipalities of Ontario (AMO) and the City of Toronto meet to review the costs submitted by municipalities and together determine a "Best Practice" cost, which is used to negotiate producer obligations to municipalities for their share of the cost for running the Blue Box program. In 2014, the net cost for managing the residential Blue Box program was approximately $216 million [15]. These costs are allocated to individual materials based on activity-based costing principles and a distribution of common costs. These costs are distributed on the basis that a material-specific net cost reflects the costs of collecting, processing and providing administrative support for that material. The PIM then calculates material-specific fee rates for packaging producers using a three-factor formula based on the net cost of material management, material-specific recycling rates, and an equalization payment, where:

1. 20% of the cost of the program is assigned to each material category based on the net cost of managing each material in the system;
2. 55% of the cost of the program is assigned based on the recovery rate achieved by that material;
3. 25% of the cost of the program is assigned based on how much it would cost to manage the material if it were recovered at a rate of 60% (only applies to materials achieving less than 60% target rate) [16].

The objective of the fee setting process is to share the cost of achieving Ontario's diversion target among packaging producers for obligated materials [16]. The intuition behind the fee setting formula is to reward materials with high recovery rates while sharing the cost among all materials participating in the program. Stated alternatively, materials with lower recycling rates partially subsidize the cost of recycling materials with higher recycling rates.

For the purposes of this study, the PIM model was used to calculate material-specific generation, recovery and cost data.

Data used in this study pertains to packaging materials found in the residential recycling stream for printed paper and packaging materials. This includes the following materials:

- Newsprint
- Magazines and Catalogs

- Telephone Books
- Other Printed Paper (e.g., Office paper)
- Corrugated Cardboard
- Boxboard
- Gabletop Cartons (e.g., milk and orange juice containers)
- Aseptic Containers (e.g., juice boxes)
- Paper Laminants (e.g., coffee cups)
- PET (Polyethylene terephthalate) Bottles (e.g., water bottles)
- HDPE (High density polyethylene terephthalate) Bottles (e.g., laundry detergent)
- Plastic Film (e.g., grocery bags)
- Plastic Laminants (e.g., chip bags)
- Polystyrene
- Other Plastics (e.g., margarine tubs and lids)
- Steel Food and Beverage Cans
- Steel Aerosols
- Steel Paint Cans
- Aluminum Food and Beverage Cans
- Other Aluminum Packaging
- Clear Glass
- Colored Glass

Figures 1–3 compare the fee rates, net cost of material management and recycling rates for the full range of Blue Box materials. Net cost of material management is calculated by taking the gross cost of material management and subtracting revenue from the sale of marketed material. Revenue for each material is calculated using the twelve-month average of the spot price received from the sale of material by provincial municipalities. Recycling rates are calculated by dividing the total quantities of material recovered by the total quantities of material generated.

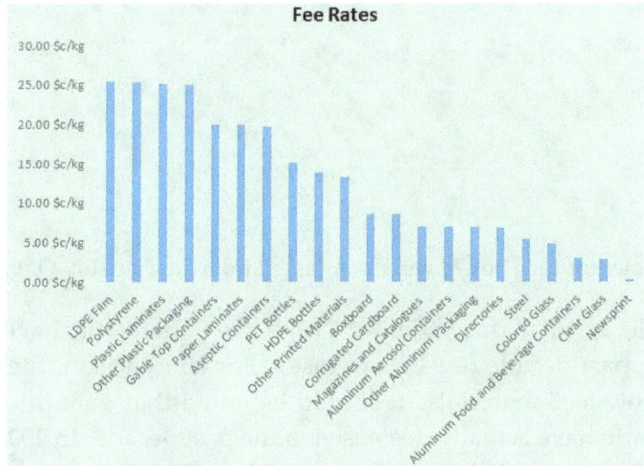

Figure 1. Fee Rates for Blue Box Materials (Adapted from Stewardship Ontario PIM Model [16]).

As shown in the above figures, on average, materials with the highest fee rates also have the highest costs of material management and lowest recycling rates. This is largely an expected result given the way in which fee rates are calculated (using the three factor formula described in Section 3.2).

Using historical data from the Stewardship Ontario PIM model, Table 2 below shows how generation of high fee rate (>10 c/kg) materials has changed over the past decade. For illustrative purposes, these figures are compared against how the generation of low fee (<10 c/kg) materials have changed during this same period.

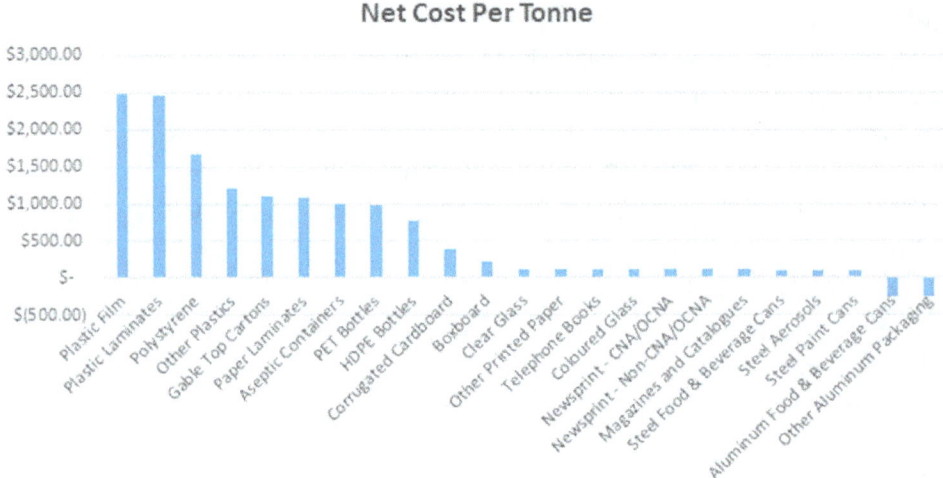

Figure 2. Net Cost Per Tonne for Blue Box Materials (Adapted from Stewardship Ontario PIM Model [16]).

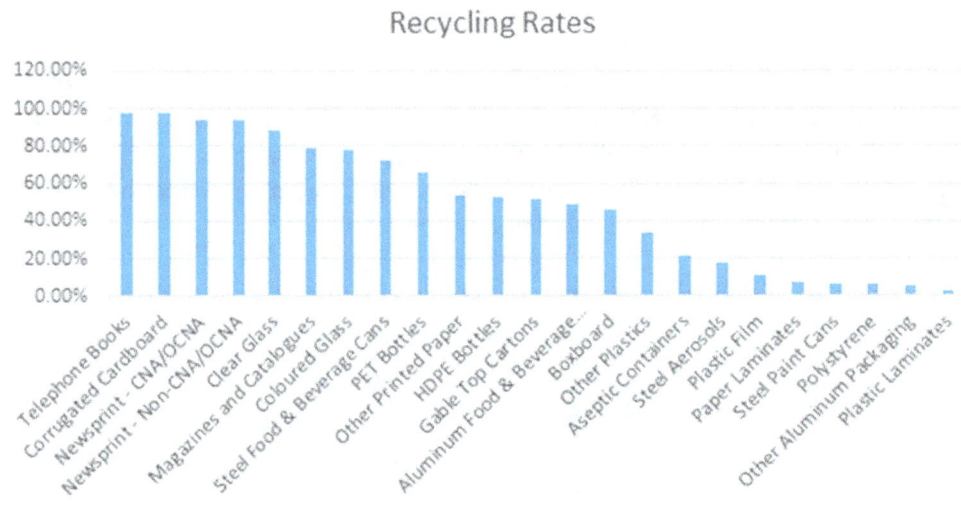

Figure 3. Recycling Rates for Blue Box Materials (Adapted from Stewardship Ontario PIM Model [16]).

As shown in Table 2, total quantities of "high fee" materials have increased by almost 151,000 tonnes over the past decade (a 94% increase). For comparison, there has been a decrease of 182,000 tonnes for "low fee" materials. It should be noted that quantities of overall household waste generation in Ontario have actually decreased in the past decade. In 2002, the average Ontarian generated 383 kg of waste per year. This is compared to 366 kg per capita/per year estimated by WDO in 2012 [17]. Some municipal officials have suggested that decreases in generation are not necessarily attributed to changes in household behavior, but due to the increasing shift towards light weight packaging by packaging producers. There is evidence to support these claims, as a review of steward sales data remitted to Stewardship Ontario indicates that the quantities of packaging waste sold into the market (expressed in terms of unit sales, not weight-based metrics, *i.e.*, tonnes) has increased over the past decade [16].

Table 2. Changes in generation of high and low fee Blue Box Packaging (Source: [16]).

Materials	2003 Quantity Generated (tonnes)	2013 Quantity Generated (tonnes)
High Fee Materials		
Gable Top Cartons	14,249 T	42,000 T
Paper Laminants	2800 T	39,205 T
Aseptic Containers	5820 T	12,800 T
Plastic Film	53,700 T	54,383 T
Plastic Laminants	35,391 T	35,391 T
Polystyrene	20,400 T	57,400 T
Other Plastics	28,300 T	70,790 T
Low Fee Materials		
Newsprint–CNA/OCNA	264,800 T	217,375 T
Newsprint–Non-CNA/OCNA	136,400 T	148,405 T
Magazines and Catalogues	95,100 T	78,908 T
Telephone Books	15,000 T	8329 T
Other Printed Paper	127,800 T	128,245 T
Corrugated Cardboard	140,000 T	169,361 T
Boxboard	130,500 T	163,988 T
PET Bottles	36,200 T	56,848 T
HDPE Bottles	23,000 T	27,598 T
Steel Food & Beverage Cans	57,800 T	45,286 T
Steel Aerosols	4300 T	4079 T
Steel Paint Cans	4800 T	5072 T
Aluminum Food & Beverage Cans	24,100 T	22,552 T
Other Aluminum Packaging	2408 T	4521 T
Clear Glass	76,200 T	74,522 T
Coloured Glass	6700 T	25,277 T

Coincidentally, this can be seen as a direct result of how packaging producers have responded to the evolution of Ontario's fee allocation model. The fees charged to packaging producers as part of the province's EPR system are calculated on a per tonne basis. As such, many packaging producers have opted to switch to light weight packaging (namely LDPE (Low Density Polyethylene Terephthalate), PET thermoforms and polystyrene crystal) to minimize the impact of the fee. While these materials traditionally have higher fee rates, fees are denominated in kilograms. As such, the total fee paid by packaging producers tends to be lower in absolute terms (due to light weighting of material). The issue with this is twofold: (1) Consumers don't readily recognize these materials as being recyclable, and (2) these items are voluminous but not very heavy. This not only results in less material being placed in the Blue Box, but lower tonnages (and thus lower recycling rates) for the material that is collected. The impact of these changes have been significant, as Ontario's recycling rate stagnated at 68% in 2010, and subsequently declined to 63% in 2014 [15]).

4. Results

A regression model was developed to ascertain what relationship (if any) exists between packaging environmental handling fees and recycling rates. Based on the intended function of Ontario's fee setting methodology, it is expected that increases in recycling rates are positively correlated with packaging fees. It should be noted that a distinction exists between high recycling rates and marginal increases in recycling rates. By definition, materials with low levels of recyclability have the highest fee rates. The purpose of the fee is to incentivize packaging producers to improve their recycling rates such that relative fee rates will decline over time.

Statistical Analysis

Using the Breusche-Pagan Lagrange (LM) multiplier test, we test to see whether a random or fixed effects regression should be used in place of a pooled ordinary least squares (OLS) analysis [18].

A Breusche-Pagan multiplier test is used to test for heteroscedasticity in a linear regression model, and is useful in helping decide between a random/fixed effects regression and a simple OLS regression. The testing reveals that the null hypothesis is rejected, as the variance across entities is greater than zero. As such, pooled OLS is dismissed as an appropriate regressive technique. To determine whether a fixed or random effects model should be used, a Hausman test was conducted to see whether the unique errors of the model (u_i) were correlated with the regressors [19]. The results show that cross-sectional variance components are greater than zero, suggesting that a fixed effects regressive model is the best available choice given the characteristics of the dataset.

However, with specific regards to this study, endogeneity poses an issue, as the independent variables (fee rates) is a function of the dependent variable (recycling rates). To correct for endogeneity of the fee rate (FR) variable, we instrument the variable FR with its one-year lagged variable. Prior year fee payments affect current year packaging recycling rates, but current year recycling rates have no bearing on prior year fee rates. An instrumental variable two-stage least squares regression is used to model our results. It should be noted that collinearity also poses a problem among the independent variables, as fee rates are a function of material-specific costs. However, using a variance inflation factors (VIF) test, it was determined that the variance of the estimated regression coefficient was sufficiently close to 1, indicating negligible multicollinearity.

The linear econometric specification of the packaging recycling rate function is:

$$RR = \beta 0 + \beta 1\,FRit-1 + \beta 2Revit + \beta 3COSTit + \beta 4TIMEit + ai + uit \tag{1}$$

RR refers to the dependent variable, packaging-specific recycling rates. Fee Rates (FR) refers to material-specific fee rates for packaging materials using the three-factor formula based on the net cost of material management, material-specific recycling rates, and an equalization payment. Revenue (REV) refers to the material-specific revenue (expressed on a per tonne basis) received from the sale of baled and marketed materials. COST refers to the cost per tonne for managing packaging materials in the recycling system. TIME is the dummy variable for each year except for the first year, and ai and uit are the components for the unobserved disturbance for packaging type i during time t.

While the specified model used in this study may seem simplistic in design, it is important to note that the emphasis of the testing is to see how fee rates affect changes in packaging recycling rates. Previous works have undertaken a more comprehensive examination of the drivers of recycling behavior, but have done so from the perspective of consumers [20,21]. Conversely, examination of municipal transfer payments showed how municipal incentivization affects municipal recycling behavior [22]. As far as can be ascertained, this is one of the few studies to specifically examine the relationship between packaging fees and recycling rates over time (in a Canadian context). Assuming that producer behavior responds to changes in fee rates, packaging producers can promote recycling by serving as both internal and external facilitators of recycling (through increased promotion and education, investing in recycling infrastructure, *etc.*). If no material relationship exists between packaging fee rates and recycling performance, we assume that changes in recycling rates are explained by factors unrelated to packaging fee rates.

Tables 3 and 4 below summarize the results from the fixed effects panel regression for packaging materials (on both a system wide and material-specific basis).

Table 3. Relationship between changes in year over year recycling rate and packaging fee rates (system wide). Instrumental Variable (2SLS) regression; Number of observations = 207; R^2 = 0.2654.

Independent Variables	Packaging Materials (RR%)	Std. Error	Z Score
FR	0.000136	0.0002	0.03
REV	0.29561	0.5868	3.90
COST	−0.214564	0.2847	4.41

Table 4. Relationship between changes in year over year recycling rate and packaging fee rates (material-specific).

Material	FR	REV	COST
Newsprint (Newspaper Association)	0.000220	0.18690	−0.14366
Newsprint–(Not part of Newspaper Association)	−0.000070	0.34266 *	0.13147
Magazines and Catalogues	0.000869	0.31110 *	−0.21833
Telephone Books	−0.000435	0.32136 *	0.27699 **
Other Printed Paper	−0.000504	0.26247 **	0.29483 **
Corrugated Cardboard	−0.000441	0.32004 *	−0.13283
Boxboard	0.000250	0.22784	−0.28680 **
Gable Top Cartons	0.000351	0.37000 *	−0.30779 *
Paper Laminates	−0.000437	0.35511*	−0.12441
Aseptic Containers	0.000594	0.22920	0.32168 *
PET Bottles	0.000251	0.13968	−0.17759
HDPE Bottles	0.000628	0.21201	−0.38966 *
Plastic Film	0.000109	0.34974 *	−0.16741
Plastic Laminates	−0.000880	0.25767 **	−0.14192
Polystyrene	0.000933	0.27153 **	−0.25416
Other Plastics	−0.000896	0.20125	0.18090
Steel Food & Beverage Cans	−0.000390	0.32253 *	−0.13927
Steel Aerosols	−0.000180	0.12530	−0.37894 *
Steel Paint Cans	0.000650	0.24550 **	−0.14989
Aluminum Food & Beverage Cans	0.000562	0.15814	−0.10016
Other Aluminum Packaging	−0.000114	0.11705	−0.10969
Clear Glass	−0.000514	0.11587	0.29892 **
Coloured Glass	−0.000592	0.27944 **	−0.32096 *

* Denotes statistical significance at the 95% interval; ** Denotes statistical significance at the 90% confidence interval.

The above results did not demonstrate an association between packaging fee rates and recycling rates (for either the system as a whole or by individual material types). For every $0.01 increase or decrease in packaging fee rates, recycling rates would change by 0.000136%. There is no evidence to suggest that packaging producers are incentivized to continue or improve their recycling performance under Ontario's existing fee model.

Of note, both cost per tonne and revenue were observed to have a statistically significant impact in our model. For every one dollar in the cost per tonne for recycling material, recycling rates would on average, decrease by approximately −0.2%. Conversely, recycling rates were positively correlated with revenue (increasing by approximately 0.3% for every one dollar increase in revenue). Model estimates did not find any discernable relationship between what types of materials are affected by changes in cost or revenue.

5. Conclusions

This study undertook a critical examination of Ontario's fee setting methodology, specifically examining the relationship between packaging fee rates and material-specific recycling rates. Using data collected for each of the 23 materials found in the residential recycling program over the past decade, a regression model was developed to gauge what relationship (if any) packaging recycling rates have with fee rates, material-specific costs and revenue. The modeling in this study indicates that packaging fee rates have no effect on packaging recycling rates. Recycling rates were positively correlated with material revenue and negatively correlated with material management costs. The findings from this study raise some serious questions regarding the efficacy of Ontario's fee setting methodology. There is no evidence that suggests the three-factor formula used to allocate fees to individual materials encourages waste diversion. The disconnect in the results and the intended function of packaging fee rates calls into question the appropriateness of Ontario's fee rate methodology.

It is the recommendation of this study that the province re-evaluate Ontario's fee incentivization model, particularly in light of the relative complexity of how fees are allocated and the lack of stakeholder awareness regarding the incentive model's functions. Partial disaggregations, in kind contributions, equalization payments, *etc.* are all critical components in determining how fees are calculated. However, comparatively few truly understand how these things work. There is a need for increased transparency with respect to the inner workings of Blue Box fee setting policies. If the province wants to improve upon these policies such that they successfully promote recycling performance, it is of paramount importance that they do away with the black box nature of the Blue Box program.

Ontario should be credited for being the first province to embrace an EPR scheme for packaging waste. However, the distribution of Blue Box costs to individual materials needs to either reflect the actual cost of managing the material in the system, or, alternatively, make the penalty for producing non-recyclable packaging sufficiently high to act as an actual deterrent. Currently, the fee incentivization model occupies a peculiar (and ineffectual) middle ground—it only partially reflects material management costs, and the penalty for low recycling rates is inconsequentially low. Packaging producers continue to switch to lighter weight, but difficult-to-recycle material, as the fee rate penalty is more than offset by savings in transportation and logistics costs. Decision-makers and policy planners need to make a conscious decision to prioritize what they want from the Blue Box system: a true cost recycling system, or one where design for the environment principles are prioritized. The issue of how fees are distributed to individual packaging materials is of particular importance to other Canadian jurisdictions. Several provinces have used Ontario's fee setting methodology as the basis for which to design their EPR programs. Given the lack of proven efficacy of Ontario's fee setting approach, it seems prudent that alternative EPR fee models be explored.

The examination of packaging fee rates (particularly those that are distributed using an incentive-based model) is still very much in its conceptual infancy, as the advent of EPR for packaging waste is a relatively new phenomenon in North America. However, as EPR systems are adopted in other provinces and states, an understanding of how packaging producers can be encouraged and incentivized to design recyclable and environmentally sound products will be of growing importance.

References

1. Lindhqvist, T.; Lifset, R. Can we take the concept of individual responsibility from theory to practice? *J. Ind. Ecol.* **2003**, *7*, 3–6. [CrossRef]
2. Gertsakis, J. Industrial ecology and extended producer responsibility. In *A Handbook of Industrial Ecology*; Ayres, R.U., Ayres, L.W., Eds.; Edward Elgar Publishers: Cheltenham, UK, 2002; pp. 27–35.
3. Mayers, K. Strategic, financial, and design implications of extended producer responsibility in Europe: A producer case study. *J. Ind. Ecol.* **2007**, *11*, 113–131. [CrossRef]
4. Deutz, P. Producer responsibility in a sustainable development context: Ecological modernisation or industrial ecology? *Geogr. J.* **2009**, *175*, 274–285. [CrossRef]
5. Mayers, K.; Lifset, R.; Bodenhoefer, K.; Van Wassenhove, L.N. Implementing individual producer responsibility for waste electrical and electronic equipment through improved financing. *J. Ind. Ecol.* **2013**, *17*, 186–198. [CrossRef]
6. Walls, M. Extended Producer Responsibility and Product Design: Economic Theory and Selected Case Studies. Available online: http://ssrn.com/abstract=901661 (accessed on 10 November 2008).
7. Gottberg, A.; Morris, J.; Pollard, S.; Mark-Herbert, C.; Cook, M. Producer responsibility, waste minimization and the WEEE Directive, Case studies in ecodesign from the European lighting sector. *Sci. Total Environ.* **2006**, *359*, 38–56. [CrossRef] [PubMed]
8. Mayers, K.; Butler, S. Producer responsibility organizations development and operations. *J. Indus. Ecol.* **2013**, *17*, 277–289. [CrossRef]

9.   Stephenson, D. Internal White Paper on EPR (For Discussion Only). 2010. Available online: www.stewardedge.ca/content/archived/2010/EPRdiscussion.doc (accessed on 17 January 2016).

10.   Sheehan, B.; Spiegelman, H. EPR in the US and Canada. *Resour. Recycl.* **2005**, *3*, 18–21.

11.   Schwartz, J.; Gattuso, D. Extended Producer Responsibility: Re-Examining Its Role in Enviornmental Progress. Available online: http://www.rppi.org/ps293.pdf (accessed on 17 January 2016).

12.   Seidel, C. Zeroing in on Waste: The Role of Extended Producer Responsibility in a Zero Waste Strategy. Recycling Council of Alberta. From Recycling Council of Alberta. 2006. Available online: http://www.gpiatlantic.org/conference/proceedings/seidel.ppt (accessed on 27 February 2006).

13.   Pro Europe. Packaging and EU Directive on Waste. 2012. Available online: http://www.pro-e.org/2009-Q&A-Packaging-and-EU-Directive-on-waste.html (accessed on 10 November 2015).

14.   Waste Diversion Ontario. About WDO. 2012. Available online: http://www.wdo.ca/content/?path=page81+item35937 (accessed on 10 November 2016).

15.   Stewardship Ontario. Blue Box Annual Report. 2003–2015. Available online: http://www.wdo.ca/content/?path=page82+item35785 (accessed on 17 January 2016).

16.   Stewardship Ontario. Pay in Model. 2005–2015. Available online: http://www.stewardshipontario.ca/stewards-bluebox/fees-and-payments/fee-setting-flow-chart/the-pay-in-model/ (accessed on 17 January 2016).

17.   Waste Diversion Ontario. 2012 Ontario Residential Diversion Rates. 2014. Available online: http://www.wdo.ca/files/8413/9040/6230/Datacall_Diversion_Rates_2012.pdf (accessed on 17 January 2016).

18.   Breusch, T.S.; Pagan, A.R. A simple test for heteroscedasticity and random coefficient variation. *Econometrica* **1979**, *47*, 1287–1294. [CrossRef]

19.   Hausman, J.A. Specification Tests in Econometrics. *Econometrica* **1978**, *46*, 1251–1271. [CrossRef]

20.   Sidique, S.F.; Lupi, F.; Joshi, S.V. Factors influencing the rate of recycling: An analysis of Minnesota counties. *Resour. Conserv. Recycl.* **2009**, *54*, 242–249. [CrossRef]

21.   Oom do Valle, P.; Reis, E.; Menezes, J.; Rebelo, E. Behavioural determinants of household recycling participation. *Environ. Behav.* **2004**, *36*, 505–540. [CrossRef]

22.   Lakhan, C. The relationship between municipal waste diversion incentivization and recycling rate performance: An Ontario Case Study. *Resour. Conserv. Recycl.* **2016**, *106*, 68–77. [CrossRef]

5

A Procedure to Transform Recycling Behavior for Source Separation of Household Waste

Kamran Rousta [1,*], Kim Bolton [1] and Lisa Dahlén [2]

[1]   Swedish Centre for Resource Recovery, University of Borås, Borås 501 90, Sweden; Kim.Bolton@hb.se
[2]   Waste Science and Technology, Luleå University of Technology, Luleå 971 87, Sweden; Lisa.Dahlen@ltu.se
*   Correspondence: Kamran.Rousta@hb.se

Academic Editor: Michele Rosano

Abstract: Household waste separation at the source is a central part of waste management systems in Sweden. Resource recovery of materials and energy increased substantially after separate collection was implemented in the 1990s. A procedure to transform recycling behavior for the sorting of household waste—called the recycling behavior transition (RBT) procedure—was designed and implemented in a waste management system in Sweden. Repeated use of this procedure, which will assist in the continual improvement of household sorting, consists of the following four consecutive steps: (i) evaluating the current sorting behavior; (ii) identifying appropriate interventions; (iii) implementing the interventions, and; (iv) assessing the quantitative effect of the interventions. This procedure follows action research methodology and it is the first time that such a procedure has been developed and implemented for the sorting of household waste. The procedure can easily be adapted to any source separation system (which may have different local situations) and, by improving the source separation, will increase the resource recovery in the waste management system. The RBT procedure, together with its strengths and weaknesses, is discussed in this paper, and its implementation is exemplified by a pilot study done in Sweden.

Keywords: recycling behavior; resource recovery; source separation; household waste

1. Introduction

Household waste separation at the source was introduced in Sweden in the early 1990s. Here household waste refers to all waste that is collected in curbside collection systems, the packaging materials in bring systems, and hazardous waste, electronics, bulky waste, *etc.* which are collected at recycling centers. There is not a single system for the entire country, but different regions provide different collection systems for the inhabitants. Source separation increases both the rate of recycling of recyclable materials and of biological treatment of food waste. Material recycling has more than doubled, and biological treatment has more than quadrupled, since source separation was implemented in Sweden. In 2014 the country generated about 4.55 million tons of household waste, of which 36% was treated by material recycling, 16% by biological methods, 47% was used for energy recovery, and 1% was landfilled [1]. This illustrates that an effective source separation scheme can increase resource recovery from household waste.

Although waste separation at the source is a common and economical way for separating the recyclable fractions, effective participation by inhabitants is required to increase the correct sorting of the recyclables and food waste [2]. Many previous studies have investigated factors that determine the sorting behavior of inhabitants to better understand this phenomenon. A wide range of factors have been studied, and the factors that are chosen for a particular study often depend on the researcher's interest and discipline, the case and situation, the scope of the study, and the method used for the study. Some studies found that intrinsic factors, such as attitudes towards recycling and environmental concern, affect sorting behavior (e.g., [3–8]). Other studies showed that convenient and easy access to recycling facilities are decisive factors (e.g., [9–12]).

Other studies developed a model and framework to explain how different factors affect recycling behavior. For example, the motivation-ability-opportunity-behavior model explains that motivation is necessary but not sufficient for environmentally friendly behavior, but that ability and opportunity to behave in the correct manner are also required [13]. For source separation of household waste, situational factors such as convenient and easy access to recycling stations are opportunities, and knowledge of why and how to sort the waste, as well as past experiences and habits of sorting, are the examples of ability. In a similar study, Barr (2002) [14] presented a conceptual framework of environmental behavior. Barr (2002) [14] explained that environmental values of inhabitants do not have a direct impact on behavior. Instead, behavioral intention is needed as a bridge between environmental values and behavior, and this bridge has a central role in his framework. Situational factors and psychological variables can influence both intention and behavior, and are included in the framework, but are not as central as behavioral intention. Context, knowledge, and experience are examples of situational factors, whereas intrinsic motivation, subjective norms, and altruism are examples of psychological variables.

The studies done by Thøgersen (1994) [13] and Barr (2002) [14] (which were previously discussed above), used recycling behavior (*i.e.,* waste sorting at the source) to exemplify how their model and framework can explain which factors are most important for environmental behavior. Both studies found that factors such as information about the waste sorting scheme and accessibility to the collection facilities played a crucial role in determining the behavior.

Pieters (1991) [15] examined recycling behavior by combining a survey study with measurements of the recyclable materials in the waste stream. The aim was to identify which factors determine inhabitants' participation in waste separation schemes. In contrast, Tucker *et al.* (1999) [16] used a mathematical model to simulate the household waste management behavior. In another study, Tucker (2001) [17] developed a hypothesized cause-effect model of recycling. Intention to recycle plays the central role in this model, and intention can be influenced by pro-recycling attitudes, social norms, and specific barriers such as perceived obstacles and perceived effectiveness. This model also emphasizes the fact that intention to recycle is not sufficient to cause recycling. Similar to Barr (2002) [14] and Thøgersen (1994) [13], this model also considers the factors, called personal difficulties, which hinder the transformation of intention into actual recycling behavior. The studies mentioned above describe the various factors and determinants that influence recycling behavior, and the importance of these different factors depends on the concept of the study, the method used in the study, and where the study was conducted.

Almost all of the above studies have been performed under specific circumstances and in a limited geographical area. It is difficult to generalize the findings to source separation schemes in different places in the world. Wilson *et al.* (2012) [18] compared waste management in 20 cities and clearly showed that there is not a single solution that is suitable for all management systems and cities. Depending on the available facilities, the length of time that the recycling scheme has been implemented, the structure of the waste management system, relevant policies, *etc.* different factors could influence the recycling scheme and its improvement in different ways. Results from previous studies provide a wide range of factors that can improve recycling behavior (*i.e.,* to create effective interventions). However, no study has introduced a procedure for designing relevant interventions and assessing their effect on recycling behavior. The procedure described in this work is based on the action research routine including *look* (gathering data and describing situation), *think* (analyzing and theorizing), and *act* (implementation action and evaluate its effectiveness) [19]. This is the first procedure that combines interventions with their assessment. In addition, the procedure can be used in any source separation scheme anywhere in the world. No matter which kind of waste collection system is implemented, procedures for continuous improvement of the system (quality management tools) are needed. Therefore, it is relevant to develop a procedure to investigate and design interventions for improvement of source separation systems and to assess their effect in an objective and quantifiable manner.

Objectives

The aim was to create, test, and evaluate a procedure for continuous improvement of any source separation system for household waste. Possible changes in sorting behavior do not depend on the procedure itself, but on the intervention that is used and how it is implemented. The procedure, called recycling behavior transition (RBT) in this paper, identifies the extent of the missorting of household waste and interventions that can improve the sorting; it develops and implements the intervention and then assesses the effect of the intervention. The objectives of the study are summarized in three research questions:

1.   How should the current sorting behavior be evaluated?
2.   How should the appropriate interventions to improve this behavior be identified?
3.   How should the effect of interventions be assessed?

2. Methods and Materials

Although the procedure is novel, it is based on established scientific methods (such as pick analysis and interview techniques). The scientific merits of these methods have been discussed in detail in the literature (e.g., [20–22]) and hence this is not a focus of the present contribution. Instead, we discuss the ways that these methods are combined in the RBT procedure, and exemplify the methods and their combination by presenting results when the procedure was implemented in a residential area in Sweden.

2.1. Design of the Study

The RBT procedure was tested in a pilot residential area in Sweden with the intention to test the method and to improve waste sorting behavior. The first step was to identify the prevailing sorting behavior, which was examined by measurement of waste composition by sampling and manual sorting (pick analysis). In the second step, factors that hindered correct sorting were identified using two types of interviews with the inhabitants: semi-structured and structured interviews. Although it is possible that these hindrances can be identified by other means, such as literature surveys, interviews have the advantage in that they include the inhabitants in the identification of appropriate interventions which increases the probability that the chosen intervention will be effective. Next, based on the interview results, interventions in the waste collection system were designed and implemented. Finally, another pick analysis was conducted to investigate the effect of the interventions. The pilot area and methods used are presented in more detail below.

2.2. Description of the Pilot Area Used to Test the Procedure

The pilot area that was selected for testing the procedure is not relevant to the procedure and, as mentioned above, the procedure can be implemented in any area that has waste sorting at the source. In spite of this, a description of the pilot area is provided since this is needed if the results presented here are to be compared with results from other areas.

The pilot area that was selected is a residential area in the city of Borås, which lies in the southwest of Sweden. This area was chosen since the waste separation scheme was not functioning as well as in other parts of the city [23]. According to Statistics Sweden [24], the pilot area had 447 inhabitants with a diverse socio-demographic background who lived in nine apartment buildings (a total of 208 apartments). Approximately 67% of the inhabitants were born outside of Sweden and immigrated to Sweden more than three years before the procedure was implemented [24]. 31% of the inhabitants were aged from 25 to 44 years and 20% were 45 to 64 years old. It was mainly inhabitants from these age groups that were involved in the interviews in Step 2 of the procedure. Approximately 63% of the citizens in this area had low income (less than 15,600 Euro per year). In addition, approximately 58% had upper secondary school or higher education, and 33% of the residents owned a car [24].

The source separation system used in the city was established in 1991. It is based on sorting food waste in black bags and combustible waste in white bags (combustible waste is all household waste which is neither food, recyclable, nor hazardous). The waste sorted in the black bags is intended for production of biofuel and that in the white bags for combustion. The black and white plastic bags are distributed free of charge to all households. Both bags are supposed to be located in the kitchen so that they are easily accessible. To maintain low costs for the logistics, both bags are collected at the same time and placed in a single container. They are subsequently separated in an optical sorting machine at material recovery facilities. This type of optical sorting is not commonly used in Sweden. Recyclable materials, such as paper packaging, plastic packaging, metal packaging, glass packaging, and newsprints, should be sorted in a bring system (*i.e.*, they should be taken to recycling stations). The nearest complete recycling station to the pilot area was one and a half kilometers away. However, newsprints and glass packaging could be sorted in separate containers that were located immediately behind the apartment buildings. Other materials, such as bulky and hazardous wastes, should be delivered to one of the five recycling centers; the nearest one being located about two kilometers from the apartment buildings. Collection of the packaging materials at recycling stations, and bulky waste and hazardous waste at recycling centers, is a common part of Swedish waste management system. In these systems, inhabitants should bring their waste to these places and sort them in the appropriate containers.

2.3. Methods

A combination of quantitative and qualitative methods is needed in the RBT procedure. These are described below.

2.3.1. Waste Composition Study (Pick Analysis)

Measuring the amount of different fractions in the waste streams (e.g., in the black and white bags) and characterizing them is called pick analysis. The method used was developed by Dahlén *et al.* (2008) [20] and has been published as a manual by the Swedish Waste Management Association [25,26]. Pick analyses were conducted in November 2011, November 2013, and November 2015 (*i.e.*, before and after interventions in the waste collection system). November was chosen since there are no special days (such as holidays) in Sweden during this month. It can therefore be assumed that the waste collected in this month was representative of ordinary waste generation during the year. The pilot area had twenty-eight 660 liter wheeled bins for collection of the white bags (intended for combustible waste) and black bags (intended for food waste). The white and black bags are collected in the same bins (*i.e.*, the bins are not separated into those dedicated for black or white bags). Seven of the bins were randomly selected each week in November, giving a total of 28 samples. Table 1 shows the details of sampling for these two pick analyses. The size and number of samples used in this study ensure the statistical certainty of the results according to the description of the method [20].

Table 1. Details of sampling for pick analysis.

Date	Unit of Sample	No. of Samples per Week	Total Samples	Total Weight (kg)
November 2011	Waste in a 660 liter bin	7	28	1732.9
November 2013	Waste in a 660 liter bin	7	28	1699.0
November 2015	Waste in a 660 liter bin	7	28	1592.5

The samples were collected for the pick analysis, and the rest of the waste in the area was weighed each week in order to estimate the total waste that was generated. Waste characterization and weighing of waste fractions in the samples were conducted separately for the white and black bags. The weight of the plastic bags themselves (white and black) was excluded from the pick analysis. Categories, subcategories, and fractions of the waste that was weighed during the pick analysis, as well as where it should be sorted, are shown in Table 2. The categories used in the pick analysis were chosen according to the municipality's instructions for waste sorting, which are similar to those used in other Swedish cities. Any other material than food waste in the black bags and combustible waste in white bags was regarded as missorted fractions.

Table 2. Waste fractions and categories used in the pick analyses.

Category	Subcategory	Fractions	Sorting Instruction
Food	Food	Leftover food, fruits, vegetables, *etc.*	Black bag
Packaging, newsprints	Paper packaging	Cardboard, paper packaging	Recycling station
	Plastic packaging	Plastic film packaging, foam plastic packaging, dense plastic packaging	
	Glass packaging Metal packaging	Colored and clear glass packaging Metal packaging	
	Newsprint	Newsprint, advertisements, paperbacks, writing/drawing papers	
	Deposit bottles [1]	PET [2], Aluminum cans and glass bottles with deposit	
Combustible	Diapers Textile fabrics	Diapers, pads Clothes, shoes, different textiles	White bag
	Combustible waste	Wood, small non-packaging plastic, garden waste, wood, non-packaging paper, cat sand, tissues, envelopes, Christmas cards, small baby tools, pens, cigarette butts, vacuum cleaner bags, *etc.*	
Other	Non-packaging plastic	Big parts	Recycling center
	Non-packaging metal	Any metal parts	
	Non-packaging glass	Broken glass	
	Other non-combustibles	Ceramics, broken mug, bricks, *etc.*	
Hazardous	Medicines [3] Batteries Small electronics Light-bulbs Other hazardous	Mobile phones, clocks, battery chargers, *etc.* Glues, chemicals, full sprays	Recycling center

[1] Deposit bottles should be collected in large supermarkets. [2] PET (polyethylene terephthalate) bottles are usually used for beverages. [3] Medicines should be collected in pharmacies.

The amounts of correctly and incorrectly sorted waste were analyzed using the MINITAB software (Minitab Ltd., Coventry, UK) to determine the average amounts of missorted materials in kg per household per week, as well as the statistical relevance of any changes in sorting behavior due to the interventions. The results from the first pick analysis revealed the prevailing waste sorting behavior in the pilot area. The results from the second and third pick analyses, which were conducted after each of the two interventions, were compared with each other and with the results from the first pick analysis using the two sample t-test with 95% confidence interval. Hence, the second and third pick analyses quantitatively assessed the effect of the interventions.

2.3.2. Interview

The results from the first pick analysis were used to design interview templates. The goals of the interviews were to identify hindrances for correct waste sorting and the changes (interventions) in the waste collection system that would enable and motivate improved sorting. To do this, two types of interviews were designed: semi-structured interviews and structured interviews. They are described in detail below.

Semi-Structured Interview

Qualitative interviews, such as semi-structured interviews, are usually performed to obtain deep insight into how people understand a phenomena, and therefore the number of interviewees is not important [21,22]. Semi-structured interviews are used to gather information about peoples' situations, to collect statements about their opinions, and to explore their motivation and experiences [27]. The initial aim of the semi-structured interviews was to understand why inhabitants sort their waste as shown by the pick analysis, if they think that they need to improve their sorting behavior, and, if so, what they think is needed to improve their sorting. To do this, eight interviewees who were comfortable with participating in the research interview were identified. This was done by sending an invitation letter to all inhabitants as well as asking inhabitants via a contact person. The interviews were conducted three to five months after the first pick analysis (over a two month period) at the University of Borås, and lasted between 50 and 65 min. A practical exercise for sorting waste was also conducted during the interview (*i.e.*, asking the interviewees how they sort different kinds of waste fractions in four alternatives: white bags, black bags, recycling stations, or recycling centers). This was designed to get a better understanding of their knowledge regarding the sorting of waste. The interviewees were asked to describe how they felt about sorting their waste and the waste management system in the city, how they understood the need for sorting and the system, and what is required for better participation in the system. The interviewees were free to discuss their ideas and comments. All the interviews were taped and transcribed. They were subsequently analyzed to identify key words, paragraphs, or themes. The results obtained from these semi-structured interviews were used to design a short (less than five minutes), face-to-face structured interview.

Structured Interview

The semi-structured interviews identified the interventions that some of the inhabitants perceived were needed in the system. In order to ascertain if these perceptions were shared by a broad range of inhabitants, a structured interview was designed. The aim of this short interview was not to perform a separate survey (via a questionnaire) but rather to complement the semi-structured interviews. The structured interview was done by the same author who conducted the semi-structured interviews. The interviews were done three months after the semi-structured interviews and over a three day period so that a larger number of people could participate. The time of the interviews was not pre-planned, but instead the interviewer met inhabitants (not children) as they were conducting their daily activities. They were asked if it was convenient for them to participate in the interview and, if so, the interviewer introduced himself and asked the age of the respondent. The respondents were then asked to give short answers to the questions provided to them. There were two types of questions. One type had "yes" or "no" alternatives and the other type questioned how and why the inhabitants sorted their waste. If an answer about "how" specific waste should be sorted was incorrect, it was followed with a "why" question in order to investigate the reason of the incorrect sorting. The questions were limited to sorting of plastic and paper packaging, since the pick analysis showed that these were the types of packaging that were missorted the most. To keep the interview as short as possible and in order to minimize the information on sorting that could be given to the interviewee (which could affect the outcome of the subsequent pick analysis), milk packaging was used as an example of paper packaging. Figure 1 shows the flowchart for the questions used in this structured interview.

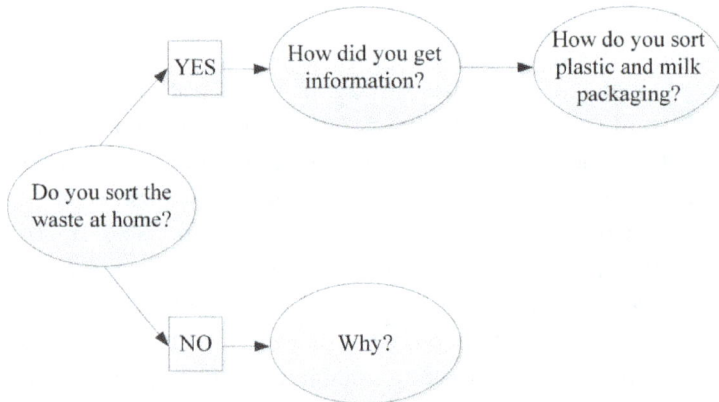

Figure 1. Flowchart of the questions in the structured interview.

2.3.3. Interventions in the Pilot Study

As discussed below, three interventions were designed. These interventions were chosen since, based on the results of the interviews, it was hypothesized that they would lead to improvements in the collection system. This hypothesis was tested in the pilot area. The interventions are described in Section 3.1.3 to clarify their relation to the interview results.

3. Results and Discussion

Implementation of the RBT procedure in the pilot area has been completed. The results obtained from the procedure are discussed below. The way that this procedure can be generalized for any source separation system is also discussed.

3.1. Pilot Area

3.1.1. First Pick Analysis—Evaluation of the Current Sorting Behavior

Figure 2 shows the average waste composition per household per week in the black and white bags.

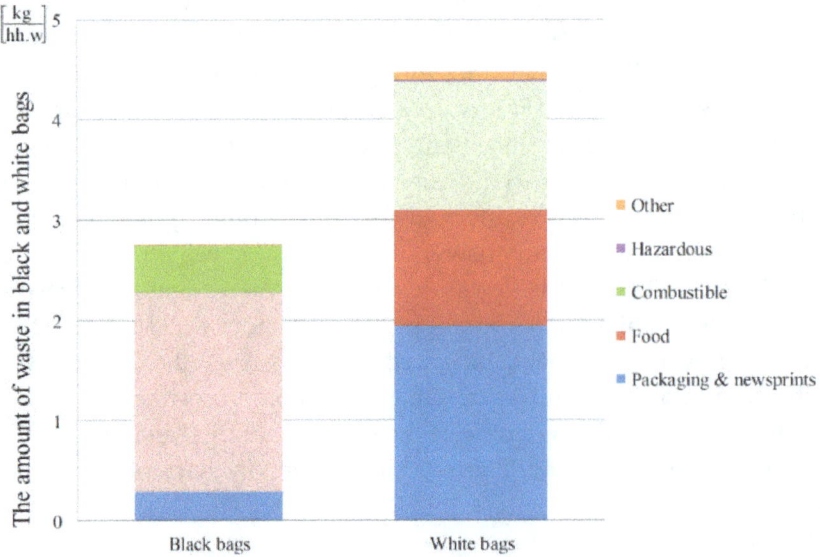

Figure 2. Composition of waste (kg per household per week) in black and white bags before the interventions (2011). The striped regions show that waste that is intended for each bag (food waste in the black bag and combustible waste in the white bag).

The total waste in the black bags, intended for food waste, was 2.8 kg per household per week (kg/hh/w) whereas in white bags (intended for combustible waste), it was 4.5 kg/hh/w. The striped areas in Figure 2 show the part of the waste that was sorted correctly. Seventy-one percent of the waste in the black bags and 29% of the waste in the white bags was sorted correctly. The total missorted ratio, obtained by summing the missorted fractions in both bags and dividing the sum by the total waste in both bags, was 55%, which is an indicator that should be decreased. The dominating waste fraction in the white bags was missorted packaging and newsprints, and was 1.9 kg/hh/w (43%). The second largest missorted fraction in the white bags was food waste (26%), which was often the food that remained in the packaging when it was discarded. This type of waste is illustrated in Figure 3.

Figure 3. Photo from the first pick analysis (2011) showing missorted food waste with packaging found in the white bags (neither food nor packaging are intended for in the white bags).

There were 0.019 and 0.004 kg/hh/w hazardous waste in the white and black bags, respectively. This missorted waste was dominated by small electronics, light bulbs, and batteries. Other missorted fractions were food waste in the white bags and combustibles in the black bags.

The results obtained from this research, and which revealed the current sorting behavior, raised the following questions: Do the households know how to sort packaging properly? If they do know, do they want to improve their sorting and, if so, what do they need to improve their sorting of packaging waste? What information would be helpful to clarify what should be sorted in the black and white bags? These questions formed the basis for the semi-structured interviews.

Hence, the results obtained from the first pick analysis answered the first research question given in Section "Objectives" (*i.e.*, how should the current sorting behavior be evaluated)?

3.1.2. Interviews for Identifying Appropriate Interventions to Improve Sorting Behavior

Semi-Structured Interview

Of the eight inhabitants that were interviewed, seven had immigrated to Sweden between 1985 and 2007 and were not Swedish citizens. Seven were between 25 and 50 years old and the eighth was 75 years old. There was a wide range in education levels—some of the inhabitants were illiterate and one had a doctoral degree. Six of them were employed, one was unemployed, and one was retired. Hence, the socio-economic background of the interviewees was typical for the pilot area (which was discussed in the Introduction). All of the interviewees could speak and understand Swedish, but sometimes a translator took part in the interview to avoid misunderstandings. The backgrounds of interviewees are listed in Table 3.

Table 3. Interviewees' backgrounds.

Person	Sex	Age	Arrival Date [a]	Years Lived in the City	Occupation	Family Members
1	F	25	2007	2	No	1
2	F	39	1990	15	Yes	5
3	F	50	1992	18	Part time	1
4	F	38	1997	10	Yes	4
5	F	40	1985	7	Yes	1
6	F	45	1997	16	Yes	2
7	M	40	2006	5	Part time	1
8	F	76	Born in Sweden	57	Retired	1

[a] Year of immigration to Sweden.

Most of the respondents had moved to the city from different countries, bringing with them different experiences of how to handle household waste. There was no organized source separation of waste in some of the immigrants' home countries. For example, one interviewee (#6) had used food waste as animal feed or fertilizer. These people had to learn how to participate in a new waste management system. In addition, there was a large variation in the motivation for participating in source separation schemes. Some of the reasons that were given for participating were resource conservation, respect for regulations, they felt that their participation was a valuable part of the waste treatment system, keeping a clean environment, their knowledge of the post-sorting processes, social norms, and that their participation gave them a sense of satisfaction. One interviewee (#1) stated: "This is the regulation for sorting the waste and I should respect it. This is what people usually do in Sweden". This means that she found waste separation as a norm in Swedish society and that this social norm can be a motivation to participate. Another interviewee (#8) said that "In the beginning, I didn't think too much about the environment. I found out that we have to sort our waste from some information leaflet. Knowing how useful it is for the environment and that I can contribute to the next generation, my children and grandchildren whom I love, means a lot to me". She continued, "At the moment, I cannot do much to protect my environment. This little job, sorting my waste, hopefully can be my part for protect environment for my children and grandchildren". Knowing what happens with the sorted waste was also a motivating factor. As expressed by one person (#6): "I always sort glass since I have learned that glass is the material that you can recycle millions of times and it motivates me to sort it." Similarly, another interviewee (#1), who had lived in Sweden for just four years, said: "When I see they produce biogas from food waste that I sort in the black bags and drive the buses, I understand my job is valuable". One respondent (#7) stated: "I do this because I want to keep my surroundings clean".

According to Thøgersen (1994) [13], these motivational factors cannot be the only reasons for correct source separation by the inhabitants. They also need situational factors such as easy access to facilities as well as knowledge of how to separate at the source. One of the aims of the semi-structured interviews was to investigate what the inhabitants perceive is needed to improve participation in source separation. Hence, once the motivational factors that influence the inhabitants sorting behavior were known, the questions focused on what actions were required from the inhabitants. This was the reason for the practical exercise (explained in the Methods section) given during the interviews, which was conducted to reveal if the interviewees knew how to correctly sort different fractions of waste. After sorting each material, whether it was correct or incorrect, they were asked to explain the reason for their choice of sorting. Most of them sorted food waste in the black bags, which was correct. Also, batteries were sorted correctly because the interviewees perceived them as being environmentally harmful. However, fractions such as newspapers, plastic-, paper-, metal-, and glass-packaging, *etc.* were often discarded in the white bags rather than being sorted and taken to recycling stations. This is in agreement with the results of the pick analysis, where about 43% of the waste in the white bags was packaging. There were some fractions, such as light bulbs, that the interviewees knew should not be in the white bags, but they were not certain where they should sort it. They were also uncertain of how to sort other fractions such as ceramics and non-packaging glass. These findings are in agreement with the studies of Henriksson *et al.* (2010) [28] who also found that people are uncertain about how to sort certain waste fractions correctly [28].

The results discussed above indicate that lack of information and knowledge may hinder the correct sorting of waste. The interviews also revealed that distance to the recycling stations was perceived as a hindrance. For example, one interviewee (#5) did not know that plastic bags such as potato chips packaging should be sorted in recycling stations. Further, even if she had been aware of how to correctly sort the waste, the distance that she needed to take the waste to the recycling station was important. She believed that improved sorting would be facilitated if there was a recycling station close to her home, preferably in the basement of the apartment building. This was also expressed by another person (#6) who put paper and plastic packaging in the white bags: "I know that I should separate them, but we don't have a recycling station close to home". In this case, the interviewee had the knowledge but still chose not to go to the recycling station. She continued: " … my child asked me why we don't sort the milk packaging. My answer was that we have no container close to our home". Some of the interviewees said that they were aware that they should sort fractions such as metal packaging, but they did not do so since the recycling station was two km away and they did not have access to a car. Even one interviewee (#8), who sorted all fractions correctly, stated: "I am able to walk to recycling station two km away to sort my waste now. I am thinking, after some years, when walking is hard for me, I may put all packaging in the white bags instead, because of this distance". Hence, it was concluded that decreasing the distance to the recycling station was a relevant intervention. Other studies have also revealed that distance to sorting facilities is a significant factor for participation in recycling systems [9,10,29].

After moving to Sweden, the respondents learned about waste management in different ways. It was not primarily through information leaflets, but by talking to and observing colleagues, neighbors, and family members. Most of the respondents could not remember whether they had received any written information. Only one interviewee (#8), who was born in Sweden, indicated that she had received written information 10 years ago. It is possible that, even though information may have been distributed, several of the interviewees may not have been able to read it due to language barriers or that it was distributed at an inappropriate time (e.g., when they were focused on adjusting to a new home, city, and country). All interviewees, apart from two respondents (#5, #8), claimed that language was still a barrier when seeking information, in spite of the fact that they had lived in Sweden for more than 15 years. In addition, five of the interviewees were illiterate when arriving in Sweden. If the information had been sent to them soon after their arrival it would not have been possible for them understand it. This is why they gained knowledge through informal communication with friends, neighbors, and colleagues. One person (#2) stated: "When I got a job five years ago I saw that they sort every waste fractions in my workplace. I found that I have to do it at home too". Another person (#5) said: "When I moved to my apartment, I saw two bags in my kitchen, one for food waste and the other combustible. It is obvious that I throw the food waste in black and all others, even packaging in the white when I don't have any other information". Lack of education can be another barrier. Many of the interviewees felt that education is essential when trying to find the right information. They also thought that the best place for learning is at school. When the information is insufficient it has a negative impact. Insufficient information can lead to incorrect sorting habits that can be difficult to change. One person (#2) said: "It is very important that one learns the right things from the beginning, because it is very hard to change the wrong habits after a while." Another interviewee emphasized that it is important to have information as early as possible in order to develop a correct habit. As expressed by one interviewee (#5): "Maybe sorting the garbage is hard in the beginning, but after a while you don't have to think about it, you do it as a norm, as a habit. It's part of my culture now. The more information you can understand and make your own the faster it will become a routine for you".

Hence, the interviews revealed that information and knowledge about how and why to sort the waste are important factors. If this is not communicated via formal authorities, it can be informal communication between friends, neighbors, family members, and colleagues. However, there is no guarantee that this informal information is correct. Some of the incorrect sorting of the respondents was due to incorrect informal information. The type of information and the time of communication are also important. For example, some interviewees sorted diapers in the black bags and they claimed that there was a sticker on the trash cans for food waste which showed that diapers should be sorted in these bags. These claims were, in fact, correct. The system for handling the black bags was changed from composting to biogas production in 2006, but the information (stickers) from the 1990s (for composting) was still on the trash cans. This indicates that simple information, such as stickers on trash cans, which is accessible at the time of disposal, has a large influence on the sorting behavior. Hence, information should be distributed at the appropriate time, and it should be continuously available or at least available at the time and place where waste is discarded.

The main findings from the semi-structured interviews are:

- The interviewees were willing to sort the waste.
- The interviewees had different motivations to sort the waste.
- Decreasing the distance to collection points for recyclables can be a relevant intervention.
- There was a lack of knowledge of how to sort the waste.
- Designing different types of information that is communicated at the appropriate time and place, via formal and informal channels, may also be a relevant intervention.

These findings are in agreement with the Thøgersen, Barr, and Tucker models [13,14,17] which show that inhabitants' intention to recycle should be accompanied with other factors in order to form the recycling behavior.

Structured Interview

The semi-structured interviews showed that (1) distance to recycling stations and (2) different types of information about how to sort waste, including the communication of this information, may be relevant interventions for improving waste source separation in this pilot area. To validate the relevance of this hypothesis for a larger number of households, a structured and short interview template was designed and conducted with 50 people (20 male and 30 female). All of the interviewees were between 20 and 50 years old. These 50 people constituted 25% of the population in this age range in this area (Swedish Statistics, 2012) and they were from 50 different households. This population therefore covered about one quarter of the total number of households in the pilot area. Figure 4 shows how respondents replied to the different questions in this short interview.

The majority of respondents, 42 out of 50 (84%), claimed that they sort their waste. We will refer to these people as recyclers. Only 8 out of 50 (16%) were non-recyclers. Among recyclers, only 24% claimed that they sorted the packaging correctly at the recycling stations. The remainder, 76%, threw packaging in the white bags, which is incorrect sorting. The main reason that was given for the incorrect sorting was the distance to the recycling station (42%), and the other reason was that they thought that their behavior was correct (*i.e.*, that had incorrect information about how to sort this type of packaging (33%)). It is not clear if the distance to the recycling station would be a barrier for the latter group once they obtain the correct sorting information. Some respondents had other reasons such as: "I don't have a car" and "I have just a little amount waste". The recyclers obtained information from informal channels (from neighbors, friends, and their workplaces) more often (90%) than from formal channels (from the municipality). This shows that this area has a strong social network to distribute this type of information. Three out of eight non-recyclers claimed that they did not sort due to a lack of information. Difficulty and lack of motivation for sorting were the reasons for the other five non-recyclers.

The result of the short, structured interviews was consistent with the findings of the semi-structured interviews. Therefore, to improve source separation in the pilot area the following interventions could be relevant: (1) decreasing the distance to the recycling station and (2) combining correct and appropriate information with effective communication channels.

Hence, the results obtained from the interviews answered the second research question given in Section "Objectives" (*i.e.*, how should the appropriate interventions to improve this behavior be identified?).

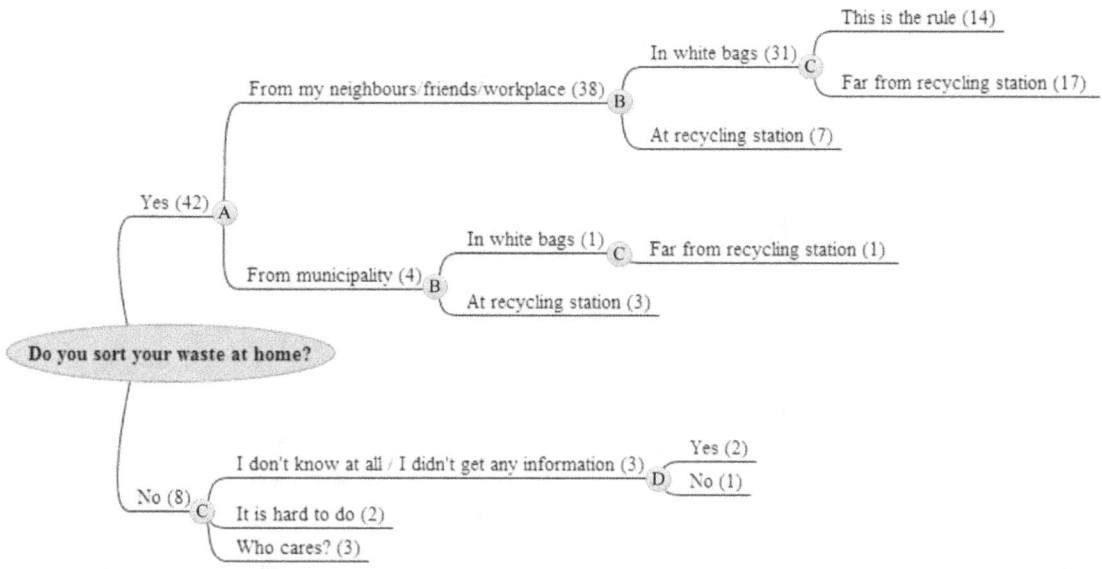

Figure 4. Results of the structured interviews with number of respondents indicated in parentheses. Nodes A to D represent questions: (**A**) How did you get information? (**B**) How do you sort plastic and milk packaging? (**C**) Why? (**D**) Do you want to sort if you get information?

3.1.3. Effect of Interventions

Based on the interview results, three interventions were implemented from 2012 to 2015. These are listed in Table 4.

Table 4. Interventions from 2012 to 2015.

Date	Intervention
June 2012	(a) Placing new stickers for black bags trash cans which clearly show that food waste should be sorted in the black bags.
June 2013	(b) Building a property close collection area for recyclables, called an environmental room, behind the apartment building to decrease the distance to the recycling station to 50 meters. The environmental room enabled the sorting of packaging, batteries, small electronics, light bulbs, and clothes.
2014–2015	(c) Communicating different types of information including: Installation of a picture on the containers for white and black bags showing where and how to sort this waste in the containers (2014). Sending written information, feedback and a "thank-you letter" three times to all households in the pilot area (2014). Having informal dialogues about using the environmental room and on how to improve the sorting in black bags (2015).

The second pick analysis was performed after implementing interventions (a) and (b), and the third pick analysis was conducted after intervention (c). The results are compared with those of the first pick analysis (2011) in Figures 5 and 6. It should be noted that two interventions were performed between the first and second pick analyses. Intervention (a) was expected to primarily influence sorting in the black bags, whereas intervention (b) was expected to influence recycling in the environmental room and in both bags. Since the white bags (intended for combustible waste) contained more recyclables than the black bags (Figure 1) intervention (b) was expected to influence the sorting in the white bags more than in the black bags.

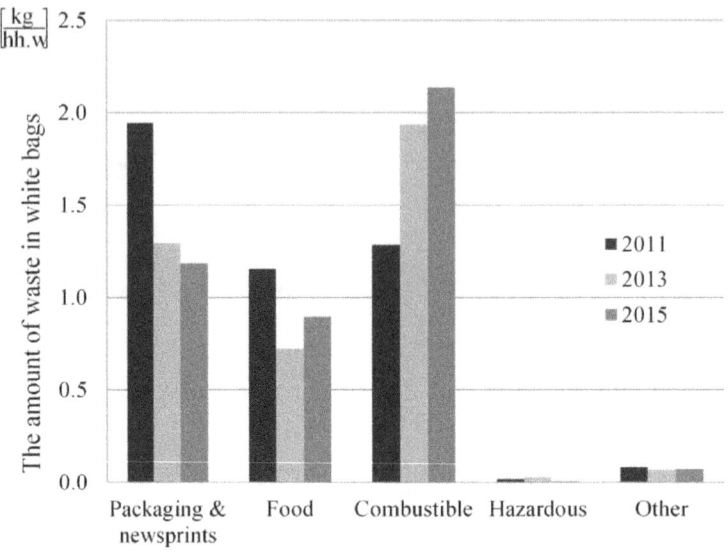

Figure 5. Comparison of waste sorting behavior before and after the interventions. The data shows waste fractions in the white bags, which are intended for combustibles (kg/hh/w).

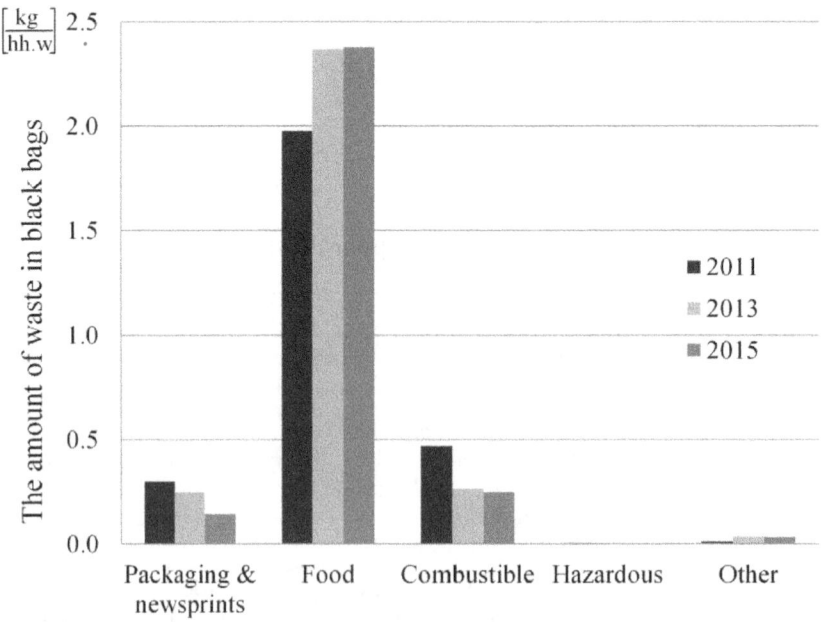

Figure 6. Comparison of waste sorting behavior before and after the interventions. The data shows waste fractions in the black bags, which are intended for food waste (kg/hh/w).

The amount of missorted packaging in the white bags decreased (33%) between 2011 and 2013, which is when the property close collection was established (intervention (b)). After intervention (c) in 2015, this decrease was 39% (Figure 5). Although there was an increase in the missorted food waste in the white bags between 2013 and 2015, there was an overall decrease between 2011 and 2015. The decrease in packaging and newsprints between 2011 and 2015 is statistically significant, as is the decrease in food waste between 2011 and 2013. The increase in food waste between 2013 and 2015, and the subsequent decrease between 2011 and 2015 are not statistically significant. The interventions did not lead to any statistically meaningful changes in the sorting of hazardous or other waste, but the hazardous waste decreased by 68% in the white bags in 2015 compared to 2011.

Missorted combustible waste in black bags (intended only for food waste) decreased (by 44%) by intervention (a), when visible information on the trash cans for black bags was provided, and by a further 3% in 2015 (Figure 6). Similarly, missorted packaging and newsprint decreased between 2011 and 2015 (52%). These two changes are statistically significant. Similarly to the white bags, there were no significant changes in the amounts of hazardous or other forms of waste.

In summary, the aim of the interventions was to decrease the missorting ratios in the pilot area. Figure 7 shows the missorted ratios in white bags, black bags, and in the total waste generated in the pilot area (*i.e.*, the average of the white and black bags).

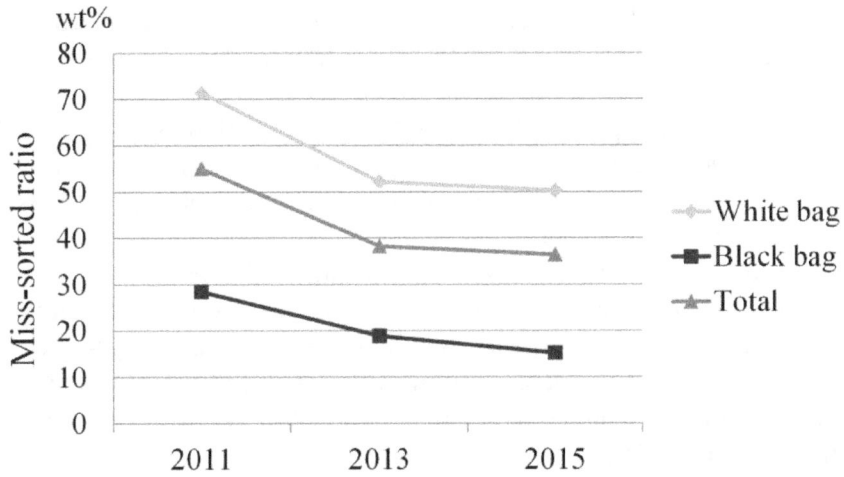

Figure 7. Comparison of missorted ratios in weight percent (wt %) before and after the interventions.

The total missorted ratio decreased from 55% to 38% after interventions (a) and (b), and decreased further to 36% after intervention (c). Figure 7 also shows that the magnitude of the reduction in the black bags between 2013 and 2015 is larger than for the white bags. Although this difference is statistically insignificant, it indicates that intervention (c) may have had the desired effect, since the focus of the informal communication was to improve sorting in the black bags. The results show that interventions (a) and (b) decreased the missorted ratio significantly. Although there was a continued decrease in the missorted ratio after intervention (c), it cannot be concluded that this is due to the intervention (the decrease may have continued even if intervention (c) had not been implemented).

Figure 7 shows that intervention (c) had less impact compared to interventions (a) and (b). There may be several reasons for this. One is that 56 of the households had inhabitants that moved into the pilot area during the project, of which 33 had inhabitants that came in 2015. Since no formal information regarding the use of the environmental room was directed towards the households after 2013, many of the new inhabitants only had access to information via informal channels. In addition, new inhabitants—especially those that moved in during 2015—may not have had time to build a social network that is necessary for informal communication between households. This means that it is possible that 27% (56 of 208) of the households did not have access to the correct information. Another reason may be the insufficient capacity of containers in the environmental room. The collection rate for the collected packaging in the environmental room was therefore limited. Observation of this room showed that the packaging containers were often overflowing on the collection days. This usually leads to poor sorting behavior which results in a mess in the room, which could negatively affect the inhabitants' participation in the sorting scheme. It is also not clear how the households perceived the new information in the form of written information, stickers, thank-you letters, and feedback.

Hence, the results obtained from the second and third pick analyses answered the third research question given in Section "Objectives" (*i.e.*, how should the effect of interventions be assessed?).

3.2. Recycling Behavior Transition (RBT) Procedure

The aim of this research was to create, test, and evaluate a procedure which can improve recycling behavior. The procedure consisted of four steps: (1) first pick analysis; (2a) a few semi-structured interviews; (2b) a larger number of structured interviews; (3) interventions based on findings from the interviews, and; (4) a pick analysis to assess the effect of interventions. The four steps, which included both quantitative and qualitative methods for data collection, were implemented and tested.

Pick analysis is a quantitative waste characterization method that measures missorted waste in a system. Structured interviews are also considered to be a form of quantitative method, whereas semi-structured interviews are a form of qualitative method; both forms of interview processes were used in this procedure. The combination of quantitative and qualitative methods (*i.e.*, using a mixed method) gives the necessary breadth and depth to understand complex phenomena [30] such as recycling behavior.

As discussed above, the first pick analysis quantifies the prevailing sorting behavior. In fact, this pick analysis identifies "*How*" inhabitants sort their waste without asking them. The output from the characterization of the collected waste is more reliable than using self-reported assessments of the participants in a source separation scheme.

In a study of the accuracy of three surveys of waste attitude/behavior undertaken in Scotland and north-west England, P. Tucker and D. Speirs (2003) [8] concluded that the results of these surveys usually give an over-optimistic picture of recycling behavior. They also concluded that it is difficult to make actions, promotions, and campaigns for improvement of recycling behavior by using results from surveys [8]. In contrast, pick analysis measures the actual waste sorting behavior of inhabitants regardless of what they answer in surveys.

The interventions that should be investigated are identified in Step 2. To do this one needs to understand "*Why*" inhabitants sort in the way that they do (as identified in Step 1), and "*What*" should be done to motivate and enable them to improve it. To do this, semi-structured interviews were designed. A semi-structured interview can provide answers to questions that cannot be checked or measured in other ways. It is difficult to get an absolute answer during an interview, but it can provide clues and indications that guide the subsequent steps. Although the results of the pick analysis give relevant data of the sorting behavior, these data require interpretation via the qualitative interview (*i.e.*, the semi-structured interview). This interpretation is needed to understand the sorting behavior.

For example, in the study conducted in the pilot area, the results of the first pick analysis showed that packaging and newsprints dominated the missorted material. Hence, the pick analysis revealed that one should focus on improving the sorting of packaging and newsprints waste. There are many possible interventions that could have been studied to achieve this goal, but the interviews identified that improved information and decreasing the distance to recycling stations were the interventions that the households perceived as being the most appropriate. The interviews also give the users, who are influenced by the interventions, a central role when developing the waste management system.

In a study in Madeline Island, Newenhouse and Schmit (2000) [31] found that qualitative approaches can add value to waste characterization studies in order to discover potential solutions to change recycling behavior. They discussed the fact that the interviews engage inhabitants in the process which, according to them, has a positive impact. The results of semi-structured interviews (Step 2a) can identify relevant interventions for improving the system, but it is important to have the opinion of a larger part of the population. The structured short interview (Step 2b) ascertains if the selected interventions are relevant for many of the inhabitants. Through these two steps, (2a) and (2b), interview results will be both qualitative and quantitative. Another reason for including Step 2b in the procedure is to involve as many inhabitants as possible. In the pilot study, the result of the structured short interviews was in agreement with those of the semi-structured interviews. If they had not been in agreement, then it would have been necessary to re-analyze the semi-structured interviews or to conduct more semi-structured interviews to identify other possible interventions.

Step 4, the second and third pick analyses, assesses the effect of the interventions. These analyses provide data on the actual recycling behavior after the interventions, which can be compared with the first analysis in order to see if the interventions were effective or not.

The RBT procedure is summarized in Figure 8 together with the methods for data collection. The procedure was called recycling behavior transition since a source separation system can be improved by a transition of the behavior of the inhabitants in sorting their waste. The dashed line in Figure 8 shows that the procedure can be repeated to enable continued development of the waste management system. After doing the last step (*i.e.*, the second or third pick analysis), there are two scenarios: (1) The interventions were effective: In this case the result of second pick analysis can be the basis for finding new interventions and further improvement, including a new round of interviews with inhabitants; (2) The interventions were not effective. In this case the factors which prevented the success of the interventions need to be identified (e.g., by a new series of interviews). It is possible to choose another qualitative approach other than semi-structured personal interviews to communicate with inhabitants (e.g., focus group or telephone interviews or combining interviews with observations). An important element with Step 2 is to understand the needs of the inhabitants for effective participation in source separation.

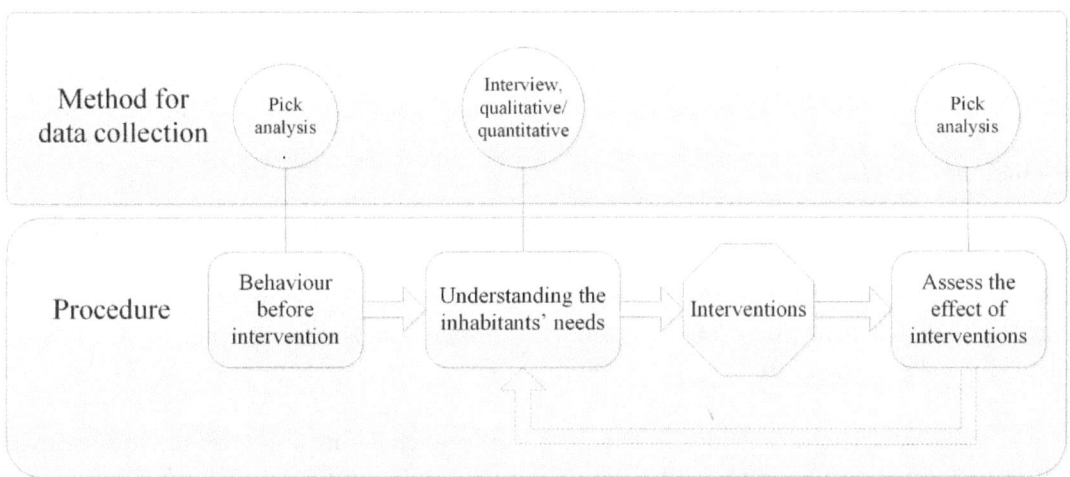

Figure 8. The Recycling Behavior Transition (RBT) procedure for continuous improvement of source separation and waste collection systems.

Recycling behavior is a complex phenomenon. The RBT procedure showed that it is possible to improve this behavior with a simple and accurate procedure. There are several advantages with the RBT procedure, and these are listed and commented upon below.

- The combination of quantitative and qualitative methods provides reliable and useful results to implement the appropriate interventions.
- The RBT procedure can be applied in any source separation system in any location, independent of local circumstances. The procedure can be used to identify the relevant intervention in other cultures, socio-economic backgrounds, and other source separation schemes.
- The RBT procedure quantifies the effect of the interventions. Actors responsible for planning waste management may hesitate to apply interventions because they do not know how to identify them and how to measure their effect. This procedure can help them not only to improve their system, but also to control the quality of the system.

- The RBT procedure focuses on user involvement in waste management systems. When the intervention is identified from the needs of the users, there is a higher probability of success.
- The continued improvement of source separation systems results in the collection of recyclable materials in larger quantities and of better quality, and consequently help in the development of a circular economy and sustainable development.

There are also several potential weaknesses of the RBT procedure that need to be recognized and, where possible, removed in future studies.

- The RBT procedure requires careful design in both time and implementation, and one often needs to wait for a specific time period for implementing interventions so that the pick analyses that are performed before and after the interventions can be compared to each other. In addition, since the procedure requires a long time, the effectiveness of the interventions is reduced if many residents move into and out of the neighborhood during the time between successive pick analyses.
- Interviews with the inhabitants to identify proper interventions are good since it identifies interventions that are relevant to the inhabitants and also increases the involvement the inhabitants in the waste management system. However, these interviews may lead to improved sorting behavior (which is good in many perspectives) which means that the effect of the intervention is difficult to isolate.
- Pick analyses, which are an integral part of the RBT procedure, require large sample sizes to be statistically meaningful. This increases the cost of the procedure.

4. Conclusions and Outlook

The waste sorting behavior of households can be quantitatively measured and evaluated by waste sampling and pick analysis (*i.e.*, waste characterization by manual sorting). Thereafter, appropriate interventions for improvement of the sorting can be identified through interviews in two steps: (a) semi-structured interviews (with a few households) where the questions are based on the pick analysis results; and (b) short, structured interviews (with a large population of households) based on the results of the semi-structured interviews. Once the interventions have been implemented, their effect can be evaluated by a new pick analysis. The missorting ratios before and after the interventions can thereby be quantified and compared.

This four step recycling behavior transition procedure enables continued improvement of waste sorting behavior in any household waste collection system that includes waste sorting at the source. The combination of quantitative and qualitative methods allows the RBT procedure to evaluate the actual sorting behavior, to understand the reasons for this behavior, to identify the needs for improvement, to design relevant interventions, and to assess the effect of the interventions. The RBT procedure can be designed and applied in any source separation system irrespective of local factors such as culture and social-economic backgrounds.

Further development of the RBT procedure is needed. Repeated use of the procedure in various types of waste separation systems will identify more strengths and weaknesses of the procedure. The RBT procedure can also be combined with other methods such as life cycle assessment and cost analysis for a wider system analysis of a waste management system.

Author Contributions: Kamran Rousta has taken part in planning the study, conducting the experimental work, analyzing the results, and has been the principal writer of the manuscript. Lisa Dahlén and Kim Bolton, supervisors, have taken part in evaluating the results and writing the manuscript.

Abbreviations

The following abbreviations are used in this manuscript:

RBT   Recycling Behavior Transition

References

1. Swedish Waste Management. *Swedish Waste Management 2015*; Rapport U2015:25; Swedish Waste Management: Malmö, Sweden, 2015. (In Swedish)
2. Rousta, K.; Dahlén, L. Source separation of household waste; technology and social aspects. In *Resource Recovery to Approach Zero Municipal Waste*; Taherzadeh, M.J., Richards, T., Eds.; CRC Press: Boca Raton, FL, USA, 2015; pp. 61–77.
3. Guerin, D.; Crete, J.; Mercier, J. A multilevel analysis of the determinants of recycling behavior in the european countries. *Soc. Sci. Res.* **2001**, *30*, 195–218. [CrossRef]
4. Jesson, J. Household waste recycling behavior: A market segmentation model. *Soc. Mark. Q.* **2009**, *15*, 25–38. [CrossRef]
5. Martin, M.; Williams, I.D.; Clark, M. Social, cultural and structural influences on household waste recycling: A case study. *Resour. Conserv. Recycl.* **2006**, *48*, 357–395. [CrossRef]
6. Meneses, G.D.; Palacio, A.B. Recycling behavior a multidimensional approach. *Environ. Behav.* **2005**, *37*, 837–860. [CrossRef]
7. Oskamp, S.; Harrington, M.J.; Edwards, T.C.; Sherwood, D.L.; Okuda, S.M.; Swanson, D.C. Factors influencing household recycling behavior. *Environ. Behav.* **1991**, *23*, 494–519. [CrossRef]
8. Tucker, P.; Speirs, D. Count me as doing my bit: An appraisal of the accuracy of waste attitude/behaviour surveys. *CIWM Sci. Tech. Rev.* **2003**, *4*, 2003.
9. Dahlén, L.; Åberg, H.; Lagerkvist, A.; Berg, P.E.O. Inconsistent pathways of household waste. *Waste Manag.* **2009**, *29*, 1798–1806. [CrossRef] [PubMed]
10. González-Torre, P.L.; Adenso-Díaz, B. Influence of distance on the motivation and frequency of household recycling. *Waste Manag.* **2005**, *25*, 15–23. [CrossRef] [PubMed]
11. Porter, B.E.; Leeming, F.C.; Dwyer, W.O. Solid waste recovery a review of behavioral programs to increase recycling. *Environ. Behav.* **1995**, *27*, 122–152. [CrossRef]
12. Refsgaard, K.; Magnussen, K. Household behaviour and attitudes with respect to recycling food waste—Experiences from focus groups. *J. Environ. Manag.* **2009**, *90*, 760–771. [CrossRef] [PubMed]
13. Thøgersen, J. A model of recycling behaviour, with evidence from danish source separation programmes. *Int. J. Res. Mark.* **1994**, *11*, 145–163. [CrossRef]
14. Barr, S. *Household Waste in Social Perspective; Values, Attitudes, Situation and Behaviour*; ASHGATE: Burlington, VT, USA, 2002.
15. Pieters, R.; Verplanken, B. Changing our mind about behavior. In *The Consumption of Time and the Timing of Consumption—Toward a New Behavioral and Socio-Economics*; North-Holland: Amsterdam, The Netherlands, 1991.
16. Tucker, P.; Smith, D. Simulating household waste management behaviour. *J. Artif. Soc. Soc. Simul.* **1999**, *2*, 31.
17. Tucker, P. Understanding recycling behaviour. *Pap. Technol.* **2001**, *42*, 51–54.
18. Wilson, D.C.; Rodic, L.; Scheinberg, A.; Velis, C.A.; Alabaster, G. Comparative analysis of solid waste management in 20 cities. *Waste Manag. Res.* **2012**, *30*, 237–254. [CrossRef] [PubMed]
19. Stringer, E.T. *Action Research*; SAGE: Los Angeles, CA, USA, 2014.
20. Dahlén, L.; Lagerkvist, A. Methods for household waste composition studies. *Waste Manag.* **2008**, *28*, 1100–1112. [CrossRef] [PubMed]
21. Myers, M.D. *Qualitative Research in Business and Management*; Sage: Thousand Oaks, CA, USA, 2009.
22. Kvale, S.; Brinkmann, S. *Den Kvalitativa Forskningsintervjun*; Studentlitteratur AB: Lund, Sweden, 2009.
23. Rousta, K.; Ekström, K.M. Assessing incorrect household waste sorting in a medium-sized Swedish city. *Sustainability* **2013**, *5*, 4349–4361. [CrossRef]
24. Swedish Statistisc. *Socio-Economic Statistics*; Swedish Statistisc: Stockholm, Sweden, 2012.
25. Swedish Waste Management. *Manual for Pick Analysis of Houshold Waste/Manual för Plockanalys av Hushållsavfall*; RVF Utveckling 2005:19; Swedish Waste Management: Malmö, Sweden, 2005. (In Swedish)

26. Swedish Waste Management. *Manual for Pick Analysis of Houshold Waste/Manual för Plockanalys av Hushållsavfall*; RVF Utveckling 2013:11; Swedish Waste Management: Malmö, Sweden, 2013. (In Swedish)

27. Drever, E. *Using Semi-Structured Interviews in Small-Scale Research. A Teacher's Guide*; ERIC: Glasgow, UK, 1995.

28. Henriksson, G.; Åkesson, L.; Ewert, S. Uncertainty regarding waste handling in everyday life. *Sustainability* **2010**, *2*, 2799–2813. [CrossRef]

29. Thøgersen, J. Facilitating recycling: Reverse—Distribution channel design for participation and support. *Soc. Mark. Q.* **1997**, *4*, 42–55. [CrossRef]

30. Tashakkori, A.; Teddlie, C. *Sage Handbook of Mixed Methods in Social & Behavioral Research*; Sage: Thousand Oaks, CA, USA, 2010.

31. Newenhouse, S.C.; Schmit, J.T. Qualitative methods add value to waste characterization studies. *Waste Manag. Res.* **2000**, *18*, 105–114. [CrossRef]

6

Recycling Approach towards Sustainability Advance of Composite Materials' Industry

Maria Cristina Santos Ribeiro [1,2,*], António Fiúza [2,3], António Ferreira [2], Maria de Lurdes Dinis [2,3], Ana Cristina Meira Castro [3,4], João Paulo Meixedo [3,4] and Mário Rui Alvim [4]

[1] Institute of Science and Innovation in Mechanical and Industrial Engineering (INEGI), Rua Dr. Roberto Frias, Porto 4200-465, Portugal

[2] Faculty of Engineering of University of Porto, Rua Dr. Roberto Frias, Porto 4200-465, Portugal; afiuza@fe.up.pt (A.Fi.); ferreira@fe.up.pt (A.Fe.); mldinis@fe.up.pt (M.L.D.)

[3] CERENA-Polo FEUP, Center for Natural Resources and the Environnent, Porto 4100-465, Portugal; ana.meira.castro@eu.ipp.pt (A.C.M.C.); jme@isep.ipp.pt (J.P.M.)

[4] School of Engineering of Polytechnic of Porto (ISEP), Rua Dr. Bernardino de Almeida, 431, Porto 4200-072, Portugal; mario.alvim@alto.pt

* Correspondence: cribeiro@inegi.up.pt

Academic Editors: Michele Rosano and Julie Hill

Abstract: Worldwide volume production and consumption of engineered composite materials, namely fiber reinforced polymers (FRP), have increased in the last decades, mostly in the construction, automobile, aeronautic and wind energy sectors. This rising production and consumption have also led to an increasing amount of FRP waste, either end-of-life (EoL) products or manufacturing rejects. Taking into account the actual and impending EU framework on waste management, in which clear targets are set with concrete measures to ensure effective implementation, landfill and incineration will be progressively unavailable as traditional end-routes for this kind of waste. Recycling techniques and end-use applications for the recyclates have been investigated over the past twenty years, but even so, more cost-effective and feasible market outlets for the recyclates should be identified that meet both the economic and the environmental points of view. This paper is aimed at enclosing and summarizing an update overview regarding all these issues: current legislation, recycling techniques and end-use applications for the recyclates. Additionally, as a case study, the assessment of the potential improvements that could be made on the eco-efficiency performance (sustainability) of a typical FRP composite materials' industry by recycling and re-engineering process approaches is also reported.

Keywords: eco-efficiency; Fiber Reinforced Polymer waste; Glass Fiber Reinforced Polymer recyclates; economic-environmental; end-of-life (EoL) products; recycling; sustainability; waste management

1. Introduction

Worldwide volume production and consumption of fiber reinforced polymers (FRP) have increased in the last decades in several fields, mostly in the construction, automobile, aeronautic and wind energy sectors [1–4]. FRP composite materials are generally made of glass (GFRP), carbon (CFRP) or aramid (AFRP) reinforcing fibers dispersed in an organic matrix, usually polyester, epoxy or vinyl-ester thermoset resins. GFRP are by far the largest group of materials in the composites industry, representing over 95% of all FRP composites [5]. According to the Lucintel market report, a leading global management consulting and market research firm, the global glass fiber market is expected to grow at a compound annual growth rate of 5.4% over the five year-period 2015–2020 [6]. Although some contraction in specific market sectors (e.g., sheet and bulk molding compounds), and in some European countries (e.g., Scandinavian countries and France), the last "Federation of Reinforced Plastics" market report of the European Composites Industry Association (EuCIA) also confirms the steady global growth of GFRP composites industry over the last four years and estimated that the overall European-GFRP production by volume increased in 2015 by 2.5% to 1069 megatons [5].

Despite all the advantages of GFRP based products over more traditional materials, the growing production and consumption also lead to an increasing amount of GFRP waste, either end-of-life (EoL) products or manufacturing rejects. Since GFRP based products present, in general, a long life-span (20–25 years), end-of life disposal was not a major concern until a few years ago. However, the waste amount resulting from EoL GFRP products will increase strongly within the next few years, and this issue has become particularly worrying. The wind energy sector only is expected to cover 15.7% of the total EU electricity demand by 2020 and 50% by 2050, and the resulting EoL wind turbine blade material, mainly constituted of GFRP based components, is estimated to reach 100,000 tons per year in Europe [4,7]. Additionally, the total amount of production waste per year of the GFRP composite industry (e.g., non-conform products and manufacturing rejects) is also following the raising production. It is estimated that the total combined volume of EoL and production waste generated by the GFRP composite market in Europe has reached 304,000 tons in 2015 [8].

Taking into account the above figures, FRP waste management has become more and more an important concern. Whereas thermoplastic based FRP materials can be easily recycled by remelting and remolding, recyclability of thermoset based FRP products, with fiber recovering, is a more difficult task due to the inherent cross-linked nature of resin matrix [9,10].

Until now, landfilling and incineration have been the most common end-routes for EoL thermoset FRP products and scrap material. However, considering the actual and impending EU framework legislation on waste management, as well as the increasing price of landfill taxes, these end-routes will be progressively unavailable. Waste management legislation focuses on dealing waste through "waste hierarchy" and will therefore put more pressure on solving FRP waste management through recycling and reuse [11]. In particular, Waste Framework Directive 2008/98/EC stipulates that *"Member States shall take the necessary measures designed to achieve that by 2020 a minimum of 70% (by weight) of non-hazardous construction and demolition waste . . . shall be prepared for re-use, recycled or undergo other material recovery"* [12]. The Directive 2000/53/EC on end-of-life vehicles, which is already in force, also follows the same principle and settled minimum limits for the amount of parts and components of EoL vehicles that should be reused, recovered or recycled: *" . . . the reuse and recovery should be equal or higher than 95% and the re-use and recycling should be equal or higher than 85% (average weight per vehicle and year) . . . "* [13]. Therefore, in the near future, due to these more restrictive and coercive EU directives, FRP producers and suppliers could lose their market share to metals and other industries if they cannot ensure that their FRP components can be reused or recycled at the end of their service life cycle [11]. Thus, at the present time, the perceived lack of economical recyclability of thermoset FRP composites is more and more important and seen as a crucial barrier to the development or even continued use of these materials in some markets.

This increases awareness of environmental matters, and the search for further sustainable materials has driven several recycling techniques to be analyzed and proposed for FRP composites, mainly for GFRP and CFRP waste materials. Although research on recycling methods is underway, related research on end-use applications for the recyclates is still at a very elementary stage; however, in order to be cost-effective, recycling approaches should always embrace both interdependent issues.

The aim of this work is to enclose and summarize an updated review regarding all these features with special emphasis on GFRP waste: available recycling techniques, end-use applications for the recyclates and market outlook. Additionally, as a case study, the assessment of the potential improvements that could be made on the eco-efficiency performance (sustainability) of a typical FRP composite materials' industry by recycling and re-engineering process approaches is also reported.

2. Waste Recycling Solutions for Thermoset FRP Wastes

2.1. Recycling Techniques for Thermoset FRP Wastes

There are three main recycling processes that can be used to get an added value from FRP thermoset waste materials: (a) incineration, with partial energy recovery from heat generated during combustion of the organic part, and co-incineration with both energy and raw material recovering; (b) thermal and/or chemical recycling, such as solvolysis, pyrolysis and similar decomposition processes, with partial recovering of energy and reinforcing fibers; and (c) mechanical recycling, involving the composite break-down by shredding, milling, comminution or other similar mechanical processes, resulting in size reduction to fibrous and/or powdered products. A detailed description of these methods can be found on Pickering [9,14] and Asmatulu *et al.* [15]. The main key points are summarized in the following sections.

2.1.1. Incineration and Co-Incineration

Incineration of FRP scrap with energy recovery is listed as a recycling method in some literature, but this feature is still up for debate. Incineration does recover part of the energy of the scrap materials whereas landfilling does not; however, air pollution resulting from incineration is a drawback of this method. On the other hand, the fiber and filler content of the materials still wind up as landfilled waste, potentially becoming hazardous waste depending on chemical analysis of the ashes [16]. According to the current legislation, limits are settled concerning levels of emissions to air, water and soil, and the residues from the incineration process should be minimized in their amount and harmfulness [17]. The benefit of energy recovering is also discussable: calorific value will depend on the organic fraction and for typical GFRP/CFRP composites that only account for 30%–40% in weight. On the other hand, incinerator operators actually charge more for accepting FRP waste in order not to overload the system. Burning plastic waste limits the amount of household waste that can be processed, which means that large volumes of domestic waste (of which there is an unlimited supply) must be sent to the landfill [11]. At present, incineration, with partial energy recovering, as the first alternative to landfilling, is less and less considered as a cost-effective end-route for composite waste.

Co-incineration in cement kilns constitutes a recent alternative end-route for GFRP waste and is thought as a slightly better and cost-effective option, as this offers combined material and energy recovering. GFRP typically contains E-glass, which is alumina-borosilicate, along with the organic resin and often calcium carbonate filler. When fed into a cement kiln, the organic resin burns providing energy and the mineral constituents provide feedstock for the cement clinker, namely Si, Ca and Al. This means that no residue is left at the end. However, there is still a significant gate fee for this process. In addition, the total amount of fuel replacement in cement kilns by GFRP waste is limited due to the presence of boron commonly found in the E-glass fiber reinforcement. More than 0.2% of boron oxide in the cement increases the setting time and reduces the early strength. In practical terms, this means that no more than about 10% of the fuel input to a cement kiln could be replaced by GFRP waste material if no significant effect on the performance of the cement is required [14]. Other drawbacks of co-incineration in cement kilns rely on the requirements that GFRP waste must comply with: fragments of composite waste should be smaller than a designated size (20 mm × 20 mm), contain low concentrations of toxic materials and heavy metals, contain no foreign material (such as metal inserts or fasteners), have a specific calorific value (higher than 5000 kcal/kg), and must not generate dust such as pulverized glass fibers [18]. At present, co-incineration in cement kilns is commercially active in Germany through CompoCycle (Zajons Zerkleinerungs GmbH, Melbeck, Germany/Holcim AG, Hamburg, Germany) and is supported by EuCIA [19].

2.1.2. Thermal/Chemical Recycling

Thermo-chemical decomposition processes have been applied for fiber and partial energy recovering, mostly for CFRP composite waste due to the inherent economic value of carbon fibers.

Although it recovers both energy and material, these recycling processes are only cost-effective in the areas where paybacks are the highest (high economic value of the fibers), and where the volume of material to be processed is large enough to justify the capital cost of the technical plant.

The most common thermal process is pyrolysis, which consists of heating the scrap material in an inert atmosphere in order to recover the polymer material as oil. This kind of atmosphere prevents combustion, and, as result, the air pollution effects are less harmful in this process than in incineration. Another advantage is that the recovered oil can be used either as fuel or be refined to regenerate resin feedstock chemicals. As a limitation of this technique, the surface fragilities induced by the thermal stress on the recovered fibers, thus reducing its original strength, have been reported [10]. Oxidation in fluidized beds is another thermal process for FRP recycling, and it consists of combusting the polymer matrix in a hot and oxygen-rich flow. Recovered fibers by this process are clean and show very little surface contamination by char deposition; however, strength and fiber length degradation also occur [9,14]. Some recent research has shown that specific etching processes can significantly recover the original strength of reinforcing fibers that have been previously damaged by a thermal treatment [20,21], but this approach is still far away from an industrial realization.

The chemical recycling methods involve dissolution of the resin by means of chemical products and are based on a reactive medium (e.g., catalytic solutions and supercritical fluids) under low temperature [22]. Being thermal stress-free and quite gentle processes, chemical methods allow the fibers to retain most of their original strength. However, some limitations of these methods have also been pointed out: they usually involve the use of hazardous solvents, they require the previous granulation of scrap material in order to improve the specific surface, which causes length reduction of recovered fibers, and, additionally, they generally lead to weak adhesion to polymer matrix in posterior applications of recycled fibers [10]. As in the case of thermal recycling plants, chemical plants for composite recycling are not yet economically viable, at least for GFRP or relatively low volumes of CFRP waste to be processed [19].

2.1.3. Mechanical Recycling

Among the recycling technologies available for thermoset FRP composite materials, the most mature technique is mechanical recycling with size reduction by shredding, crushing or milling processes. The resultant recyclates, a mix of powdered and fibrous material, can be incorporated as filler or reinforcement replacement into new composite materials or as a closed-loop recycling process. This technique usually involves three steps: (a) initial size reduction of scrap material in some primary crushing process to pieces in the order of 50–100 mm in size; (b) final size reduction in jaw crushers, hammer or knife mills where the waste material is ground into a finer product ranging to 10 mm in size down to particles less than 50 μm; and (c) sorting and classifying operations to grade the resultant recyclates into fractions of different size (through cyclones or air zig-zag separators combined with sieving techniques) [23]. Typically, the finer grade fractions are mostly of a powdered nature with a high proportion of filler and resin particles, whereas the coarser fractions tend to be of a fibrous nature where the particles have a high aspect ratio and fiber content [9]. Although mechanical recycling has been considered mostly for GFRP composites, in which reinforcing fibers have a relatively low economic value, it can also be applied to process CFRP waste with environmental benefits as demonstrated by Howarth et al. [24].

Mechanical recycling shows significant environmental and economic advantages when compared to the previous recycling routes. In fact, mechanical size reduction: (a) does not produce atmospheric pollution by gas emission or water pollution by chemical solvents effluents; (b) does not require sophisticated, and expectably expensive, equipment like the ones that are required in the other processes; and (c) allows the processing of larger amounts of waste at higher throughputs. As drawbacks, two less attractive features have been pointed out: (a) safety issues due to risk of ignition during the shredding process; and (b) the lower value of the final product hardly competitive with homologous virgin raw materials such as calcium carbonate or shopped glass fibers. Nevertheless, ensuring that economically viable end-use applications for the recyclates exist, mechanical recycling at industrial scale processing is so far considered the most suitable recovery technique, at least for relatively low cost and promoter-free FRP materials.

2.2. End-Use Applications for GFRP Recyclates

Over the last 25 years, several end-use applications were investigated for mechanically recycled thermoset GFRP waste or recovered glass fibers, either as raw material for new composites or into a closed-loop recycling process for the same source-material. In the envisioned applications, GFRP recyclates were applied as filler, reinforcement or core material replacement as follows [25,26]: (a) filler material for artificial wood, high density polyethylene plastic lumber, rubber pavement blocks, dense bitumen macadam and bulk or sheet (BMC/SMC) molding compounds; (b) reinforcement for wood particleboard and soils; and (c) core material for textile sandwich structures. Most of the foreseen applications have not succeeded for one or both of the following reasons: (a) tendency of the recyclate addition to negatively affect the mechanical properties of final composite; and (b) negative cost balance, where mechanical recycling and sorting operational costs outweighed the market value of the virgin product.

Among the several potential applications of mechanically recycled FRP waste in new composite materials, a significant amount of research work has been carried out on Portland cement concrete in which the effect of GFRP recyclates, and more rarely CFRP recyclates, has been analyzed and assessed either as reinforcement, aggregate or filler replacement [27–41]. In the analyzed studies, a wide-range of replacement amounts was assessed: between 1% up to 20% in weight of total aggregates (after conversion of volume content to weight content). The applied FRP waste size fraction also differs widely, to relatively large pieces of GFRP or CFRP waste (5–30 mm square by 0.02–10 mm depth) [27,28] down to very fine grade fractions with particle average diameter less than a few microns [29–33,41]. However, in the biggest part of research studies, GFRP waste addition consisted of fluffy mixtures of powdered and fibrous particulate material with different lengths of glass fibers [34–40]. In addition, recovered glass fibers through pyrolysis recycling process were investigated for short reinforcement in cement mortars [42].

Besides the environmental benefits, and as a function of specific mix design formulation, reported added values due to FRP recyclate incorporation in cement based materials include slight to strong decreases of permeability with subsequent improved durability [29,31–33,35], less drying shrinkage [29,35], better workability [27,30–32], reduced risk of cracking induced by restrained shrinkage [31,35], improved fracture and tensile behavior [28,31,40], higher thermal insulation [32,41], and a global cost reduction of raw materials. In some particular cases, for lower sand replacement ratios, slender increases on compressive [34–39], splitting tensile [27,35], and/or flexural strengths [28,32,34,37–39] were observed. However, most of the time, undesirable features were noticed such as significant losses in the mechanical properties (in most of the cases due to high water–cement ratio required to achieve the desirable workability) [27–31,33,34,36–39,41,42], higher wear loss [33], higher setting times [30,31], potential incompatibility problems derived from alkalis–silica reaction (depending upon glass fiber nature and content) [29–39], higher susceptibility to chloride ion penetration [42], and weak adhesion at recyclate-binder interface [27]. This last issue, commonly found in the design process of composite materials modified with recycled plastics, was also addressed in some research works through the combined incorporation of GFRP recyclates and chemical coupling agents [32].

The global outputs of part of the above research works were addressed by Yazdanbakhsh and Bank (as 2013) in their revision study [43]. The resultant main highlights, benefits and drawbacks of FRP waste incorporation into Portland cement concrete materials are still valid today, even considering the most recent advances in that field. As they state in their conclusions, in general, the partial replacement of aggregates in Portland cement concretes and mortars by mechanically recycled FRP *"do not notably affect the durability of the final cementitious materials, but significantly reduce their mechanical properties*" [43]. In addition, the partial replacement of mineral aggregates by GFRP recyclates generally leads to minor decays in mechanical properties of final composite if fibrous GFRP waste fractions are applied instead of finely powdered GFRP recyclates.

Recently, some experiments were also carried out undertaking the incorporation of GFRP recyclates into polymer concretes (PC) and polymer mortars (PM) [25,26,44–47]. The effect of different replacement ratios of sand aggregates by both fine and coarse GFRP waste fractions on final mechanical properties of polyester based PM was assessed, as well as the effect of the incorporation of silane coupling agents. The obtained results showed that the partial replacement of sand aggregates by either of both GFRP waste fractions (up to 15% in weight of total aggregates) improves the compressive and flexural behavior of resultant PC/PM materials. Lately, the *Global Fiberglass Solutions Inc.* group (GFSI) also started to investigate this recycling route [48]. Compared to the end-use applications in cementitious based concrete materials already reported, the proposed solution overcomes some of the problems found, namely: (a) the possible incompatibilities problems due to alkalis–silica reaction; (b) the decrease in the mechanical properties; and (c) the poor bond between GFRP recyclates and matrix binder. Taking into account the obtained results, this last approach seems to be a very promising alternative end-route for mechanically recycled GFRP waste in concrete materials.

3. Sustainability Improvement of FRP Composite Materials' Industry: A Case Study

The sustainability of a business, company or industry is closely related to its eco-efficiency performance. Eco-efficiency is a management philosophy that encourages the companies to search for environmental improvements that also yield parallel economic benefits. The term was aimed at summing up, in a single expression, the business end of sustainable development: *"doing more with less"*, which means delivering more value while using fewer resources. Its focus is on business opportunities allowing companies to be more environmentally responsible and more cost-effective. Hence, implementing eco-efficiency is first and foremost about navigating for opportunities, and such opportunities can be found through four main approaches: (a) the re-engineering process approach, in order to reduce the consumption of resources and reduce pollution while at the same time saving costs; (b) the recycling approach, re-valorizing by-products and production waste through cooperation with other companies, promoting recycling and the reuse of recyclates into new added value products; (c) the re-designing approach according to ecological design rules that lead to less environmental impact, higher rate of recyclability and dissemble facility; and (d) the re-thinking market approach in order to find new ways of meeting customer needs [49].

In the present case study, the sustainability improvements that can be made in a composite materials industry were assessed by measuring the eco-efficiency performance of the company before and after the implementation of certain measures related to both the re-engineering process and recycling approaches. A pultrusion manufacturing company with headquarters in Maia (Portugal),—*ALTO, Perfis Pultrudidos Lda.*—,was the subject of this case study, and the analysis was restricted to the main business branch of this small/medium enterprise: the production and selling of standard GFRP pultrusion profiles.

3.1. Methods

3.1.1. Measurement of Eco-Efficiency Performance

The quantification of eco-efficiency performance of a company or business is a complex process that involves the measurement and control of several relevant parameters or indicators, globally applied to all companies (*Generally Applicable Indicators*), or specific according to the nature and specificities of the business itself (*Business Specific Indicators*). The indicators fall into two main groups based on the eco-efficiency formula represented by the ratio of the two *"eco"* dimensions of economy and ecology relating products or service values to environmental influence.

The *Generally Applicable Indicators* for product/service value are: quantity of goods produced or quantity of services provided to costumers (i) and net sales (ii). Those relating to the environmental influence in product/service creation are linked to the consumption of energy (i); raw materials (ii) and water (iii); emission of greenhouse gases (iv); and ozone depleting substances (v). The *Business Specific Indicators* are also discriminated according to their economic or ecological nature, but they are not global and must be individually defined from one business to another. A complete company's eco-efficient profile will include both types of indicators, value profile and environmental profile, and additionally, the eco-efficient ratios given by the previous two elements as "numerator" and "denominator" data.

In this particular study, the framework recommended by "The World Business Council for Sustainable Development" (WBCSD) was adopted [50] and the guidelines of the ISO 14301:1999 standard [51] were followed and applied. The main pertinent *Generally Applicable Indicators* for this case study, as well as the *Business Specific Indicators*, were defined and determined according to the above standard recommendations. With basis on indicators' figures, the value profile, the environmental profile and the pertinent eco-efficiency ratios were established and analyzed. The analysis was restricted to the main business branch of the company (ALTO): the production and selling of GFRP pultrusion profiles. The time-scale of the analysis was 75 working days and enclosed the production of seven different standard GFRP profiles illustrated in Figure 1. The main inputs and outputs of the pultrusion production process of ALTO are specified in Table 1.

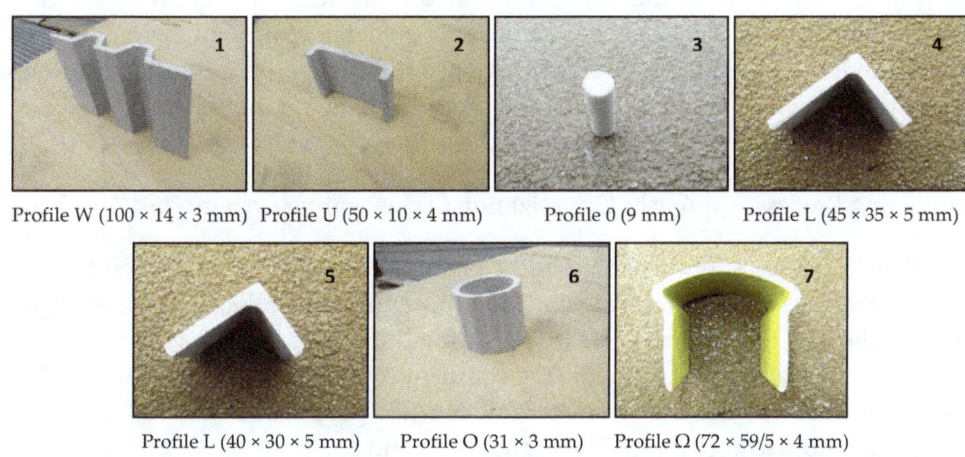

Profile W (100 × 14 × 3 mm)   Profile U (50 × 10 × 4 mm)   Profile 0 (9 mm)   Profile L (45 × 35 × 5 mm)

Profile L (40 × 30 × 5 mm)   Profile O (31 × 3 mm)   Profile Ω (72 × 59/5 × 4 mm)

Figure 1. Cross-sections of the GFRP pultrusion profiles analyzed in this case study.

Table 1. Main inputs and outputs of pultrusion manufacturing process of ALTO.

Main Inputs	Main Outputs
Electric Energy	GFRP pultrusion profiles
Virgin Raw Materials:	Pultrusion Wastes:
• Thermoset polyester resin; • Glass reinforcing fibers; • Calcium carbonate, pigments, catalyst systems and other additives	• Non-conform profiles; • By-products and manufacturing rejects; • Scrap material derived from cutting and assembly processes of GFRP profiles

Three *Generally Applicable Indicators* (two for product value and one for environmental influence), and one *Business Specific Indicator* of environmental influence were selected for eco-efficiency assessment. The specifications of each indicator are detailed in Table 2.

For each pair of "product value" and "environmental influence" indicators, the respective six eco-efficiency ratios were computed for the analyzed framework period (75 days). The same indicators and eco-efficient ratios were then predicted for an equivalent time period taking into account the implementation of improvement strategies.

Table 2.   Selected *Generally Applicable* and *Business Specific Indicators* for eco-efficiency performance assessment.

Generally Applicable Indicators	Category	Aspect/Unit
Quantity of Product: Total amount of GFRP profiles sold.	Product Value	Mass/kg
Net Sales: Total recorded sales less sales returns and allowances.	Product Value	Monetary/€
Energy Consumption: Total amount of electric energy consumed in pultrusion process.	Environmental Influence	Energy/kWh
Materials Consumption: Sum of weight of all raw materials required for GFRP profile production: polyester resin, glass reinforcing fibers (roving, mat and veil), calcium carbonate, pigments, catalyst system and additives.	Environmental Influence	Mass/kg
Business Specific Indicators	**Category**	**Aspect/Unit**
Total Waste to Landfill: Total amount of production waste for disposal (non-conform products, by-products, manufacturing rejects and leftovers derived from cutting and assembly processes of GFRP profiles).	Environmental Influence	Mass/kg

3.1.2. Improvement Strategies: Re-Engineering Process and Recycling Approaches

After analyzing all the procedures involved in the production process of GFRP profiles, it was concluded that it would be possible to improve the sustainability and eco-efficiency ratios of the company by reducing the environmental influence indicators: energy consumption, materials consumption and total waste to landfill. This can be possible by taking action on two key fronts, as described in the following two items.

• Re-engineering Process Approach: Optimization of die heating system

In the pultrusion process implemented in the company (ALTO), dry glass reinforcing fibers are pulled through a thermoset polyester resin bath for impregnation, and after the wetting process, the reinforcement is allowed to enter into a heated forming die where it attains the cross-section shape of the die and cures. Finally, outside the die, the composite profile already consolidated is pulled by a continuous pulling system and then a cut-off saw cuts the profile at a desired length. A schematic representation of the pultrusion process is presented in Figure 2.

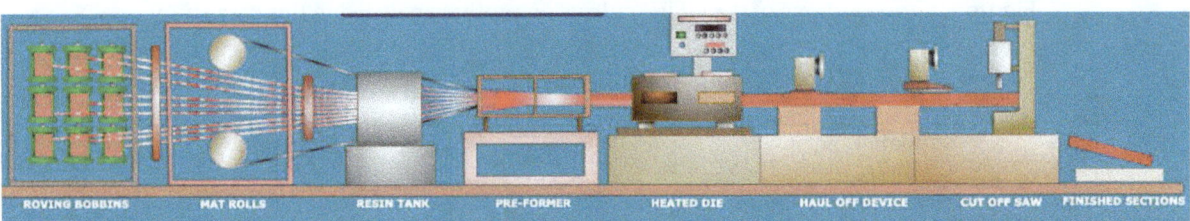

Figure 2. Schematic diagram of pultrusion manufacturing process.

Typically in the pultrusion process, and also in this case, the die is heated by external planar heaters as the most common heating system. However, this type of external heating system leads to a significant loss of heat in the surroundings of the die.

Earlier studies conducted by the authors showed that significant savings on energy consumption of the pultrusion process could be achieved using embedded cylindrical heaters into the die instead of external planar resistances [52]. Experiments were conducted in a 900 mm length die during the manufacturing process of a standard pultrusion profile (Profile U: $50 \times 10 \times 4$), keeping all the other process parameters constant—pulling speed, pulling force, total resistance power and temperature profile (TP)—along the die. These process parameters were already fine-tuned by the large experience of the manufacturer, and conducted to a high standard of quality of pultruded part. Temperature profile was first experimentally obtained by thermography techniques for the external heating system, and then numerically simulated by finite element analysis (FEA). After validation of FEA simulation, energy consumption with an internal heating system was estimated using the same technique. The obtained results showed that internal resistances significantly enhance the energetic performance of pultrusion process, leading to a 57% decrease of energy dispended in the die heating process, which represents a reduction of 17% of total energy consumed in the pultrusion process. The warm-up time is also reduced up to 50%, which reduces significantly the lead-time of each order and increases the production time. Moreover, in posterior studies, it was also found that the optimized position of the internal heaters throughout the die could even additionally reduce the energy consumption linked to the heating process 8% more [53]. More details of conducted research studies can be found in Silva *et al.* [52,53].

- Recycling Approach: Mechanical recycling of production waste and reuse of recyclates in new composite materials

In the actual framework of the pultrusion sector and, in general, in that of the composite materials' industry, production waste, non-conform and end-of-life products are usually landfilled due to their limited recycling ability, even when thermoplastic-based products are considered [54]. Currently, by-products, non-conform profiles and production waste of ALTO are also landfilled (Figure 3), with subsequent negative environmental impacts and supplementary added costs to this company. Waste to landfills constitute around 7% of total annual production of 40 tons and leads to an estimated cost for the company of 4 M€ per year. However, mechanical recycling of GFRP waste materials, with reduction to powdered and fibrous particulates (Figure 3), constitutes a recycling process that can be easily attained in heavy-duty cutting mills. The posterior reuse of obtained recyclates, either into a closed-loop process as calcium carbonate replacement for resin matrix of GFRP profiles (which represents an average of 20% in weight of total raw materials applied in the manufacturing process), or as reinforcement into new composite materials, will lead to both cost reduction in raw materials and landfill process and minimization of total amount of waste to landfill.

(a)   (b)

Figure 3. (a) typical waste of GFRP pultrusion process; (b) samples of obtained recyclates after mechanical recycling in a heavy-duty cutting mill using different sized-meshes inside the grinding chamber.

As referred in Section 2.2, mechanically recycled GFRP waste remains, however, mired by the scarceness of cost-effective end-use applications and clearly developed recycling routes (logistics, infrastructures and recycling facilities) between waste producers and potential consumers for the recyclates. Presently, new outlets and end-markets with added value for the GFRP recyclates are required. Regarding this subject, the use of the recyclates as raw materials in the production of polymer based concrete materials (precast industry) seems to be a very promising end-use application.

Polymer concrete (PC) materials are high performance resin based concretes, in which a polymer acts as binder matrix for the mineral aggregates. High mechanical strength, improved resistance to chemical and frost attack, high damping characteristics, very fast curing time and excellent bond to several substrates are the main advantages of these materials over cement based concretes [55–58]. Nevertheless, at present, the main asset of PC materials over conventional concretes is their great ability for incorporating recycled waste products, mainly due to the hermetic nature of resin matrix. Recycling and waste encapsulation constitute nowadays a new and emerging branch market for PCs. Most of the successful applications involve either industrial by-products or end-of-life products [59–63].

The previous and impending experimental works carried out by the present research team have already demonstrated that GFRP recyclates can be successfully incorporated into PC materials as reinforcement and partial replacement of aggregate components, leading to both flexural and compressive strength increase of modified concrete materials [25,26,44–47]. The main outcomes of these studies show that the upcycling of GFRP waste is possible resulting in added value PC products. It was demonstrated that aggregate replacement amounts up to 15% in weight by mechanically recycled GFRP waste are viable and cost-effective. Larger replacement amounts are also technically possible but lead to progressive drops in mechanical strength of the final products. Compared with related end-use applications in conventional Portland cement concrete materials, the proposed solution overcomes some of the limitations found (see sub-chapter 2.2). These limitations, by and large resultant from the use of a cementitious binder as matrix, are avoided using a cementless concrete as host material for the recyclates.

The above recycling approach highlights a viable technological option for improving the quality of GFRP filled PC materials, thus opening a door to selective recycling of GFRP waste. It is expected that around 80% of actual production waste of ALTO can be mechanically recycled and reduced to fibrous/filler material, and posteriorly reused either as reinforcement for polymer based concrete materials (pre-casting industry) or as partial calcium carbonate replacement of resin matrix in the pultrusion process, into a closed-loop recycling process.

3.2. Results and Discussion

3.2.1. Current and Predicted Value and Environmental Indicators

In the present analysis, it was assumed that the value indicators, namely the total amount of products sold and net sales, would not be affected by the implementation of the new improvement strategy approaches. This means that the current and the predicted value indicators will be the same, whereas the current and the predicted environmental influence indicators will be differentiated, taking into account the turnovers of those measures.

In what concerns the prediction of the new environmental influence indicators, the following assumptions were assumed:

- The replacement of the die heating system (external planar resistances by internal cylindrical heaters) leads to 17% savings on the total consume of electric energy due to the pultrusion process, irrespective of the type of die/GFRP profile production;
- The savings on electric energy due to the optimization of heater position along the die are not taken into account;
- The reduction on warm-up periods of the die at the beginning of each run/order is disregarded and is not reflected in an eventual increase of production rate (the value indicators were kept equal);
- Eighty-percent of the current amount of production waste to landfill is able to be mechanically recycled. This percentage corresponds to the average production waste fraction constituted of relatively unaltered and clean GFRP material—non-conform products and left-overs resulting

from cutting and assembly processes of GFRP profiles on site. The remained 20% of production waste is mainly constituted of roving, veil and mat scrap, not properly consolidated in a resin matrix, and other manufacturing rejects not able for mechanical recycling;

- Twenty-five percent of the total amount of calcium carbonate applied in the production process of GFRP profiles is the maximum amount that could be replaced by fine-ground GFRP recyclates into a closed-loop recycling process. Higher amounts of replacement, as experimentally tested by the producer, lead to slight decays in final mechanical properties and fire reaction performance of standard profiles.

The obtained value indicators are presented in Figure 4, whereas the measured and predicted environmental indicators are depicted in Figure 5. Presented values are discriminated according to the seven types of pultrusion profiles produced during the framework time. They include four *Generally Applicable Indicators* of product value and environmental influence indicated in Table 2 and one *Business Specific Indicator* of environmental influence (total of production waste to landfill).

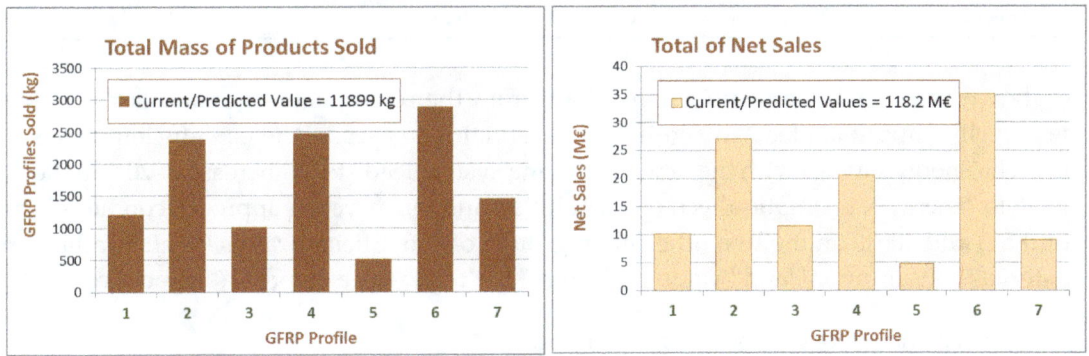

Figure 4. Value indicators according to the type of GFRP profile produced by the company ALTO.

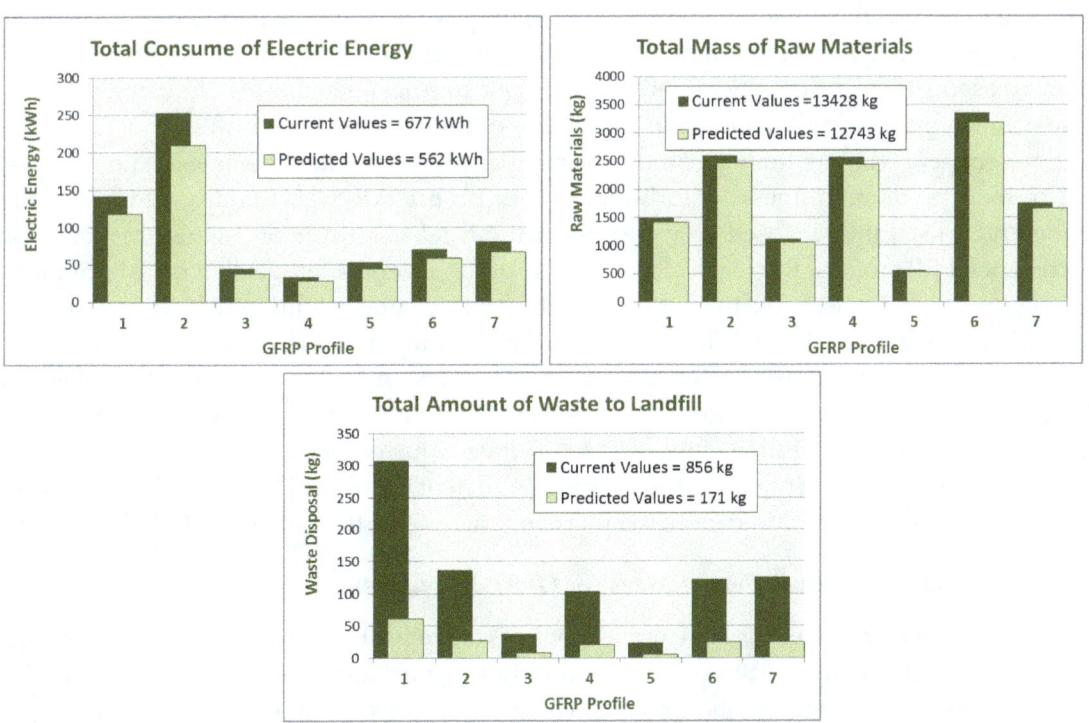

Figure 5. Current and predicted environmental influence indicators according to the type of GFRP profile produced by the company ALTO.

3.2.2. Current and Predicted Eco-Efficient Ratios

With basis on the above indicators, current eco-efficient ratios were determined and compared with those that could be obtained implementing the improvement strategy approaches. The obtained results are presented in Table 3.

Table 3. Current and predicted eco-efficiency ratios (before and after implementation of improvement strategies).

Eco-Efficiency Ratios	Mass of Product Sold per:		
	Energy Consumption	Materials Consumption	Total Waste Disposal
Current	17.58 kg/kWh	0.89 kg/kg	13.91 kg/kg
Expected	21.17 kg/kWh	0.93 kg/kg	69.58 kg/kg

Eco-Efficiency Ratios	Net Sales per:		
	Energy Consumption	Materials Consumption	Total Waste Disposal
Current	174.59 €/kWh	8.80 €/kg	138.08 €/kg
Expected	210.32 €/kWh	9.28 €/kg	691.22 €/kg

As shown in Table 3, the turnovers resultant from the re-engineering manufacturing process and the recycling approach lead to significant potential increases in the eco-efficient ratios of the company. The energy savings with the new die heating system leads to an increase of 20% in the "value indicators" to "energy consumption" eco-efficient ratios, and the recycling approach conducts increases of around 5% and 400% on the two other main groups of eco-efficient ratios ("value indicators" to "raw materials consumption" and "value indicators" to "total waste disposal", respectively).

3.2.3. Case Study: Main Conclusions and Revenues

The implementation of a new die heating system and, especially, mechanical recycling approach, with partial waste reuse of scrap material derived from manufacturing, cutting and assembly processes of GFRP profiles, lead to both minimization of waste disposal and cost reduction on raw materials, electric energy and landfill process. These features lead to significant improvements in the subsequent assessed eco-efficiency ratios of the present composite materials' industry, yielding to a more sustainable product and manufacturing process of pultruded GFRP profiles.

At present, as a result of this case study, the company implemented the new die heating systems in four of their production lines. In addition, part of production waste is diverted to mechanical recycling and reused into a closed-loop process as a partial substitute of calcium carbonate in the production of GFRP profiles. Regrettably, and in spite of the several attempts that have been made, the remaining production waste able to be mechanically recycled and intended to be diverted to the PC precast industry is still sent to landfill, jointly with the other manufacturing rejects. Eighty-percent of the total production waste of ALTO (which corresponds to around 2.2 tons per year) is not enough to challenge or attract the PC precast industry: the relatively small quantity does not justify the necessary adjustments that would be required in the production lines. Clusters and cooperative initiatives should be first created and implemented in order to create efficient "bridges" between the industries that generate GFRP waste in a certain region, the recycling plants and the potential PC precast industries.

4. Market Outlook and Future Perspectives for GFRP Recyclates

The intended perspectives for final applications of concrete and/or composite materials modified with GFRP recyclates include, among others, molding compounds, precast slabs, paving blocks, railroad sleepers, wall panels, manhole covers, valve chambers, cement floor screeds, valley gutters, roofing sheets and flat sheets for signage; however, few of these products came out of the investigation field and had an industrialized expression. One of the few successful applications (manhole covers, utility boxes and urban furniture made of thermoset and glass fiber flakes waste aggregated in a resin under high pressure cold molding) was developed by *Reprocover*, in Belgium, and it has been commercialized since 2011 [64]. Nevertheless, apart from some in-house recycling (such as the above example), until now, attempts to commercialize these products as a recycling route for GFRP waste have failed.

Recently, some progress made by *Extreme EcoSolutions* (Nijkerk, The Netherlands) and *Hambleside Danelaw* (Daventry, UK) in this field, reported by Job in 2014 [48], seemed to indicate that they were close to commercializing some end-use applications for mechanically recycled GFRP waste, namely, as filler for polyethylene film products and as reinforcement for construction elements, respectively. However, as far as it has been publicized, these products have not yet arrived to the market [65,66].

Regarding the application of GFRP recyclates to PC materials, the investigation line that was started by Ribeiro and co-workers in 2010 [25,26,44–47] also gained the attention of *Global Fiberglass Solutions Inc.* group. Over the last two years, this company has invested significant efforts on research and product development and expects to commercialize final precast products for rail and roadways infrastructures under the trademark of "*Ecopolycrete*" [67].

Even so, and despite all of the efforts that have been done on developing cost-effective recycling routes, the industrial applications of GFRP recyclates still remain hindered by the lack of clearly developed recycling paths (logistics, infrastructures and recycling facilities) between GFRP waste producers and potential consumers for the recyclates, and this seems to be the main barrier to a more generalized use of these recyclates in commercial applications. However, it is foreseen that this scenario will change in the next few years as strong investments are being made in this field. This will meet the EU circular economy policy on waste management with objectives and targets to improve waste management, stimulating innovation in recycling, limiting the use of landfilling, and creating incentives to change consumer behavior.

Acknowledgments: The financial support of "*Fundação para a Ciência e a Tecnologia*" (FCT), under the "*Programa Operacional Potencial Humano*" (POPH/ESF), Operational Program funded by the European Social Fund (SFRH/BPD/98869/2013 grant), as well as the technical support of ALTO, Perfis Pultrudidos, Lda., are gratefully acknowledged.

Author Contributions: Maria Cristina Santos Ribeiro, António Fiúza and Maria de Lurdes Dinis conceived and designed the experiments; Mário Rui Alvim performed the experiments; Maria Cristina Santos Ribeiro, Ana Cristina Meira Castro and João Paulo Meixedo analyzed the data; Mário Rui Alvim and António Ferreira contributed with analysis tools and resources; Maria Cristina Santos Ribeiro, Maria de Lurdes Dinis and António Fiúza wrote the paper.

References

1. Hollaway, L.C. A review of the present and future utilisation of FRP composites in the civil infrastructure with reference to their important in-service properties. *Constr. Build. Mater.* **2010**, *24*, 2419–2445. [CrossRef]

2. Zaman, A.; Gutub, S.A.; Wafa, M.A. A review on FRP composites applications and durability concerns in the construction sector. *J. Reinf. Plast. Compos.* **2013**, *32*, 1966–1988. [CrossRef]

3. GWEC, Global Wind Turbines Council. Market forecast for 2015–2019. Available online: http://www.gwec.net/global-figures/market-forecast-2012-2016/ (accessed on 9 May 2016).

4. Beauson, J.; Tilholt, H.; Brøndsted, P. Recycling solid residues recovered from glass fiber-reinforced composites—A review applied to wind turbine blade materials. *J. Reinf. Plast. Compos.* **2014**, *33*, 1542–1556. [CrossRef]

5. EuCIA—European Composites Industry Association. Composites Market Report 2015: Market developments, trends, challenges and opportunities (The European GRP Market/The Global CRP Market). Available online: http://www.eucia.eu/news/composites-market-report-2015 (accessed on 9 May 2016).

6. Global Glass Fiber Market 2015–2020: Trends, Forecast, and Opportunity Analysis. Available online: http://www.lucintel.com/glass_fiber_market_2020.aspx (accessed on 9 May 2016).

7.   Recycling of Wind Turbine Rotor Blades—Fact or Fiction? Available online: http://www.dewi.de/dewi_res/fileadmin/pdf/publications/Magazin_34/05.pdf (accessed on 9 May 2016).

8.   Jacob, A. Composites can be recycled? *Reinf. Plast.* **2011**, *55*, 45–46. [CrossRef]

9.   Pickering, S.J. Recycling thermoset composite materials. In *Wiley Encyclopedia of Composites*, 2nd ed.; Nicolais, L., Borzacchiello, A., Lee, S.M., Eds.; John Wiley & Sons: New York, NY, USA, 2012; Volume 4, pp. 2599–2614.

10. Pimenta, S.; Pinho, S. Recycling carbon fiber reinforced polymers for structural applications: Technology review and market outlook. *Waste Manag.* **2010**, *31*, 378–392. [CrossRef] [PubMed]

11. Conroy, A.; Halliwell, S.; Reynolds, T. Composite recycling in the construction industry. *Compos. Part A* **2006**, *37*, 1216–1222. [CrossRef]

12. 2008/98/EC. Directive 2008/98/EC of the European Parliament and of the Council of 18 November of 2008 on Waste and repealing some Directives. Available online: http://eur-lex.europa.eu/oj/direct-access.html (accessed on 9 May 2016).

13. 2000/53/EC. Directive 2000/53/EC of the European Parliament and of the Council of 18 September of 2000 on end-of-life vehicles—Commission Statemets. Available online: http://eur-lex.europa.eu/oj/direct-access.html (accessed on 9 May 2016).

14. Pickering, S.J. Recycling technologies for thermoset composite materials—Current status. *Compos. Part A* **2006**, *37*, 1206–1215. [CrossRef]

15. Asmatulu, E.; Twomey, J.; Overcash, M. Recycling of fiber-reinforced composites and direct structural composite recycling concept. *J. Compos. Mater.* **2014**, *48*, 539–608. [CrossRef]

16. Bartholomew, K. Fiberglass reinforced plastics recycling, technical report – Minnesota technical assistance program, December 2004. Available online: http://mntap.umn.edu/fiber/resources/report12-04.pdf (accessed on 9 May 2016).

17. 2000/76/EC. Directive 2000/76/EC of the European Parliament and of the Council of 4 December of 2000 on the incineration of waste. Available online: http://eur-lex.europa.eu/oj/direct-access.html (accessed on 9 May 2016).

18. Nomaguchi, K.; Hayashi, S.; Abe, Y. A Solution for Composites Recycling Cement Process. In Proceedings of the COMPOSITES 2001: Convention and Trade Show Composites Fabricators Association, Tampa, FL, USA, 3–6 October 2001.

19. Job, S. Recycling glass fiber reinforced composites—History and progress. *Reinf. Plast.* **2013**, *57*, 19–23. [CrossRef]

20. Yildirir, E.; Miskolczi, N.; Onwudili, J.A.; Németh, K.E.; Williams, P.T.; Sója, J.S. Evaluating the mechanical properties of reinforced LDPE composites made with carbon fibers recovered via solvothermal processing. *Compos. Part B* **2015**, *78*, 393–400. [CrossRef]

21. Yang, L.; Sáez, E.R.; Nagel, U.; Thomason, J.L. Can thermally degraded glass fiber be regenerated for close-loop recycling of thermosetting composites? *Compos. Part A* **2015**, *72*, 167–174. [CrossRef]

22. Morin, C.; Loppinet-Serani, A.; Cansell, F. Near and supercritical solvolysis of carbon fiber reinforced polymers (CFRPs) for recycling carbon fibers as a valuable resource: State of the art. *J. Supercrit. Fluids* **2012**, *66*, 232–240. [CrossRef]

23. Palmer, J.; Ghita, O.R.; Savage, L.; Evans, K.E. Successful closed-loop recycling of thermoset composites. *Compos. Part A* **2009**, *40*, 490–498. [CrossRef]

24. Howarth, J.; Mareddy, S.R.; Mativenga, P.T. Energy intensity and environmental analysis of mechanical recycling of carbon fiber composite. *J. Clean. Prod.* **2014**, *81*, 46–50. [CrossRef]

25. Ribeiro, M.C.S.; Meira-Castro, A.C.; Silva, F.G.; Santos, J.; Meixedo, J.P.; Fiúza, A.; Dinis, M.L.; Alvim, M.R. Re-use assessment of thermoset composite waste as aggregate and filler replacement for concrete-polymer composite materials: A case study regarding GFRP pultrusion waste. *Resour. Conserv. Recycl.* **2015**, *104*, 417–426. [CrossRef]

26. Meira Castro, A.C.; Carvalho, J.P.; Ribeiro, M.C.S.; Meixedo, J.P.; Silva, F.J.G.; Fiúza, A.; Dinis, M.L. An integrated recycling approach for GFRP pultrusion waste: recycling and reuse assessment into new composite materials using Fuzzy Boolean Nets. *J. Clean. Prod.* **2014**, *66*, 420–430. [CrossRef]

27. Alam, M.S.; Slater, E.; Billah, A.H.M. Green concrete made with RCA and FRP scrap aggregate: Fresh and hardened properties. *J. Mater. Civil Eng.* **2013**, *25*, 1783–1794. [CrossRef]

28. Ogi, K.; Shinoda, T.; Mizui, M. Strength in concrete reinforced with recycled CFRP pieces. *Compos. Part A* **2005**, *36*, 893–902. [CrossRef]

29. Tittarelli, F.; Moriconi, G. Use of GFRP industrial by-products in cement based composites. *Cem. Concr. Compos.* **2010**, *32*, 219–225. [CrossRef]

30. Tittarelli, F.; Kawashima, S.; Tregger, N.; Moriconi, G.; Shah, S.P. Effect of GRP by product addition on plastic and hardened properties of cement mortars. In Proceedings of the 2nd International Conference on Sustainable Construction Materials and Technologies, Ancona, Italy, 28–30 June 2010.

31. Tittarelli, F.; Shah, S.P. Effect of low dosage of waste GRP dust on fresh and hardened properties of mortars: Part 1. *Constr. Build. Mater.* **2013**, *47*, 1532–1538. [CrossRef]

32. Tittarelli, F. Effect of low dosage of waste GRP dust on fresh and hardened properties of mortars: Part 2. *Constr. Build. Mater.* **2013**, *47*, 1539–1543. [CrossRef]

33. Correia, J.R.; Almeida, N.M.; Figueira, J.R. Recycling of FRP composites: Reusing fine GFRP in concrete mixtures. *J. Clean. Prod.* **2011**, *19*, 1745–1753. [CrossRef]

34. Asokan, P.; Osmani, M.; Price, A.D.F. Assessing the recycling potential of glass fiber reinforced plastic waste in concrete and cement composites. *J. Clean. Prod.* **2009**, *17*, 821–829. [CrossRef]

35. Asokan, P.; Osmani, M.; Price, A.D.F. Improvement of the mechanical properties of glass fiber reinforced plastic waste powder filled concrete. *Constr. Build. Mater.* **2010**, *24*, 448–460. [CrossRef]

36. Osmani, M.; Pappu, A. An assessment of the compressive strength of glass reinforced plastic waste filled concrete for potential applications in construction. *Concr. Res. Lett.* **2010**, *1*, 1–5.

37. Osmani, M.; Pappu, A. Utilization of glass reinforced plastic waste in concrete and cement composites. In Proceedings of the 2nd International Conference on sustainable Construction Materials and Technologies, Ancona, Italy, 28–30 June 2010.

38. Osmani, M. Innovation in cleaner production through concrete and cement composite recycling. In Proceedings of the 3rd International Workshop Advances in Cleaner Production, S. Paulo, Brazil, 18–20 May 2011.

39. Garcia, D.; Vegas, I.; Cacho, I. Mechanical recycling of GFRP waste as short-fiber reinforcements in microconcrete. *Constr. Build. Mater.* **2014**, *64*, 293–300. [CrossRef]

40. Sebaibi, N.; Benzerzour, M.; Abriak, N.E. Influence of the distribution and orientation of fibers in reinforced concrete with waste fibers and powders. *Constr. Build. Mater.* **2014**, *65*, 254–263. [CrossRef]

41. Corinaldesi, V. Lightweight plasters containing plastic waste for sustainable and energy-efficient buildings. *Constr. Build. Mater.* **2015**, *94*, 337–345. [CrossRef]

42. Criado, M.; Garcia-Dáz, I.; Bastidas, J.M.; Alguacil, F.J.; López, F.A.; Monticelli, C. Effect of recycled glass fiber on the corrosion behavior of reinforced mortar. *Constr. Build. Mater.* **2014**, *64*, 261–269. [CrossRef]

43. Yazdanbakhsh, A.; Bank, L.C. A critical review of research on reuse of mechanically recycled FRP production and end-of-life waste for construction. *Polymers* **2014**, *6*, 1810–1826. [CrossRef]

44. Castro, A.C.M.; Ribeiro, M.C.S.; Santos, J.; Meixedo, J.P.; Silva, F.J.G.; Fiúza, A.; Dinis, M.L.; Alvim, M.R. Sustainable waste recycling solution for the glass fiber reinforced polymer composite materials industry. *Constr. Build. Mater.* **2013**, *45*, 87–94. [CrossRef]

45. Ribeiro, M.C.S.; Fiúza, A.; Castro, A.C.M.; Silva, F.G.; Dinis, M.L.; Meixedo, J.P.; Alvim, M.R. Mix design process of polyester polymer mortars modified with recycled GFRP waste materials. *Compos. Struct.* **2013**, *105*, 300–310. [CrossRef]

46. Ribeiro, M.C.S.; Fiúza, A.; Castro, A.C.M.; Meixedo, J.P.; Dinis, M.L.; Costa, C.; Ferreira, F.; Alvim, M.R. Recycling of Pultrusion Production Waste into Innovative Concrete-Polymer Composite Solutions. *Adv. Mater. Res.* **2011**, *295–297*, 561–565. [CrossRef]

47. Ribeiro, M.C.S.; Meixedo, J.P.; Fiúza, A.; Dinis, M.L.; Castro, A.C.M.; Silva, F.J.G.; Costa, C.; Ferreira, F.; Alvim, M.R. Mechanical Behavior Analysis of Polyester Polymer Mortars Modified with Recycled GFRP Waste Material. *World Acad. Sci. Eng. Technol.* **2011**, *75*, 365–371.

48. Job, S. Recycling composites commercially. *Reinf. Plast.* **2014**, *58*, 32–38. [CrossRef]

49. Lehni, M. Eco-efficiency: Creating more value with less impact. In *WBCSD Report*; World Business Council for Sustainable Development: Geneva, Switzerland, 2000.

50. Verfaillie, H.A.; Bidwell, R. *Measuring Eco-Efficiency—A Guide to Reporting Company Performance*; World business Council for Sustainable Development: Geneva, Switzerland, 2000.

51. International Organization for Standardization. ISO 14301:1999. Environmental Management—Environmental performance evaluation—Guidelines. ISO Standard, International Organization for Standardization: Geneva, Switzerland, 1999.

52. Silva, F.J.G.; Ferreira, F.; Costa, C.; Ribeiro, M.C.S.; Castro, A.C.M. Comparative study about heating systems for pultrusion process. *Compos. Part B* **2012**, *43*, 1823–1829. [CrossRef]

53. Silva, F.J.G.; Ferreira, F.; Ribeiro, M.C.S.; Castro, A.C.M.; Castro, M.R.A.; Dinis, M.L.; Fiúza, A. Optimising the energy consumption on pultrusion process. *Compos. Part B* **2014**, *57*, 13–20. [CrossRef]

54. Halliwell, S. *End of Life Options for Composite Waste: Recycle, Reuse or Disposal? National Composites Network Best Practice Guide*; National Composites Network: Chesterfield, UK, 2006.

55. Bhutta, M.A.R.; Ohama, Y. Recent status of research and development of concrete-polymer composites in Japan. *Concr. Res. Lett.* **2010**, *1*, 125–130.

56. Fowler, D.W. State of the art in concrete polymer materials in the US. In Proceedings of the 12th International Congress on Polymer Concrete, Chuncheon, Korea, 27–28 September 2007.

57. Ribeiro, M.C.S.; Nóvoa, P.R.; Ferreira, A.J.M.; Marques, A.T. Flexural performance of polyester and epoxy polymer mortars under severe thermal conditions. *Cem. Concr. Compos.* **2004**, *26*, 803–809. [CrossRef]

58. Suh, J.D.; Lee, D.G. Design and manufacture of hybrid polymer concrete bed for high-speed CNC milling machine. *Int. J. Mech. Mater. Des.* **2008**, *4*, 113–121. [CrossRef]

59. Barrera, M.G.; Campos, M.C.; Gencel, O. Polyester polymer concrete: Effect of the marble particle sizes and high gamma radiation. *Constr. Build. Mater.* **2013**, *41*, 204–208. [CrossRef]

60. Bignozzi, M.C.; Saccani, A.; Sandrolini, F. New polymer mortars containing polymeric waste. Part 1: microstructure and mechanical properties. *Compos. Part A* **2000**, *31*, 97–106. [CrossRef]

61. Garbacz, A.; Sokolowska, J.J. Concrete-like polymer composites with fly ashes–comparative study. *Constr. Build. Mater.* **2013**, *38*, 689–699. [CrossRef]

62. Nóvoa, P.J.R.O.; Ribeiro, M.C.S.; Ferreira, A.J.M.; Marques, A.T. Mechanical characterization of lightweight polymer mortars modified with cork granules. *Compos. Sci. Technol.* **2004**, *64*, 2197–2205. [CrossRef]

63. Reis, J.M.L.; Jurumenha, M.A.G. Experimental investigation on the effects of recycled aggregate on fracture behavior of polymer concrete. *Mater. Res.* **2011**, *14*, 326–330. [CrossRef]

64. Reprocover. Available online: http://reprocover.com (accessed on 9 May 2016).

65. Extreme EcoSolutions. Available online: http://extreme-ecosolutions.com (accessed on 9 May 2016).

66. Hambleside Danelaw, Building Products. Available online: http://www.hambleside-danelaw.co.uk/ (accessed on 9 May 2016).

67. Ecopolycrete. Available online: http://www.ecopolycrete.com (accessed on 9 May 2016).

7

Coal Mining Waste as a Future Eco-Efficient Supplementary Cementing Material: Scientific Aspects

Moisés Frías [1],*, Rosario García [2], Raquel Vigil de la Villa [2] and Sagrario Martínez-Ramírez [3]

[1]   Eduardo Torroja Institute for Construction Sciences (IETcc-CSIC), Madrid 28033, Spain

[2]   CSIC-UAM Associated Unit, Department of Geology and Geochemistry, Autonomous University of Madrid, Madrid 28049, Spain; rosario.garcia@uam.es (R.G.); raquel.vigil@uam.es (R.V.d.l.V.)

[3]   Institute for the Structure of Matter (IEM-CSIC), Madrid 28006, Spain; sagrario@iem.cfmac.csic.es

*   Correspondence: mfrias@ietcc.csic.es

Academic Editors: Michele Rosano, Katerina Adam and Maria Menegaki

Abstract: The stockpiling of tailings around coal mines poses a major environmental problem. Nonetheless, this clay mineral (kaolinite)-based waste can be reused as a supplementary cementitious material (recycled metakaolinite) in the manufacture of future eco-efficient cements. This paper explores the most significant scientific questions posed in connection with the conversion of this waste into pozzolans, such as the variation in product mineralogy depending on the sintering temperature and its effect on reaction kinetics in the pozzolan/$Ca(OH)_2$ system over a period of 365 days. The findings show that the optimal sintering temperature is 600 °C, such that the cementitious properties of the activated product are determined solely by the conversion of kaolinite into metakaolinite and are unaffected by the other clay minerals (micas). The presence of 20% activated coal waste favors the formation of larger amounts of aluminous phases such as C_4AH_{13} and C_4AcH_{12} than in the reference paste and enhances C–S–H gel polymerization.

Keywords: coal mining waste; mineralogy; blended cements; hydrated phases

---

1. Introduction

Today's society is undergoing a change from a linear (produce-use-discard) to a recycling-based circular economic model that pursues two ends: the generation of less waste and its reuse as raw materials in industrial processes. This sustainable approach is one of the mainstays of Europe's 2020 strategy [1].

Cement manufacturing is a benchmark in this regard, in light of its use decades-old use of industrial by-products and waste in different stages of production: as alternative fuels (worn tires), raw materials (waste water treatment sludge) or supplementary cementitious materials (SCMs) to produce eco-efficient cements [2–4]. Metakaolinite, a material well known for its high pozzolanicity that can be obtained by thermally activating natural kaolinite, is one of the standardized SCMs [5–11]. Products with a high metakaolinite content can also be obtained by activating high-kaolinite industrial waste, a more eco-friendly approach compliant with the fundamental premises of the circular economy [12–14].

The stockpiling of kaolinite-based coal mining waste (CW) has been posing severe environmental problems for decades [15–17]. These tailings are one of the primary sources of pollution in China [18], where around 4.5 billion tons have accumulated in over 1700 dumps, together occupying an area of $150\,km^2$. In light of its kaolinitic nature, if recycled, this carbon waste may be an alternative source of MK [19–21]. Prior research conducted by Frías et al. [22,23], García et al. [24] and Vegas et al. [25] has shown that this inert waste can be converted by thermal activation into a high added value pozzolan (activated coal waste, ACW) apt for use in the manufacture of innovative eco-efficient cements. The need has thus arisen to determine the short- and long-term behavior of these pozzolanic materials in blended cements.

This study expands on the existing scientific knowledge of both preliminary thermal activation and the subsequent hydration of these materials when blended with cement. A number of instrumental techniques (XRD, SEM-EDX, TG/DTA, micro-Raman and NMR) are deployed to analyze the phases initially present in the thermally activated MK and in 365 days hydrated blends containing 80% cement. The findings are compared to the results for a reference Ordinary Portland Cement (OPC) paste to determine the effect of the pozzolanic reaction between the ACW and the portlandite in the blended cement system.

2. Experimental Section

2.1. Materials

Raw coal mining waste (CW) was collected in the vicinity of an open-pit mine owned by Sociedad Anónima Hullera Vasco-Leonesa and located in the Spanish province of Leon. The waste was crushed, sieved under 90 microns and thermally activated at temperatures ranging from 600 °C to 900 °C for 2 h to produce ACW. Laser granulometry results showed that 10%, 50% and 90% of the ACW particles are less than 1.7, 8.11 and 32.63 microns, respectively.

The cement used was a commercial ordinary portland cement classified as CEM I 52.5 N [2]. The potential composition of the Portland clinker, computed as stipulated in the existing Spanish legislation [26], is given in Table 1. The blended cements contained 20% ACW, the maximum replacement ratio allowed for the manufacture of type II/A cements (6%–20%). After removal from the molds, the 1 cm × 1 cm × 6 cm prismatic specimens were cured in water at 20 °C until tested at 1, 28 or 365 days.

Table 1. Potential composition of clinker.

(%)	C_3S	C_2S	C_3A	C_4AF
OPC	43.61	27.71	9.29	9.70

2.2. Analytical Methods

The accelerated chemical method (40 °C) was applied to assess the pozzolanicity of the carbon waste activated for 2 h at 600 °C in an electrical muffle furnace. Further to that method, 1 g of pozzolan was immersed in 75 mL of saturated lime solution for a given reaction time. The solution was subsequently filtered and the CaO concentration was calculated in both the problem and control solutions. The sole difference between this method and the standardized pozzolanicity test for pozzolanic cements lies in the material, which was a pozzolan rather than a cement.

2.3. Instrumental Techniques

The mineralogical composition of the bulk samples was determined by X-ray powder diffraction (XRD) and the <2 μm fraction by the oriented film method, both on a Siemens D-5000 (Munich, Germany) X-ray diffractometer fitted with a Cu anode. Semi-quantitative mineralogical composition was calculated with the powder reflection method. The areas under the reflections for each mineral were found with Gaussian fitting, in which conventional software was applied to subtract the baseline intensity. The relative error for this method was calculated at 10%.

Sample morphology and microanalysis were conducted on an FEI Inspect (Hillsboro, OR, USA) scanning electron microscope equipped with a tungsten (W) source DX4i dispersive X-ray analyzer and an Si/Li detector. The chemical composition was determined as the mean of ten scans per sample, shown along with the standard deviation. The chemical formulas for the minerals were determined with EDS semi-quantitative analysis and the results expressed in oxides (wt.%).

TG/DTA trials were run on a Stanton Equipment Inc. STA 781 thermogravimetric analyzer. Powder samples weighing 30 mg to 35 mg were heated at a rate of 10 °C/min in an N_2 atmosphere.

Dispersive Raman spectra were recorded at 532 nm (Nd:YAG) on a Renishaw InVia system fitted with a Leica confocal Raman microscope and an electrically cooled CCD camera. Frequency was calibrated to silicon (520 cm^{-1}) and the spectral resolution was set at 4 cm^{-1}.

The Bruker AVANCE-400 (9.4 T) spectrometer used for ^{29}Si NMR analysis operated at a ^{29}Si frequency of 79.4 MHz. Kaolin (d = −91.5 ppm) referenced to TMS (d = 0 ppm) was used as the external standard for chemical shift. The delay time was 60 s and the spinning speed 4 kHz.

3. Results and Discussion

3.1. Comparison of Mineralogy after Activation at 500 °C to 900 °C

According to the XRD findings, the CW sample contained phyllosilicates (micas and kaolinite), quartz, calcite, dolomite and feldspars (Figure 1). Pozzolanic activity was induced by the phyllosilicates, for thermal activation in air at 600 °C to 900 °C is known to prompt dehydroxylation in many clay minerals [27,28], with the concomitant total or partial breakdown of their crystalline lattice structure and the formation of a highly reactive transition phase.

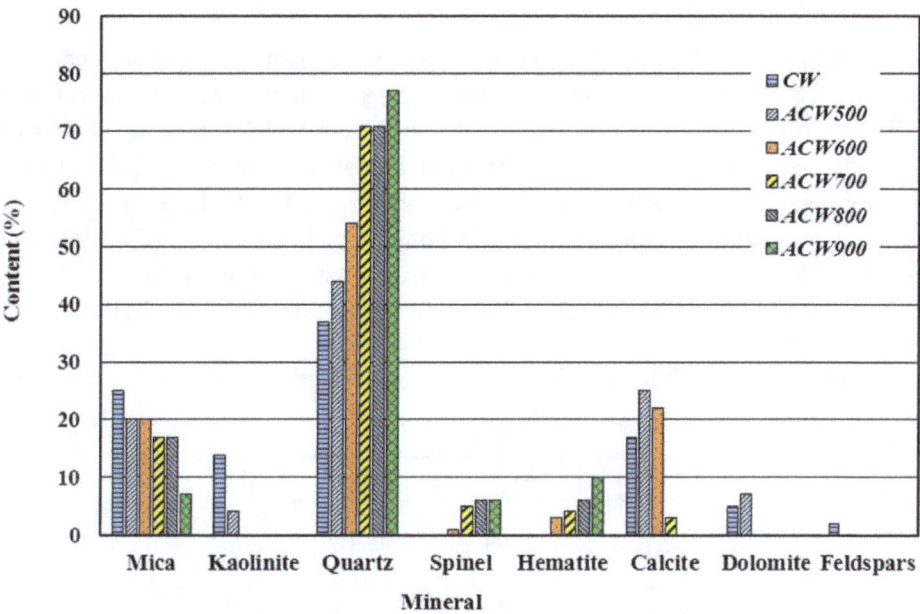

Figure 1. X-ray powder diffraction (XRD)-determined mineralogical composition (%) for the starting (CW) and thermally treated (ACW 500/900) coal waste.

Clay is most reactive when the activation-induced loss of hydroxyl ions leads to structural collapse and disarrangement. The sintering temperature required to convert inert kaolinite into activated metakaolinite (MK) usually ranges from 600 °C to 800 °C. Mica, in contrast, must be heated to over 930 °C to be activated and even then yields weak pozzolans.

The waste was heated at 500 °C, 600 °C, 700 °C, 800 °C and 900 °C to determine the optimal temperature for phyllosilicate dehydroxylation and new phase formation. The XRD results showed that full kaolinite dehydroxylation took place at 600 °C, with spinel-like phase neoformation occurring at 900 °C (Figure 1). The metakaolinite clusters visible in the SEM micrographs of ACW-600 were spongy and had blurred edges, whereas in the ACW-900 product the MK clusters were compact and had very sharply defined edges (Figure 2).

Figure 2. Micrographs of ACW produced by sintering CW at 600 °C (**left**) and 900 °C (**right**).

Consequently, 600 °C was the temperature required to convert the kaolinite in coal waste into metakaolinite.

3.2. Pozzolanic Activity

The amount of lime fixed by the activated waste over time (found with the accelerated chemical method described in the experimental section) was expressed as the ratio between the CaO content (in mmol/L) in the problem solution and the content in the control solution to obtain the pozzolanic index (PI) at each test age. According to the data in Table 2, waste ACW-600 exhibited high pozzolanicity, because 80% of the lime had been consumed in the first 28 days. The reaction rate declined after that age, with a PI of only 89% after 365 days. The behavior of ACW-600 was comparable (except in the first 24 h) to that of other types of industrial waste such as silica fume (SF) and bamboo leaf ash (BLA), regarded in the literature to be highly reactive pozzolans [29–31].

Table 2. Pozzolanic index (%) for ACW-600, bamboo leaf ash (BLA) and silica fume (SF) by hydration time.

PI (%)	ACW-600	BLA	SF
1 day	44.6	81.9	83.1
28 days	80.3	90.9	88.2
90 days	86.7	91.4	88.2
365 days	89.9	92.1	90.1

3.3. Hydrated Phase Behavior

3.3.1. XRD Analyses

XRD analysis detected portlandite, ettringite, tetracalcium aluminate hydrate (C_4AH_{13}) and tetracalcium aluminate carbonate hydrate (C_4AcH_{12}) in both the OPC and the 20% ACW blended cement (Figures 3 and 4, respectively). The proportions of each crystalline hydrated phase varied with hydration time and the presence or absence of the ACW.

Portlandite was the predominant hydrated phase in OPC at all test ages, followed in the first 28 days by ettringite. Tetracalcium aluminate carbonate hydrate (C_4AcH_{12}) formation was clearly visible after the first 28 days and the content of this phase remained flat throughout the period studied. Tetracalcium aluminate hydrate (C_4AH_{13}) was detected at 365 days. Portlandite declined substantially after 365 days and ettringite only slightly, while calcite grew.

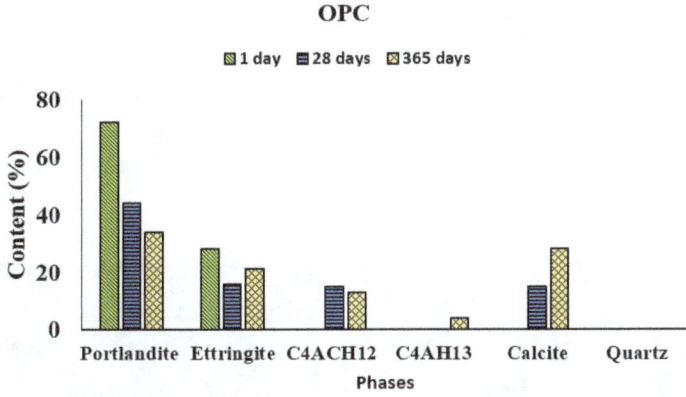

Figure 3. Mineral composition of OPC paste determined by XRD.

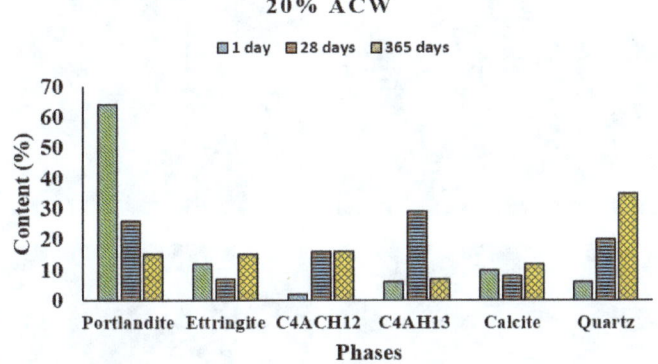

Figure 4. Mineral composition of 20% ACW determined by XRD.

As in OPC, portlandite was the majority phase in the 1 day 20% ACW. Quartz, absent in OPC, was detected in the blended cement. The pattern of variation with reaction time of ettringite, tetracalcium aluminate carbonate hydrate (C_4AcH_{12}) and tetracalcium aluminate hydrate (C_4AH_{13}) also differed in 20% AWC relative to OPC.

Ettringite was the second-most abundant phase in the latter paste only up to day 1, when C_4AcH_{12} and C_4AH_{13} formation was clearly visible. The inclusion of 20% ACW favored the formation of C_4AH_{13}, which was the majority phase at 28 days. Its appearance was attributed to the supersaturation of the aqueous phase with calcium hydroxide. The high concentrations of Ca^{2+} and OH^- in the solution due to the presence of $Ca(OH)_2$ favored the precipitation of C_4AH_{13}, although the content of this phase declined in the 365 days samples.

Calcium monocarboaluminate is normally detected immediately after hydration. The reaction between C_3A and $CaCO_3$ is governed by a solid-state mechanism, while the presence of the latter modifies the initial swift hydration of the former as a result of the barrier quickly generated by the reaction product, calcium carboaluminate hydrate. As hydration progresses, hemicarboaluminate gradually converts into monocarboaluminate. That process was not observed in the present study, however, because the first analysis was conducted 24 h after the materials were mixed with water. Calcite content grew with time.

3.3.2. SEM/EDX Analyses

The combined use of these two techniques afforded information on the morphology, texture and chemical composition of the hydrated phases, and particularly about the phases that could not be identified with XRD. SEM identified C–S–H gels (Figure 5). In the first 365 days, their CaO/SiO_2 ratio ranged from 1.6 to 2.5, therefore constituting Taylor classification type II gels. The ratio was closer to 2.5 in the OPC gels and lower in the gels present in the blended cement. Structural order rose with higher Ca/Si moduli and longer reaction times. In the early age pastes (1 day), the C–S–H gels comprised flaky clusters (Figure 5B), whereas at later ages (28 days) they exhibited thin layers of C_4AcH_{12}, C_4AH_{13} and portlandite intergrowths (Figure 5A). All three products were consistently laminar although their sizes varied. At longer hydration times (365 days), thin layers of portlandite aggregated into more compact clusters (Figure 5D), while ettringite prisms grew in the voids, together with C–S–H gel and thin layers of C_4AcH_{12} and C_4AH_{13}.

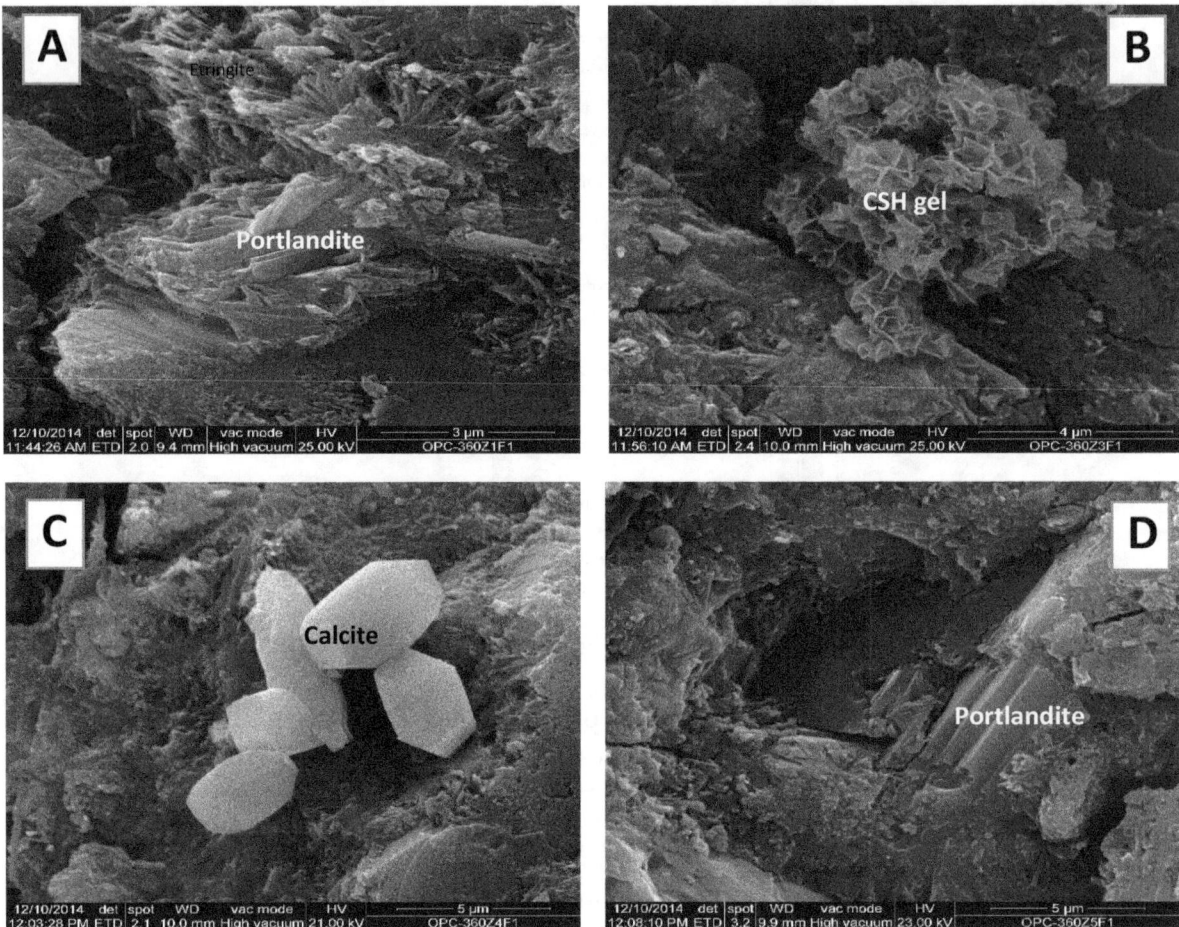

Figure 5. SEM micrographs of OPC and 20% ACW cement pastes at different curing times: (**A**) portlandite, (**B**) CSH gel, (**C**) calcite and (**D**) portlandite.

The inclusion of 20% ACW induced a more compact microstructure in the blended cement cured at 365 days, which exhibited larger and thicker layers of C_4AH_{13} and portlandite. Moreover, the C–S–H gel and prismatic ettringite clusters were less porous in the blended paste, in which rhombohedral calcite crystal content was likewise significant (Figure 5C).

3.3.3. TG/DTA

According to the thermal differential analyses (Figure 6), both cements exhibited three main endothermal signals. The wide band at 20 °C to 200 °C was attributed to water loss (45–50 °C); the band at 95 °C to 100 °C to ettringite and C–S–H gel dehydroxylation; and the band at 130 °C to 200 °C to the dehydroxylation of silicoaluminate (C_2ASH_8) and tetracalcium aluminate (C_4AH_{13}) hydrates, characteristically found in metakaolinite-based pozzolans [32,33]. Further to the XRD and SEM findings, the peak at 140 °C can be attributed primarily to C_4AH_{13} formation. The content of this hydrated phase rose with reaction time, particularly where carbon waste was present in the cement paste. No losses were identified at temperatures of around 200 °C, which would be attributed to carbo-aluminate phase (C_4AcH_{12}) decomposition.

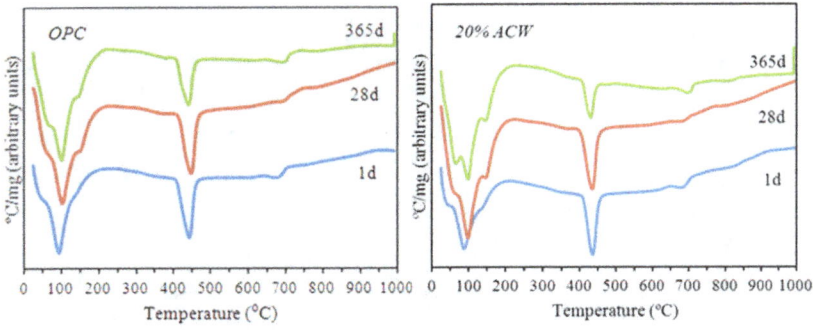

Figure 6. DTA curves for OPC and 20% ACW by hydration time.

Portlandite dehydroxylation was observed at 400 °C to 500 °C. The $Ca(OH)_2$ contents from TG curves and the portlandite content (fixed lime: $100 - (20\%\ ACW/OPC \times 100)$) in the pastes containing 20% ACW compared to OPC is given in Table 3 for the three reaction times studied.

Table 3. $Ca(OH)_2$ contents (%) and difference in both cements.

Cements	1 Day	28 Days	365 Days
OPC	13.7	17.9	19.4
20% ACW	12.5	15.7	11.3
Difference	−8.5	−12.1	−41.8

Pursuant to these data, the ACW clearly exhibited pozzolanic activity after 28 days due to the pozzolanic reaction between the metakaolinite and the portlandite generated during portland cement hydration, with the formation, primarily, of additional amounts of C–S–H gels and C_4AH_{13}, as discussed earlier.

The endothermal band at 650 °C to 700 °C was attributed to decarboxylation of the calcite present in both the starting OPC and the partially carbonated samples (an effect of curing and storage).

3.3.4. Micro-Raman

The 365 days Raman spectra for the OPC and 20% ACW pastes are shown in Figure 7. Both exhibited signals generated by characteristic cement hydration products: C–S–H gel (668 cm^{-1}), ettringite and monosulfoaluminate (999 cm^{-1} to 980 cm^{-1}). Whereas the signal peaking at 1084 cm^{-1} and assigned to calcite was narrow on the OPC spectrum, in the spectrum for the sample containing 20% waste it was wider (1070–1050 cm^{-1}) due to the presence of carboaluminate-like compounds. The OPC paste spectrum also contained a signal attributable to the OH groups in unreacted portlandite.

The ^{29}Si NMR spectra for the 365 days samples recorded to determine the characteristics of the C–S–H gel formed are reproduced in Figure 8. The weak signal observed at −71 ppm indicated that, despite the time elapsing, both pastes contained anhydrous phases. The main difference between the two spectra was that the Q1 unit signal was much more intense in the OPC than in the 20% ACW cement paste, denoting a smaller proportion of end of chain units and hence a higher degree of polymerization in the latter. Other authors have reported similar results for the pozzolanic reaction in fired clay-based waste [34]. Furthermore, the amount, distribution and microstructural characteristics of the C–S–H gel formed have a significant effect on the mechanical strength of the cement. Studies conducted by Tobón et al. [35] showed that strength rises with the amount of C–S–H gel. The presence of over 5% pozzolanic materials (which generate more polymerized gels), in turn, raised blended cement compressive strength substantially.

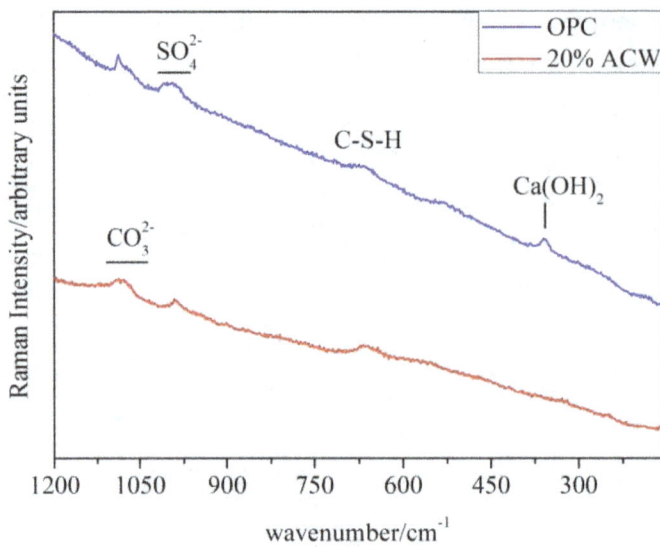

Figure 7. Micro-Raman spectra for 365 days OPC and 20% ACW pastes.

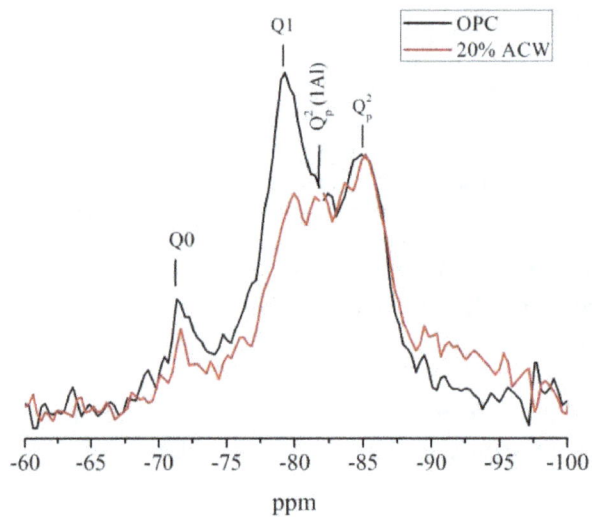

Figure 8. ^{29}Si NMR spectra for 365 days OPC and 20% ACW pastes.

4. Conclusions

As a result of its kaolinitic nature, coal mining waste affords an alternative avenue for obtaining metakaolinite, with substantial environmental and social benefits, one of the main pillars of the European circular economy.

The optimal sintering temperature for thermally activating coal waste, 600 °C, induces phyllosilicate dehydroxylation, giving rise to pozzolanic metakaolinite. When the temperature reaches 900 °C, in contrast, spinel-like phases form, lowering pozzolanicity. The amorphism and porosity of the metakaolinite clusters formed afford the material a high pozzolanic index, comparable to the indices for other highly reactive pozzolans.

Further to the mineralogical, chemical and microstructural findings discussed above, ACW is apt for use as a pozzolanic material in blended cements. The addition of 20% ACW to OPC intensifies C_4AcH_{12} and C_4AH_{13} formation. In the presence of the activated waste, C_4AH_{13} is the majority reaction product. Its higher content induces more compact microstructures with a greater abundance of thin layers of C_4AH_{13} and portlandite and lesser amounts of porous C–S–H gel and ettringite clusters. The C–S–H gel forming in OPC may exhibit fairly high Ca/Si ratios.

Economically, energetically and mineralogically speaking, coal tailings are apt for manufacturing eco-efficient cements, for at 600 °C kaolinite is fully converted to metakaolinite, whose hydration generates the same amorphous and crystalline compounds as ordinary cement.

Acknowledgments: This research was funded by the Spanish Ministry of the Economy and Competitiveness under coordinated projects MAT2012-37005-CO3-01, BIA2015-65558-C3-1-2-3R (MINECO/FEDER).

Author Contributions: The first author and head researcher for this study, Moises Frías, planned the experiments, conducted the TG/DTA analyses and drafted the manuscript. Rosario García and Raquel Vigil helped plan the paper and conducted the XRD and SEM-EDX trials. Sagrario Martínez conducted the micro-Raman and NMR studies and revised the manuscript.

References

1. Environment. Available online: http://ec.europa.eu/environment/circular-economy/index_en.htm (accessed on 25 April 2016).
2. Cement: Compositions, Specifications and Conformity Criteria for Common Cements. Available online: http://www.din.de/en/getting-involved/standards-committees/nabau/projects/wdc-proj:din21:204548497 (accessed on 17 March 2014).
3. Stark, J. Recent advances in the field of cement hydration and microstructure analysis. *Cem. Concr. Res.* **2011**, *41*, 666–678. [CrossRef]
4. Juenger, M.C.G.; Siddique, R. Recent advances in understanding the role of supplementary cementitious materials in concrete. *Cem. Concr. Res.* **2015**, *78*, 71–80. [CrossRef]
5. Soriano, L.; Monzó, J.; Bonilla, M.; Tashima, M.M.; Payá, J.; Borrachero, M.V. Effect of pozzolans on the hydration process pof portland cement cured at low temperatures. *Cem. Concr. Compos.* **2013**, *42*, 41–48. [CrossRef]
6. Nezerka, V.; Slízková, Z.; Tesárek, P.; Plachý, T.; Franeová, D.; Petráñová, V. Comprehensive study on mechanical properties of lime based pastes with additions of MK and brick dust. *Cem. Concr. Res.* **2014**, *64*, 17–29. [CrossRef]
7. Pavlík, V.; Uzáková, M. Effect of curing conditions on the properties of lime, lime-MK and lime-zeolite mortars. *Constr. Build. Mater.* **2016**, *9*, 14–25. [CrossRef]
8. Cassagnabere, F.; Escadeillas, G.; Mouret, M. Study of the reactivity of cement/MK binders at early age for specific use in steam cured precast concrete. *Constr. Build. Mater.* **2009**, *23*, 775–784. [CrossRef]
9. Frías, M. The effect of MK on the reaction products and microporosity in blended cement pastes submitted to long hydration time and curing temperature. *Adv. Cem. Res.* **2006**, *18*, 1–6. [CrossRef]
10. Frías, M. Study of hydrated phases present in a MK-lime system cured at 60 °C and 60 months of reaction. *Cem. Concr. Res.* **2006**, *36*, 827–831. [CrossRef]
11. Martínez, S.; Frías, M. Micro-raman study of stable and metastable phases in $MK/Ca(OH)_2$ system cured at 60 °C. *Appl. Clay Sci.* **2011**, *51*, 283–286. [CrossRef]
12. Frías, M.; García, R.; Vigil, R.; Ferreiro, S. Calcination of art paper sludge waste for the use as a supplementary cementing material. *Appl. Clay Sci.* **2008**, *42*, 189–193. [CrossRef]
13. Rodríguez, O.; Frías, M.; de Rojas, M.I.S. Influence of the calcined paper sludge on the development hydration heat in blended cement mortars. *J. Therm. Anal. Calorim.* **2008**, *92*, 865–871. [CrossRef]
14. Frías, M.; Vigil, R.; García, R.; de Rojas, M.I.S.; Baloa, T. Mineralogical evolution of MK-based drinking water treatment waste for use as pozzolanic material. The effect of activation temperature. *J. Am. Ceram. Soc.* **2013**, *96*, 3188–3195.
15. Prasad, S.; Byragi, T.; Vadde, R. Environmental aspects and impacts its mitigation measures of corporation coal mining. *Procedia Earth Planet. Sci.* **2015**, *11*, 2–7.
16. Zhang, Y.; Ge, X.; Nakano, J.; Liu, L.; Wang, X. Pyrite transformation and sulfur dioxide release during calcination of coal gangue. *RSC Adv.* **2014**, *4*, 42506–42513. [CrossRef]
17. Bian, Z.; Dong, J.; Lei, S.; Leng, H.; Mu, S.; Wang, H. The impact of disposal and treatment of coal mining wastes on environment and farmland. *Environ. Geol.* **2009**, *58*, 625–634. [CrossRef]

18. Liu, H.; Liu, Z. Recycling utilization patterns of coal mining waste in China. *Resour. Conserv. Recycl.* **2010**, *54*, 1331–1340. [CrossRef]

19. Beltramini, L.; Suarez, M.; Guillarducci, A.; Carrasco, M.; Grether, R. Aprovechamiento de residuos de la depuración del carbón mineral: Obtención de adiciones puzolánicas para el cemento Portland. *Tecnol. Cienc.* **2010**, *N.4*, 7–18.

20. Li, D.; Song, X.; Gong, C.; Pan, Z. Research on cementitious behaviour and mechanism of pozzolanic cement with coal gangue. *Cem. Concr. Res.* **2006**, *36*, 1752–1759. [CrossRef]

21. Zhang, T.; Gao, P.; Gao, P.; Wei, J.; Yu, Q. Effectiveness of novel and traditional methods to incorporate industrial wastes in cementitious materials—An overview. *Resour. Conserv. Recycl.* **2013**, *74*, 134–143. [CrossRef]

22. Frías, M.; Vigil, R.; de Rojas, M.I.S.; Medina, C.; Valdés, J. Scientific aspects of kaolinite based coal mining wastes in pozzolan/Ca(OH)$_2$ system. *J. Am. Ceram. Soc.* **2012**, *95*, 386–391. [CrossRef]

23. Frías, M.; de Rojas, M.I.S.; García, R.; Valdés, J.; Medina, C. Effect of activated coal mining wastes on the properties of blended cement. *Cem. Concr. Compos.* **2012**, *34*, 678–683. [CrossRef]

24. García, R.; Vigil, R.; Frías, M.; Rodriguez, O.; Martínez, S.; Fernández, L.; de Soto, I.S.; Villar, E. Mineralogical study of calcined coal waste in a pozzolan/Ca(OH)$_2$ system. *Appl. Clay Sci.* **2015**, *108*, 45–54. [CrossRef]

25. Vegas, I.; Cano, M.; Arribas, I.; Frías, M.; Rodríguez, O. Physical-mechanical behaviour of binary cements blended with thermally activated coal mining waste. *Constr. Build. Mater.* **2015**, *99*, 169–174. [CrossRef]

26. AENOR. *Spanish Standard UNE 80 304: 2006*; Cements. Calculations of potential composition of Portland Clinker; AENOR: Madrid, Spain, 2006.

27. Ambroise, J.; Murat, M.; Pera, J. Investigations on synthetic binders obtained by middle-temperature thermal dissociation of clay minerals. *Silic. Ind.* **1986**, *7*, 99–107.

28. Sayanam, R.A.; Kalsotra, A.K.; Mehta, S.K.; Singh, R.S.; Mandal, G. Studies on thermal transformations and pozzolanic activities of clay from Jammu region (India). *J. Therm. Anal.* **1989**, *35*, 99–106. [CrossRef]

29. Villar, E.; Valencia, E.; Santos, S.F.; Savastano, H.; Frías, M. Pozozlanic behaviour of bamboo leaf ash: Characterization and determination of the kinetic paprameters. *Cem. Concr. Compos.* **2011**, *33*, 68–73. [CrossRef]

30. De Sánchez Rojas, M.I.; Rivera, J.; Frías, M. Influence of the microsilica state on pozzolanic rate. *Cem. Concr. Res.* **1999**, *29*, 945–949. [CrossRef]

31. Hewllet, P.C. *Lea's Chemistry of Cement and Concretes*, 4th ed.; Arnold: London, UK, 1998.

32. Frías, M.; Rodríguez, O.; de Rojas, M.I.S. Paper sludge, an environmentally sound alternative source of MK based cementitious materials. A review. *Constr. Build. Mater.* **2015**, *74*, 37–48. [CrossRef]

33. Frías, M.; Martínez, S.; Blasco, T.; Frías-Rodríguez, M. Evolution of mineralogical phases by ^{27}Al and ^{29}Si NMR in MK-Ca(OH)$_2$ system cured at 60 °C. *J. Am. Ceram. Soc.* **2013**, *96*, 2306–2310. [CrossRef]

34. Medina, C.; del Bosque, I.F.S.; Asensio, E.; Frías, M.; de Sanchez Rojas, M.I. Mineralogy and microstructure of hydrated phases during the pozzolanic reaction in the sanitary ware waste/Ca(OH)$_2$ system. *J. Am. Ceram. Soc.* **2016**, *99*, 340–348. [CrossRef]

35. Tobón, J.I.; Payá, J.J.; Borrachero, M.V.; Restrepo, O.J. Mineralogical evolution of Portland cement blended with silica nanoparticles and its effect on mechanical strength. *Constr. Build. Mater.* **2012**, *36*, 736–742. [CrossRef]

8

The EU Circular Economy and Its Relevance to Metal Recycling

Christian Hagelüken [1,*], Ji Un Lee-Shin [2], Annick Carpentier [3] and Chris Heron [3]

[1]   Umicore, Hanau 63457, Germany
[2]   Umicore, Brussels 1000, Belgium; JiUn.LeeShin@eu.umicore.com
[3]   Eurometaux, Brussels 1000, Belgium; Carpentier@eurometaux.be (A.C.); Heron@eurometaux.be (C.H.)
[*]   Correspondence: christian.hagelueken@eu.umicore.com

Academic Editors: Michele Rosano and Julie Hill

Abstract: This paper provides an overview of ongoing European policy actions to improve the circular management of non-ferrous metals. After explaining why metals are at the center of the European Union's circular economy initiative, the authors outline a number of issues that still need tackling to "close the loop", and prevent Europe's metals from being landfilled, incinerated, or exported without guarantee of high-quality treatment. Electronic waste is focused on in detail during this analysis, because of the special challenges in environmentally sound recovery of smaller quantities of valuable and precious metals. In particular, the authors find that a mandatory certification scheme for recyclers of electronic waste, in or out of Europe, would help to incentivize high-quality treatment processes and efficient material recovery. More generally, the article finds that the European Commission's waste legislation proposals and Action Plan begins to address key challenges, provided the requirements are implemented strongly and consistently across Member States. In particular, it is crucial that EU policy establishes level playing field conditions for European metals recyclers

Keywords: metals; recycling; circular economy; electronic waste; WEEE (waste electrical and electronic equipment); EU; policy; resource efficiency

---

1. The EU Circular Economy Package December 2015

After the financial crisis, in 2010 the European Commission put forward its 10-year "Europe 2020" strategy [1] for improving and boosting EU competitiveness and employment. A resource efficient Europe was identified as a main engine for sustainable growth, by bringing major economic opportunities, improvement of productivity, cutting costs and increasing competitiveness. Since then, the Commission has published the Roadmap to a Resource Efficient Europe [2], giving long-term policy guidance on increasing resource productivity and decoupling growth from resource use, taking into account environmental impacts. From the Roadmap, the most central policy action proposal to date on resource efficiency has been the circular economy package tabled on 2 December 2015 [3].

The European Commission had initially adopted an earlier proposal in July 2014 [4] (Figure 1), which was withdrawn to make means for a more ambitious motion to include more than just waste management.

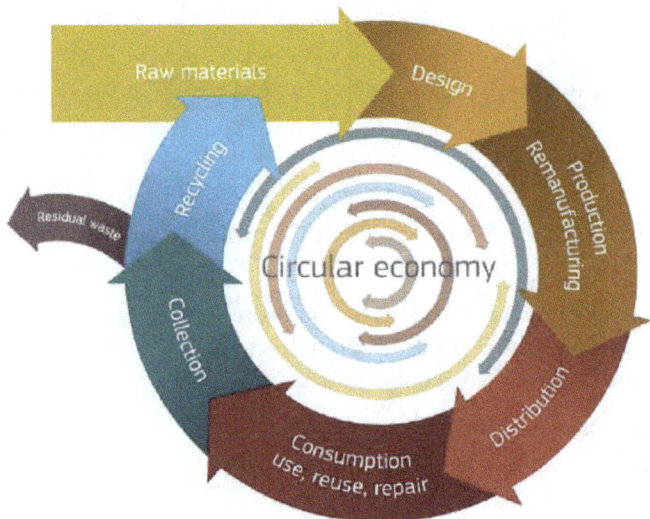

Figure 1. Towards a circular economy: A zero waste programme for Europe.

Therefore, on 2 December 2015, the European Commission presented its new Package proposal, consisting of: (1) a communication with an action plan, establishing measures with a timeline covering the whole cycle: From production, and consumption to waste management and market for secondary raw materials [3]; and (2) the revised legislative proposals on waste including the general Waste Framework Directive [5], the Landfilling Directive [6], and the Packaging and Packaging Waste Directive [7].

The European Commission defines circular economy as a transition "where the value of products, materials and resources is maintained in the economy for as long as possible, and the generation of waste minimized" [3]. Metals, including Critical Raw Materials, are a central element in the package and an ideal candidate for a circular economy as they are eternally recyclable, and secondary metals do not face downcycling or quality issues. Their comprehensive recycling and reintroduction into new lifecycles helps to secure Europe's access to metals; a key building block for a competitive economy. There are many examples where metal recycling rates are already very high: precious metals in jewelry and industrial process catalysts, steel, and base metals such as aluminum, copper and lead. However, there is significant room for improvement for many other non-ferrous metals, especially precious and specialty metals (the latter sometimes also referred as "minor metals"), which could be recovered more effectively from industrial residue streams, and end-of-life consumer goods, such as vehicles, electronic applications, or rechargeable batteries. This does not mean to strive for a 100% recycling of all metals, as technical and economical limits need to be considered. As discussed in Reference [8] an optimum mix of secondary and primary metals supply exists. Recovering the last bits of metals from complex and low-grade materials can become more energy intensive than supplying these from primary sources. However, in most cases, the recovery of secondary metals from products/materials is far less energy intensive than their mining, as the metal concentration in many products is higher than in ores, especially in the case of precious and specialty metals (e.g., in electronic products). As mining conditions are expected to become more difficult (lower ore grades, more complexity, greater depths, etc.) the optimum mix will move towards an increasing share of secondary metals [8].

The focus of this paper will be on the existing challenges to improve the circular management of metals, and the measures needed to create the necessary framework conditions to support this. It will also identify how the circular economy package can support high-quality recycling of metals under level playing field-conditions, and incentivize material recovery.

2. A Circular Economy for Metals-Benefits, Opportunities, Challenges

Unlike other raw materials, such as fossil fuels or food, metals are not consumed. Since they do not lose their intrinsic properties during recycling, metals can be used and re-used multiple times, maintaining their quality and functionality. In this sense, metals are a material with permanent characteristics that can be qualified as a permanently available resource, as long as they remain within the planet boundaries, and are not dissipated into environments where their recycling is not feasible for technical-economic reasons [8]. Metals are essential components in key sustainable innovations, including low-carbon transport, renewable energies and digital communications.

Demand for raw materials will increase alongside higher market penetration of these applications, as well as the increasing global population and its middle class. This has created two major challenges within the EU: Securing cost-efficient and sustainable access to raw materials, and increasing resource efficiency. Recycling provides a highly-efficient way of reintroducing valuable materials back into the economy, and, by doing so, tackling the key strategic challenges, while lowering environmental impacts and energy intensity of materials supply.

Metals recycling has significant benefits, as are summarized below [9]:

- Substituting primary raw materials. Europe produces only about 3% of the primary raw materials it needs for metals production, while Europe's urban mine offers a great potential to recycle more, especially technology and critical metals.
- Reducing environment and CO_2 impacts of the production of secondary raw materials compared to primary material. Recycling saves up to 20 times the energy needed to produce metals and reduces the impact on water, air, soil, and biosphere.
- Reducing dependency on imported materials and secured supply of valuable materials, some of which are critical materials (metals that are important to the EU economy and show a supply risk).
- Avoiding landfill and incineration of metals, which is not only a loss of valuable raw materials, but also generates impacts on the environment.
- Supporting economic activities in Europe at the different stages of the metals recycling value chain (collection, pre-processing and end-processing), and in downstream industries thanks to the security of supply.

The growing demand for metals, however, cannot be met by recycling alone. Primary (mining) and secondary (recycling) supply will remain complementary in the future. Due to the permanent nature of metals and the long lifetime of some metal bearing products and infrastructures (which can stay in stock for tens or hundreds of years), we have been building up a significant anthropogenic stock, creating a potential future urban mine. Setting up a circular economy means that at the end of these products' lives—whenever and wherever this will take place—they need to be properly and efficiently recycled. Typically, consumer goods, such as electrical and electronic applications, and rechargeable batteries have relatively short lifecycles. They also contain a number of valuable and critical metals.

Today, advanced metallurgical recycling technologies exist for such complex products, so technically circularity can be achieved for a high number of their metals. The circular economy package can play an important role to support quality circular management.

The objective of a "circular economy" is quite clear: A circle will only be closed if materials physically find their way into new product lifecycles. In line with the waste hierarchy, the priority should be to maintain products and materials in the economy as long as possible through waste prevention and reuse. However, products will eventually reach their end-of-life and access to these for recycling, prevention of illegal or dubious waste exports, and use of best available recycling technologies need to be ensured.

There are various challenges that need to be addressed. Collection of end-of-life appliances or residue streams is a necessary prerequisite to allow recycling, but it is not sufficient. The next step is to ensure that products and materials are treated in high quality processes along the entire value chain, so as to be able to recover a wide range of metals with good yields and with good environmental performances, including safe elimination of hazardous substances. The growing complexity of products makes this increasingly challenging and there are some limitations to recover "all" metals from complex products [10]. As metal cycles are complex and interdependent a good managing of the overall system is crucial. For example, base metals such as lead and zinc solve key functions in certain products but, increasingly, they are linked in concert with other base metals, copper, nickel and cobalt, and with a multitude of special and precious metals in today's products and infrastructure. At the products' end-of life, in most cases, the recovery of precious and special metals is closely linked to the extractive metallurgy for base metals, i.e., the latter playing an enabling role to close the loop for the former. The understanding of this "Web of Metals" and the existing of a state-of-the-art metallurgical recycling infrastructure, hence, is a key prerequisite to achieve a circular economy [11].

In addition to the logistical, technical and environmental constraints, there are also economic challenges. High quality recycling entails costs that need to be covered. Ideally, the value of the recoverable materials matches the costs of the entire recycling treatment chain. If this is not the case, the gap needs to be bridged through other means (e.g. through recycling fees or innovative business models). Hence, especially at times of depressed raw material prices, the legislator needs to establish supportive framework conditions and a level playing field to allow high quality recycling of complex products containing valuable materials, and close the economic gap where needed.

3. The Recycling Value Chain—The Case of Consumer Electronics

The recycling value chain can be defined as the sequence of operations leading to the recovery of materials from waste. These operations include (1) collection, the beginning of any waste management process; (2) preparation for material recovery, which covers manual and/or mechanical operations and physical sorting; and (3) material recovery, which consists of chemical, physical and/or metallurgical operations, but does not include incineration for energy recovery and the reprocessing into materials that are to be used as fuel. The recycling value chain ends when the waste is reprocessed into products or materials, which do not require any further processing whether for the original or other purposes [12]. In other words, final outputs of a recycling chain are metals and materials in a sufficiently pure quality that are capable of replacing primary metals (i.e., originating from mining chains) as input raw materials for the manufacturing of new products.

The recycling value chain is illustrated in Figure 2 for the case of electronics recycling, which requires a complete process chain from collection to sorting and dismantling into separate components containing valuable metals (pre-processing), and then to a subsequent final metallurgical processing (end-processing). In this last end-processing step, valuable metals are extracted usually in a combination of pyrometallurgical and hydrometallurgical processes and purified to pure metals, which are then delivered back to the market for new product lives. The actual physical metal recycling takes place at the very end of the chain, but the preceding steps are crucial for directing the fraction containing the various metals to quality final recovery processes. Figure 2 illustrates this recycling chain with an estimate on order of magnitude on number of actors worldwide.

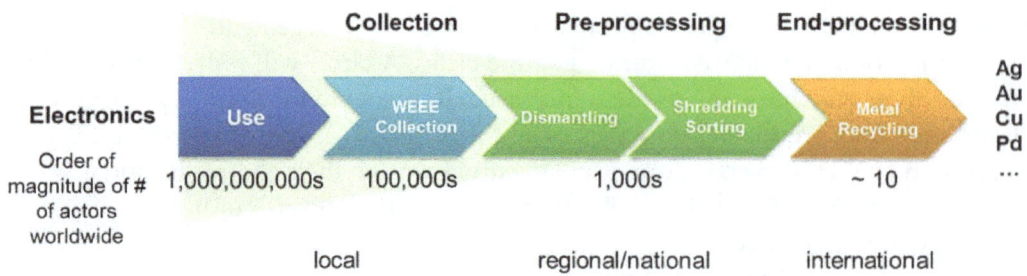

Figure 2. Recycling value chain for electronic waste.

"Mono-substance" waste streams, such as glass or steel scrap, are generally available in large quantities and contain little or no hazardous substances. Recycling of these streams is technically relatively simple. To the contrary, complex "poly-substance" end-of-life products, such as consumer electronics, contain a wide range of different materials, including hazardous substances. These streams require a different and more complex recycling chain treatment. Whilst for mono-substance streams, the focus is on cost optimization and mass and volume recovery, for complex poly-substance streams, it is essential to focus on environmentally sound value recovery of materials that are often present in low concentrations. For example, the mass of precious metals contained in electronics is negligible, but they represent a significant share of the intrinsic material value and of the ecologic footprint [9].

Recent developments in electronic products further increase these challenges. While sales numbers for many consumer products are still rising, their rapid change in design and composition, progressive miniaturization and complexity makes recycling of these products increasingly difficult. Advanced recycling technology can to a certain extent cope with such developments but for the case of a multitude of different elements, often in combinations that do not exist in nature and partly intensely interconnected and with low concentrations, the laws of thermodynamics set technical limits [10]. At the same time, substitution and thrifting in precious metal content have led to a decrease of intrinsic material value per device [13].

An ideal recycling value chain for consumer electronics, firstly, needs to ensure separate high collection, notably thanks to incentives and consumer awareness, combined with appropriate infrastructures that facilitate access of end-of-life products for recycling. However, one issue is that high priced small electronics (e.g., mobile phones or laptops) are often hoarded in households [14].

It is also essential that collection and recycling systems ensure traceability to prevent waste streams being diverted into the wrong channels, dumps or poor quality treatment processes, where their valuable metals will be lost and the environment damaged [12]. Here, pre-processors also have an important role in dismantling and separating such complex products, ensuring they are transferred to high-quality end-processors. A set of technical guidelines and standards has been developed to define effective and quality pre-processing (see Section 5).

The challenging trade-off is between a cost-effective recovery of large volume materials and a gradual recovery of critical and valuable materials present in small quantities. Nevertheless, valuable and critical metals are spread across various fractions and partial loss is inevitable in case of highly complex products, 100% recovery of "all" metals is not achievable [10] (see Section 1). However, just cherry picking the few most valuable metals—e.g., gold and copper in case of electronic waste—while emitting hazardous substances is certainly not the right approach from a sustainability and circular economy perspective.

Finally, subsequent final metallurgical processing needs to take place at state-of-the-art end-processors. High-quality processes are required to recover high metal yields, and to safely eliminate hazardous components and emissions. It is also important to note that while collection is conducted on a local and regional scale, pre-processing mostly makes use of regional and interregional operations (see Figure 2). This differs from the final step, which requires sophisticated large-scale integrated metallurgical plants, requiring significant investments and sufficient feeds to operate, especially for the recovery of precious and special metals. Economies of scale are crucial to ensure that the costs associated with recycling of consumer waste electronics are kept at a reasonable level. Here, only a limited number of such plants can treat materials sourced internationally. Therefore, a good cooperation and coordination of stakeholders along the recycling chain is essential to be overall successful in closing the loop. The weakest step in the chain determines the overall performance, i.e., highly efficient end-processing will be of little relevance if collection is weak or pre-processing is not capable of channeling the fractions with valuable and critical metals into the most appropriate metallurgical recycling processes. Of specific relevance are technical and economical optimization at the interfaces between pre-processing and end-processing and the facilitation of shipments along the chain to the downstream quality treatment plants.

4. Recycling Drivers and Success Factors-Specific Challenges for Resource Relevant Consumer Products

The case of consumer electronics shows the challenges society faces to close the loops. Nevertheless, there are other product examples where circularity already works well. Thus, in order to identify the appropriate frame conditions for a broader circular economy approach, it is worth having a closer look at recycling drivers and success factors.

Probably, the most outstanding example of circularity is gold in jewelry, artifacts, coins or ingots, where the high metal and emotional value provides sufficient incentives that prevent losses. Based on various data sources and deep insights into the precious metals markets, the authors estimate that, of the approximately 180,000 tons of gold so far mined in mankind's history, more than 90% is still in use, respectively, in stock.

However, not all precious metals are recycled efficiently. Overall, average recycling rates for precious metals are above 50%, with huge differences across applications [15]. For example, from chemical and oil refining process catalysts, over 90% of the precious metals contained therein are recovered, even in cases of long lifecycles of over 10 years [9,15]. Such a closed cycle is typically taking place in a business-to-business (B2B) environment, with no private consumers involved in its different steps. The user of the metal-containing product (e.g., the chemical plant) returns the spent product directly to a refiner who recovers the metals and returns them to the owner for a new product cycle. In most cases, the metals remain the property of the user for the entire cycle and the metal-refiner conducts recycling as a service. With such a set-up, the whole cycle flow becomes very transparent and professionally managed by industrial stakeholders, resulting in very small metal losses. Recycling is from the beginning planned in as an integral part of the business model, which is purely market driven.

Another example is automotive catalysts, which contain platinum group metals (PGM). From composition and metal concentration, they are very similar to chemical process catalysts (0.2%–0.3% PGM on a ceramic matrix), but PGM recycling rates are 60%–70% [9,15]. Compared to many other metal applications this is still a very high recycling rate, driven by a high intrinsic value and easy disassembly from a car. However, metallurgical recovery rates for PGM from automotive catalysts may be over 95%. The gap to the 60%–70% effective recycling rate is caused among other factors by exports of end-of-life (EoL) cars to regions with insufficient recycling infrastructures, and by a long and opaque recycling value chain.

Less promisingly, electronic wastes have average precious metal recycling rates below 15% [9,15] due to the reasons elaborated in the previous chapter. Precious metals recovery rates would again be over 95% if all waste materials were efficiently transferred to state-of-the art metallurgical refineries.

In summary, in open cycles taking place in a business to consumer (B2C) environment, metal losses are significantly higher due to other reasons than what is caused by the technical recycling process itself. In B2C applications, ownership of the metals changes each time a transaction occurs, transparency is low, businesses often do not follow an industrial logic, and, hence, results in higher metal losses than in B2B closed loop systems. To improve overall recycling rates, we need frame conditions that contribute to move from open B2C business models towards closed B2B cycles [16].

Beyond the intrinsic value and the business models applied, a number of other factors impact successful circularity. These can be grouped in product-intrinsic and external factors:

Intrinsic factors comprise of material value (example jewelry), complexity/heterogeneity of a product (incl. product design and ease to access/disassemble key components), the presence or absence of hazardous substances, the business model type/lifecycle type (B2C, B2B), and the product's transferability between users (e.g., mobile phone versus coffee machine).

External factors comprise of collection infrastructure (ease of handing in an end of life product) and external incentives (e.g., leasing, deposit systems), appropriate waste legislation (including the effectiveness of monitoring and enforcement), and stakeholder behavior and motivation (awareness and cultural habits, emotional links to a product, etc.). The latter is of significance, not only for consumers, but for all other stakeholders involved: How serious do manufacturers and retailers strive to close the loop, how well are EPR systems set up and maintained, to which extent do authorities engage in monitoring and enforcement, and how seriously do municipalities, waste management and recycling companies along the chain strive for comprehensive collection and high quality repair and recycling? Business ethics also play an important role, as sound, effective and environmentally compliant recycling is usually more costly than "quick and dirty" approaches.

Figure 3 compares qualitatively to which extent these different factors impact the overall success of circularity in Western Europe for some product examples. It needs to be noted that interdependencies between these factors exist and usually the factor combination is decisive for the overall impact. Take, as example, chemical catalysts, WEEE (waste electrical and electronic equipment) and lead acid car batteries, all of which contain hazardous elements. For PGM chemical catalysts, the B2B business model type, in combination with the high material value, drives a well-managed cycle within industrial players, capable of coping with the hazardous characteristics. In the case of car batteries, the hazardous properties triggered legislation on proper battery handling and recycling, in some countries supported by deposit fees. For a private consumer, a spent car battery is valueless (without losing the deposit he paid), hence, will handle it as waste if infrastructure is accessible. This has led to a widely closed loop for car batteries—as long as they are not exported outside Europe along with EoL vehicles. This is different with some small WEEE that has an emotional value to consumers. A classic example is a mobile phone: It is small and thus can be easily stored or thrown in a waste bin; due to its purchased value and personal attachment (personal data, photos, etc.), it is somehow regarded as "precious"; it can be sold as a second hand good (high transferability). Therefore, establishing a closed loop for mobile phones becomes very difficult as long as no incentives, such as leasing business models or deposit systems, exist. However, not all WEEE are the same, and size plays a difference/role—a non-functioning fridge is just a burden for the consumer—as well as transferability linked with attractiveness, e.g., in Europe, the market for second-hand coffee machines is small, and, once a machine is dysfunctional, most consumers will dispose of it.

	Glass, paper, PET	precious-metal jewellery	PGM-chemical catalysts	Ge bearing industrial residues	WEEE, (ELV)	Pb car batteries (Germany)
intrinsic factors						
Material value	o	++	++	+	(+)	(+)
complexity / heterogeneity	-	-	o	o	++	o
hazardous substances contained?	-	-	o	o	+	++
business model / lifecycle type	B2C	B2C	B2B	B2B	B2C	B2C
Product transferability between users	o	o	-	-	++	o
external factors						
Collection-infrastructur	++				o	o
external collection incentives	(deposits)					deposit
legislation / monitoring / enforcement			+		o	+
stakeholder behavior & motivation		X			X	
grade of achievement CE*	++	++	++	+	--	++

* in relation to entity of resource relevant materials contained in a product

- low / no   o medium / partly   + high / yes

B2C consumer goods, open product cycles, low transaprency on material flows

B2B industrial users, closed product cycles, relatively high transparency on material flows

Figure 3. Qualitative assessment of factors impacting the circularity for some selected products [17].

As expected, circularity is well achieved for the B2B examples of chemical process catalysts with PGM or germanium bearing industrial production scrap (e.g. from Ge wafer production), and, for B2C, precious metal jewelry. Products in B2C lifecycles, as well as certain B2C products with—from a consumer's perspective—have relatively low value, such as paper, glass and PET (polyethylene terephthalate) bottles, are very well recycled, mainly due to low product complexity, a very well established collection infrastructure, and the absence of any "emotional links" to these products. In some countries (e.g., Germany), collection of PET bottles and certain glass bottles is incentivized by deposit systems. WEEE shows the worst results due to challenges from almost all factors. The intrinsic and external factors described above, and their relative scoring in the table, are derived from over 20 years' experience of the main author in the field of metals recycling. It would be interesting to further analyze these empirical observations, and to elaborate on interdependencies, as well on the interplay of physical/technical, economical and behavioral/cultural influences. However, this will be (research) work of its own and goes beyond the scope of this paper.

Summing up, the degree of circularity of products depends on a complex range of impact factors, as illustrated in Figure 3. In some cases—e.g., jewelry, valuable products in a B2B environment—market forces alone are sufficient for success. However, especially for complex consumer products that contain a mix of valuable and hazardous materials market forces alone are not sufficient, but appropriate, policy and legislation is crucial to improve the current status of low circularity.

5. Quality Recycling Based on Process Standards and Certification—A Cornerstone for Closing the Metals Loop

To create a global level playing field on recycling activities and support Europe's competitiveness in the field, high quality recycling processes and practices are needed. "Quality recycling" may be understood as a minimum level of quality of the recyclate/output material and/or a minimum level of quality of the treatment process, including the existing depollution requirements at EU level under the Waste Electrical and Electronic Equipment Directive [18].

As mentioned previously, metals have intrinsic properties during recycling, thus, can be used multiple times and re-used maintaining their quality and functionality. Therefore, for metals we need quality recycling through proper treatment using state-of-the-art processes and facilities. For complex waste streams, such as electronic waste, due to the content of hazardous components, the EU Directive on WEEE outlines minimum selective depollution treatment and mandates the development of state-of-the-art minimum standards for its treatment [18]. Furthermore, the directive allows that the treatment operation may be undertaken outside the respective Member State or the EU, provided that the shipment of WEEE is in compliance with existing rules, and treatment takes place in conditions that are equivalent to EU requirements [18]. However, depollution performance, the extent to which raw materials are recycled, the level of safety, health and environmental measures and control procedures differ considerably not only on a global scale but also within the EU.

The report on "Countering WEEE Illegal Trade" (CWIT) that was financed by the European Commission and conducted by Interpol found that in Europe only 35% (3.3 million tons) of all the e-waste discarded in 2012, ended up in the officially reported amounts of collection and recycling systems. The other 65% (6.15 million tons) was either exported (1.5 million tons), recycled under non-compliant conditions in Europe (3.15 million tons), scavenged for valuable parts (750,000 tons) or simply thrown in waste bins (750,000 tons), meaning that there is a serious economic loss of materials and resources directed to compliant e-waste processors in Europe [19]. Hence, non-compliance of the WEEE Directive, as well as the Waste Shipment Regulation, which also sets the conditions for exports of waste to non-OECD countries, notably with view to ensuring environmentally sound management of waste, becomes problematic. Dumping or low quality sub-standard treatment ("backyard recycling"), not only has negative impacts on human health and the environment, but also leads to significant pressure on recycling prices and disturbs a fair level playing field, as such operations externalize environmental and social costs. This makes it difficult for compliant recyclers to achieve a stable capacity utilization and to invest in new processes or capacity expansion, which can delay or even prevent the medium- and long-term harvesting of the EU's potential on secondary raw materials.

Under mandate M/518, the European Commission requests that European standardization organizations develop European standards for treatment, including recovery, recycling and preparing for re-use of WEEE [20]. CENELEC, the European Committee for electrotechnical standardization, is currently developing a series of such European standards and technical specifications (EN 50625 series), reflecting the state-of-the-art in recycling covering the whole value chain, as well as environment, health, governance and process efficiency provisions. Nevertheless, there is no obligation or legal requirement for the mandatory implementation of finalized standards and technical specifications of the EN 50625 series, and only a few Member States, such as France, Ireland and the Netherlands, have decided to comply with these standards. Therefore, this situation not only creates leakage of WEEE to Member States, where standards remain optional, but distort a significant part of the WEEE stream, as it continues to be treated in sub-optimal conditions. Additionally, the existing WEEE Directive does not set sufficient incentives for the recovery of precious metals from WEEE, as recycling targets are weight based, and the existing point of measurement for recycling does not account for the output of recovered materials.

One solution would be to make it compulsory for WEEE recyclers to comply with the EN 50625 series of standards and technical specifications. Recyclers outside of Europe should also be required to treat EU WEEE exports under equivalent conditions.

Putting in place a certification scheme with a third party audit would verify harmonized enforcement across the EU and outside. Such legal requirements are necessary to ensure that the circular economy for complex waste streams, such as WEEE, will be closed physically, where recycled materials are re-injected in new product lifecycles, and, economically, where costs match recycling treatment.

6. The Essential Role of Policy and Legislation

When tackling the challenges and barriers to a more circular use of metals in Europe, a systemic approach is required, using a combination of complementary measures across the value chain. As elaborated in Section 4, especially for products with lower intrinsic material value or other missing attractiveness to recycle, policy and legislation play key roles. Without a WEEE directive, presumably the bulk of our electric and electronic wastes would still go to landfills or to incineration. The same is valid for, e.g., batteries or plastics. Paper and glass are only recycled at high rates due to the comprehensive infrastructure that has been set up in most European countries, financed by various fee systems. However, one-size-fits-all policy requirements will not be effective in realizing the circular economy model across all European sectors. The regulator can adopt measures that will help address the challenges identified. The circular economy package is a step in the right direction, as it addresses some of the issues and promotes the recovery of materials from waste. Below is an overview of the main and general key measures that could address the above-mentioned challenges for metals, which have been put together by the European Non-Ferrous Metals Association (Eurometaux) [8]. The table (Table 1) aims to pinpoint several issues in the legislative package, without going into an in-depth policy analysis or discussion on legal wording, as this is beyond the scope of this paper. Several measures are already included in the present circular economy package proposal from the European Commission, while others require further development or are under discussion, such as the point of measurement for the recycling targets (see Figure 4). At present, the European Commission's proposal is undergoing co-decision procedure between the European Parliament and the Council.

Table 1. Existing challenges and possible solutions in the Circular economy Package.

Challenge	Solution
Inconsistent measurement of recycling rates Varying interpretation of what "recycling" means and how it should be measured. Currently, Member States interpret "recycling" and calculate their recycling rates differently, creating inconsistencies. Some base their recycling rates on collected waste, even though it covers waste that will be exported after collection or sorting, incinerated or landfilled, which does not incentivize the recovery of materials.	**A single method for measuring recycling rates, at the input to final recycling process** The Commission proposal on the rules to calculate the attainment of the recycling targets is understood as the "weight of the input waste entering the final recycling process", which further incentivises the recovery of materials from waste reflecting reality of what recycling should be. "Final recycling process" is defined as "the recycling process which begins when no further mechanical sorting operation is needed and waste materials enter a production process and are effectively reprocessed into products, materials or substances" (See Figure 4 below table) This means that a final recycling process generates output in a sufficiently pure quality, capable of replacing primary materials. In addition, independent of where finally the point of measurement will take place (statistical data availability plays here a role as well) it is crucial to keep the definition of final recycling process in order to underline that the circle can only be closed if products/materials have passed the recycling chain to the very end.
Suboptimal end-of-life collection schemes. Collection is the first step of the recycling value chain. If collection is not performed efficiently the recycling rate cannot increase significantly. The efficiency of the collection schemes vary widely across Europe.	**Minimum requirements for Extended Producers Responsibility** The Commission's proposal to support separate collection and define minimum requirements for extended producer responsibility (EPR) will help improve collection. The EPR schemes should also cover the entire cost of waste management, including the final recycling process
Landfill and incineration Landfilling and incineration of post-consumers goods. Too many end-of-life products which contain valuable materials, sometimes even critical, are landfilled or incinerated.	**Target to reduce landfill to 10%** The objective to reduce the landfilling of recyclable waste to 10% is essential to support recycling. Enforcement at Member State level is crucial. Incineration with energy recovery is a complementary option that is lower than recycling in the waste hierarchy but that has its merit in cases where recycling is not feasible.
Illegal waste exports, both legal and illegal (Illegal) Exports of waste due to high intrinsic value of certain scrap and embedded energy content. Exports of hazardous waste are often labelled as second-hand goods, and waste for disposal as waste going to recovery.	**Stronger controls and a requirement for waste exports to be recycled under 'equivalent conditions'** Different measures can be instrumental in fighting against illegal shipments, including • Harmonised control of shipments at harbours to avoid "port hopping" • Identification of second hand goods in customs declarations to facilitate targeted controls • In the case of complex waste streams (e.g., WEEE and waste batteries), introduce a requirement that secondary materials may only be exported if a certified final processor is duly identified. The Commission recognises the need to combat illicit shipment and step up enforcement of the revised waste shipment regulation.
No level playing field for quality recyclers Recycling processes must meet minimum quality criteria and these need to apply across the recycling value chain in Europe and elsewhere	**Certification of treatment facilities** The Circular economy package is not sufficiently ambitious as it merely proposes the "promotion of industry-led voluntary certification of treatment facilities". A mandatory certification scheme would ensure equivalent and fair conditions. (see Chapter 5)
A focus on mass not value As EU waste policies traditionally focus on volume and weight, certain valuable materials in end-of-life products are not yet sufficiently targeted in collection and recycling.	**A product-centric approach** The proposed Action Plan rightly identifies the need to pay increased attention to the recovery of valuable and critical raw materials from end-of-life products. It is important to treat each product according to its specificities, including material composition and the techno-economic opportunities from recycling.
Barriers to industrial symbiosis & recycling There are a number of barriers to intra-EU shipments of waste, end-of-life products and by-products for recycling. This includes the non-harmonised status of waste and by-products across Member States, the use of national waste codes or the lack of appropriate codes for given waste, the overly complex and lengthy procedures (notification and transit) and in some cases the inappropriate implementation of the proximity principle.	**Measures to facilitate waste shipments in Europe** Harmonised definitions of waste and by-products and the use harmonised EURAL codes (no separate national codes) across Member States is key. New codes should be created for waste for which no waste codes exist. The Waste Shipment regulation provides a simplified procedure for "pre-consented recovery facilities", but does not provide much benefit in reality. It should be improved to allow fast track and immediate shipment of waste from and to pre-consented recovery facilities once the competent authorities have been notified. Control can take place at any time thanks to an easy tracking of shipments through their identification number. An electronic system would also facilitate the procedure.
Effective regulation of hazardous substances. Because they are naturally occurring substances a metal free environment is by nature impossible. A sheer ban of hazardous substances would decrease the amount of waste recycled in Europe, and be an incentive to landfill, export materials and import more materials and products	**Effective, smart & proportionate legislation** End-of life products containing hazardous substances should be properly treated i.e., in respect of the legislation in place. Legislation should be effective, smart and proportionate. It should not unnecessarily discourage the use of secondary resources, especially within the context of the EU Circular economy and global competition. We need to secure a sound risk management and avoid penalizing EU recyclers unnecessarily.
Technical and economic challenges when recycling complex products The trend in changes in the composition and design of products raises challenges in terms of economic viability of the recycling process. In addition, an ever increasing variety of element combinations in components and products cause technical challenges and require the development of new recycling approaches.	**Targeted innovation support** A targeted support to innovation supporting the recycling of complex products coupled with an increased dialogue across the value chain is needed to address this challenge. The Commission has committed to support Circular economy developments through its research and innovation financing programme, Horizon 2020, and Cohesion Policy funs.

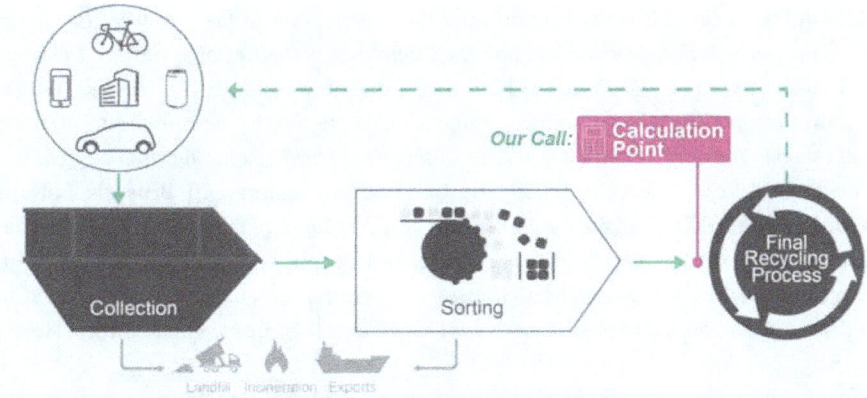

Figure 4. Final recycling process and measurement of recycling rate [21].

7. Conclusions and Way Forward

To conclude, metals are the perfect material for a circular economy, as they can be recycled again and again without losing their properties. However, there are a number of challenges that need to be addressed in a systemic way to ensure that Europe maintains its leading position in the recycling of base metals, and to further develop the recycling of valuable and critical metals that are present in small quantities in complex products.

The main challenges relate to accessing the waste and end-of-life products, and to ensure that quality operators recycle them. Addressing these challenges would ensure that more metals are recycled and at higher yields. It would also support the competitiveness of quality recyclers and, hence, their investments and the related job opportunities.

The European Commission's circular economy package is a step in the right direction, focusing more than ever, on the recovery of materials embedded in waste.

For the waste proposals, the Commission's ambition now needs to be maintained through the co-decision procedure, and Member States need accept the challenge of reaching ambitious targets (notably through the new point of measurement of recycling targets). Without harmonized implementation of the waste legislation, an effective European circular economy will not be possible.

It is still too early to tell whether the ambition of the European Commission's Action Plan will be translated into reality. The next three years will be crucial to implement strong regulations on areas, including resource-efficient eco-design, illegal waste exports and intra-EU waste trade.

Without comprehensive actions to address the different challenges identified in this paper, Europe will be unable to capture the full economic benefits of a circular economy. The European metals industry has invested in becoming a global recycling leader, and is committed to work with policy makers to achieve this ambition.

Author Contributions: Christian Hagelüken has taken part in conceptualizing, writing and checking the paper. Ji Un Lee Shin has taken part in writing and checking the paper. Annick Carpentier has taken part in writing and checking the paper. Chris Heron has taken part in reviewing the paper structure and checking the English.

References

1. European Commission (EC). *Communication from the Commission: Europe 2020, A Strategy for Smart, Sustainable and Inclusive Growth, COM (2010) 2020 Final*; European Commission (EC): Brussels, Belgium, 2010.
2. European Commission (EC). *Communication from the Commission to the European Parliament, the Council, the European Economic and Social Committee and the Committee of the Regions: Roadmap to a Resource Efficient Europe, COM (2011) 0571 Final*; European Commission (EC): Brussels, Belgium, 2011.

3.   European Commission (EC). *Communication from the Commission to the European Parliament, the Council, the European Economic and Social Committee and the Committee of the Regions: Closing the Loop-An EU Action Plan for the Circular economy. COM (2015) 614/2*; European Commission (EC): Brussels, Belgium, 2015.

4.   European Commission (EC). *Communication from the Commission to the European Parliament, the Council, the European Economic and Social Committee and the Committee of the Regions: Towards a Circular economy: A Zero Waste Programme for Europe. COM (2014) 0398*; European Commission (EC): Brussels, Belgium, 2014.

5.   European Commission (EC). *Proposal for a Directive of the European Parliament and of the Council Amending Directive 2008/98/EC on Waste. COD 2015/0275*; European Commission (EC): Brussels, Belgium, 2015.

6.   European Commission (EC). *Proposal for a Directive of the European Parliament and of the Council Amending Directive 1999/31/EC on the Landfill of Waste. COD 2015/0274*; European Commission (EC): Brussels, Belgium, 2015.

7.   European Commission (EC). *Proposal for a Directive of the European Parliament and of the Council Amending Directive 94/62/EC on Packaging and Packaging Waste. COD 2015/0276*; European Commission (EC): Brussels, Belgium, 2015.

8.   Wellmer, F.W.; Hagelüken, C. The Feedback Control Cycle of Minerals Supply, Increase of Raw Materials Efficiency, and Sustainable Development. *Minerals* **2015**, *5*, 815–836. [CrossRef]

9.   Hagelüken, C. *Recycling of (Critical) Metals: In Critical Metals Handbook*; Gunn, G., Ed.; Wiley: Oxford, UK, 2014.

10. UNEP. *Metal Recycling-Opportunities, Limits, Infrastructure—A Report of the Working Group on the Global Metal Flows to the International Resource Panel*; United Nations Environment Program: Paris, France, 2013.

11. Reuter, M.A.; Matusewicz, R.; Van Schaik, A. Lead, Zinc and their Minor Elements: Enablers of a Circular economy. *World Metal.-ERZMETALL* **2015**, *68*, 132–146.

12. Eurometaux.   Proposed Measures to Ensure an Effective Circular Economy.   Available online: http://www.eurometaux.org/DesktopModules/Bring2mind/DMX/Download.aspx?Command=Core_Download&EntryId=8461&PortalId=0&TabId=57 (accessed on 6 June 2016).

13. Van Kerckhoven, T. Optimizing (precious) metals recovery out of electronic scrap. In Proceedings of the Electronics Recycling, Singapore, 13 November 2014.

14. Nokia Corporation. *Global Consumer Survey Reveals that Majority of Old Mobile Phones Are Lying in Drawers at Home and Not Being Recycled*; Nokia Corporation: Helsinki, Finland, 2008.

15. UNEP. *Recycling Rates of Metals-A Status Report: A Report of the Working Group on the Global Metals Flows to the International Resource Panel*; United Nations Environment Program: Paris, France, 2011.

16. Hagelüken, C.; Buchert, M.; Ryan, P. Materials Flow of Platinum Group Metals in Germany. *Int. J. Sustain. Manuf.* **2009**, *1*, 330–346. [CrossRef]

17. Hagelüken, C. Kreislaufwirtschaft 2015-wo Stehen wir Heute im Vergleich zu vor zehn Jahren? In Proceedings of Berliner Recycling und Rohstoffkonferenz, 16–17 March 2015.

18. European Commission (EC). *Directive 2012/19/EU of the European Parliament and the Council of 4 July 2012 on Waste Electrical and Electronic Equipment (Recast). L 197/38*; European Commission (EC): Brussels, Belgium, 2012.

19. Huisman, J.; Botezatu, I.; Herreras, L.; Liddane, M.; Hintsa, J.; Luda di Cortemiglia, V.; Leroy, P.; Vermeersch, E.; Mohanty, S.; van den Brink, S.; et al. Countering WEEE Illegal Trade (CWIT) Summary Report, Market Assessment, Legal Analysis, Crime Analysis and Recommendations Roadmap, 30 August 2015, Lyon, France. Available online: http://www.cwitproject.eu (accessed on 6 June 2016).

20. M/518 EN Mandate to the European Standardisation Organisation for Standardisation in the Field pf Waste Electrical and Electronic Equipment (Directive 2012/19/EU (WEEE)). Available online: http://ec.europa.eu/environment/waste/weee/pdf/m518%20EN.pdf (accessed on 6 June 2016).

21. Waste Framework Directive: European Material Industries call for Measurement of Real Recycling Rates. Available online: http://www.cepi.org/node/20494 (accessed on 6 June 2016).

9

The Importance of Specific Recycling Information in Designing a Waste Management Scheme

Adekunle Oke [1],* and Joanneke Kruijsen [2]

[1]   Aberdeen Business School, Robert Gordon University, Aberdeen AB10 1RT, UK
[2]   Energy and Sustainability, Robert Gordon University, Aberdeen AB10 GJ, UK; J.Kruijsen@rgu.ac.uk
*   Correspondence: a.o.oke@rgu.ac.uk

Academic Editor: Michele Rosano

Abstract: Recycling information can be complex and often confusing which may subsequently reduce the participations in any waste recycling schemes. As a result, this research explored the roles as well as the importance of a holistic approach in designing recycling information using 15 expert-based (in-depth) interviews. The rationale was to offer a better understanding of what constitutes waste, recycling, and how recycling information should be designed and presented to make recycling more attractive/convenient. Based on the research participants' perceptions with supports from the existing studies, this research sub-categorised recycling information into three different themes, termed the "WWW" (what, when, and where) of recycling information components. As a result, these components (or attributes) were extensively described (using findings of semi-structured interviews) to elicit pragmatic guidance for practitioners, policy-makers, and other stakeholders in designing structured communication or information strategies that may simplify and subsequently increase waste recycling practices. The policy implications of holistic information in enhancing recycling are further discussed.

Keywords: attitudes; behaviour; communication; information; prompts; policy; recycling; waste

1. Introduction

The depletion of natural resources, and its associated waste production, has been linked to unsustainable human attitudes and behaviours [1,2]. Nevertheless, an understanding of the thought processes and activities behind the generation of waste may offer new perspectives on how to encourage waste prevention, including resource conservation efforts, without a dramatic change to human behaviours and lifestyles. Waste production is a complex issue [3] confronting local, national, and international governments [4]. Its management may require the integration of inter-disciplinary worldviews while its understanding may be further enhanced using various socio-cultural perspectives [3]. As a result, numerous studies (such as [4,5]) have been conducted within the realms of waste management, many of which focused on socio-demographic and psychological aspects of waste production and management. Findings from these studies have inspired different environmental policies, including legal frameworks that instigated the design of many waste management strategies around the world. Nevertheless, a survey of 2000 households in England suggests that a considerable amount of people (about 30% of the survey participants) are still confused about what and where to recycle [6].

Policy makers and other stakeholders are therefore confronted with the task of appealing to the subjective and cautious reasoning of individuals in order to instill a waste prevention, reuse, recycling, and/or upcycling ethos. In practice, one of the challenges confronting waste management policy makers and planners is to establish whether recycling information would achieve its intended objectives. Another challenge includes the extent (in terms of format, structure, and frequency) at which recycling information should be provided in order to influence behaviours.

As a result, this research was designed to provide a pragmatic guidance for the design of a well-informed communication strategy by exploring the roles and importance of recycling information in modifying recycling behaviour using people's perceptive. On the one hand, the intention was to contribute to the existing knowledge on the effects of information on recycling behaviours and to make recycling more accessible and convenient for people to perform. On the other hand, the research was to encourage a more pro-environmental consciousness and deliberate decision making [2,7] that could impact on the existing consumers' culture and its associated throw-away attitudes.

2. Information and Recycling Behaviour

Although the efficacy of using the "individual" as a unit of analysis have been recently challenged [8], the decisions to participate in recycling scheme irrespective of the contexts still rest on the "individual". As a result, the "individual" has a liberty of invoking their subjective judgement when to (or when not to) perform pro-environmental behaviours based on different factors underpinning such a decision-making process. In this respect, different factors such as demographics [9,10]; socio-economics [11,12]; scheme design [4,13]; and identity [14] that are likely to influence recycling behaviour were previously identified. While these factors were well articulated in the literature on recycling at home, factors influencing recycling at work (as well as other contexts) are still sketchy (see [1] for a review). However, the importance of information on recycling schemes in encouraging public involvement have been documented (see [15,16] for a review) albeit no guidance on the components as well as how to design a communication strategy. For example, [17] observed that adequate knowledge of what to recycle through recycling information and provision of feedback were positively significant to recycling behaviour at home. In a similar study, [18] reported a positive influence of publicity and promotion on household recycling behaviour while [10,19] also observed a positive association between recycling knowledge and household waste recycling behaviour. This suggests that recycling information in terms of feedback, publicity, promotion, or a well-designed communication strategy is an effective tool to engage and to enhance recycling behaviour at home [16].

However, behaviour has been reported [19] to return back to the baseline soon after the intervention was withdrawn, suggesting inconsistencies in behaviour over a prolonged period of time. While factors influencing recycling at home are well established and documented, little is known about factors that may likely promote recycling at work (see [1,20] for a review). Nonetheless, the influence of information on recycling at work has been previously investigated (see [15,21–23] for a review). Findings from these studies suggest that information has profound effects on how recycling is practiced at work. Information on recycling schemes can therefore serve as a motivation as well as a barrier to recycling, not only at work but also at home.

Whilst research has attempted to attribute recycling performance to available information on recycling, no known research has established how and in what format including the frequency at which this recycling information should be presented. Consequently, specific information on what (recyclables and non-recyclables), where (location of recycling containers), and how to recycle would enhance recycling behaviour [15,23,24]. Although the influence of information is mixed and observed to be behaviour specific [25], it may probably differentiate perceived recyclers from non-recyclers [13]. Further, the importance of a well-designed information to recycling participation is often over-looked and our present knowledge on how recycling information should be designed and presented is still lacking. This includes relevant and important information about the recycling process, recycling scheme, and available recycling facilities that may likely encourage recycling. As a result, the understanding of activities that generate waste (through a holistic information on the dynamics of waste, from production to consumption of materials) is required. This research was designed not only to enhance scheme design but also to promote recycling behaviour.

3. Research Methodology

3.1. Data Collection

Recall that the main purpose of this research was to design a "grounded model" of information and/or communication aiming at improving recycling practices by exploring the self-expressing views and perceptions of the participants. However, this current study formed a part of the larger and ongoing study on waste recycling at home and at work in the UK. Using a pragmatist perspective [26–28], the research addressed people's lived experiences and practices in relation to waste recycling in general and assisted in conveying the meanings of waste and recycling. Previous studies on recycling have focused on two major strands representing recycling practices at home and recycling at work. While it is tempting or appropriate to contextualise this research, it is of no practical use to disenfranchise one particular context. As a result, this research addressed the development of information/communication strategies that can be practically adaptable to recycling schemes in any contexts.

In order to understand people's perceptions on the influence of information on their recycling behaviour, qualitative research approach was adopted in this research. Although this study reported on the qualitative interviews, two different perhaps complementary approaches were adopted for data collection: interviews and visual (site) observation. The rationale was to understand the contributions of information to recycling practices using the research participants' worldviews [26]. The first approach involves an in-depth (expert-based) inductive exploration of the participants' subjective views so as to understand how information could be designed and framed [27] within the domain of waste recycling. For this research, experts are not people who only work within waste management sector but also include those who could provide required information in terms of recycling practices at home and at work. This involved gathering specific data relevant to recycling activities and logically making sense of the data using the participants' lived experiences without any theoretical lens. The second approach was conducted to confirm the contributions of existing recycling facilities (such as bin locations and signage/prompts) in the visited sites.

The participants that could provide appropriate data to achieve the research goal were purposefully selected and contacted from different organisations in the UK. A total of 15 semi-structured face-to-face interviews were conducted in person and guided by the interview protocol that was designed for the interview purpose. Rather than the sample size, the richness as well as the appropriateness of the collected data [29] was the central focus of this research. According to Miles and Huberman [30], sample size in explorative, qualitative studies evolves during data collection processes and is not predetermined. Although there is a lack of agreement on the appropriate sample size for qualitative studies [31], the concept of saturation [32] has been adopted in qualitative studies [33]. The sample size as used in this research was sufficient considering that multiple data collection methods were adopted [34]. In addition, about 10 interviews were recommended by Creswell [35] for phenomenology study while 12 participants were considered to be adequate for interview-based study [36]. As a result, the average time for the one-on-one interview was 45 minutes and covered different aspects of waste management and particularly recycling at home and at work. The intention was to understand recycling behaviour (in terms of the motivations and barriers) as well as the participants' views on waste management practices in the UK. Each interview session was digitally recorded following the participants' active consent, transcribed using NVivo 11, and inductively analysed by identifying key themes in terms of recycling information/communication using inductive thematic analysis. In addition, the participants were ascribed with different pseudonyms after been assured of anonymity and confidentiality in line with research ethics and data protection procedure.

The second data collection approach involves visual inspection (or observation) of the existing recycling facilities and to assess the effectiveness of the schemes available in the visited organisations. Using this approach, seven different organisations and two private residential flats where selected interviewees work and reside respectively were visited for visual observations. While the interviews were used to espouse the research participants' perception on recycling information, the site observation was undertaken to complement the participants' subjective views and to understand the existing facilities including how recycling is structured within the visited sites. The two approaches were initiated and implemented concurrently between September 2015 and March 2016.

3.2. Data Analysis

As previously mentioned, the collected data were inductively analysed without pre-defined themes in order to convey the meanings of the participants' views in a logical and coherent format. Although no distinct themes were generated prior to data collection and analysis, findings from existing studies on recycling behaviour including the researcher's epistemological perspective may bias the identification of themes. In order to identify the emerging themes from the data, the interview transcripts were read through many times for a better understanding of the data and familiarisation.

As a result, thematic analysis [26,37], was performed using NVivo 11—a qualitative data analysis software package. In this approach, the transcripts' contents were categorised and labels were assigned to the emerging themes (categories). For ease of data analysis and synthesis however, the data analytical procedure was informed by Ritchie et al.'s [37] recommendations which included data management (for example data preparation, data labelling, and data itemizing/sorting), abstraction, and interpretation.

4. Result and Discussion

4.1. Participants' Characteristics

Table 1 shows the socio-demographics of the participants (gender, age, education, employment status, income, and ethnic background).

Table 1. Participants' Socio-demographic Information.

Participants	Gender	Age	Education	Employment Status	Income (£)	Ethnic Background
001	Male	56–65	HNC	Full-time	25,000–49,999	Scottish
002	Male	36–45	Higher Education	Full-time	50,000–99,999	British
003	Male	46–55	A/AS	Full-time	50,000–99,999	British
004	Female	26–35	Higher Education	Full-time	25,000–49,999	British
005	Male	46–55	Higher Education	Full-time	50,000–99,999	British
006	Male	46–55	Diploma	Full-time	>100,000	Scottish
007	Male	>65	Other	Part-time	<24,999	Scottish
008	Female	16–25	A/AS/higher or equivalent	Part-time	<24,999	Asian
009	Female	26–35	Higher Education	Full-time	<24,999	Scottish
010	Male	36–45	GSCE or Equivalent	Full-time	25,000–49,999	Scottish
011	Female	26–35	Higher Education	Part-time	<24,999	Any other background
012	Male	56–65	A/AS/higher or equivalent	Full-time	25,000–49,999	Scottish
013	Male	36–45	Higher Education	Full-time	25,000–49,999	African
014	Male	46–55	Higher Education	Full-time	25,000–49,999	British
015	Male	26–35	GSCE or Equivalent	Full-time	25,000–49,999	Scottish

Source: Author

The gender of the participants showed that 4 females and 11 males participated in the study, the youngest participant was between 16 and 25 while the oldest participant was over 65. Although there was a huge gender difference and skewed toward male participants, the gender difference was not pre-determined and was based on those who responded to the invitation to participate in the research. Considering that e-mail was sent out to invite participants, the gender differences supported the early studies (such as [38,39]) that observed that males (men) were more likely to respond and participate in a web-based survey compared to their females (women) counterparts. On the other hand, it may reflect the dynamics of waste recycling [40] in a household, given that all the participants claimed to be responsible (in charge of) for recycling at home. As a result, it was contrary to [40] who reported that females were more likely to initiate and sustain recycling at home.

Nevertheless, all the participants had one formal education or the other while the highest qualification was a higher degree (Bachelor and above). Although the level of education has no influence on the participants' income (as reflected in Table 1), there is only one extreme case of income at above £100,000. In addition, the participants included one African, one Asian, while others were White (British and Scottish); this was based on the availability (and accessibility) of the participants rather than the racial landscape of the study context.

4.2. Recycling Information Components

Based on the participants' accounts, information aiming at enhancing recycling was broadly categorised into three different themes (see Figure 1). The model (themes) in Figure 1 included what, when, and where (including how) to recycle although the contexts of waste generation were not considered for this analysis. It was assumed in this study that a provision of adequate and correct information is necessary in enhancing recycling irrespective of the waste generation contexts.

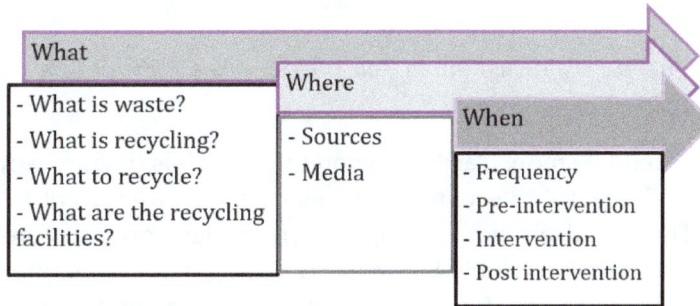

Figure 1. Components of Recycling Information (Source: Author).

As a result, an understanding of these components would not only assist waste planners to design attractive schemes but would also provide an opportunity for people to engage and participate. Rather than focusing on attitude and other socio-psychological attributes of social actors, a well-designed scheme with sufficient information in terms of these components may activate other intrinsic attributes as well as enhancing recycling participation.

4.2.1. "What" of Waste Recycling

The "what" of waste-recycling information (see Figure 1 above) as briefly described, analysed, and presented in this study includes: What is waste? What is recycling? What to recycle? What are the recycling facilities?

Waste: What Does It Mean?

Over the years, waste has been construed as the by-product of human activities, particularly resource-intensive and consumer-based economic lifestyles [41], and often disposed of in landfills. This has led to a misconception that waste constitutes materials with no socio-economic and environmental values. This misconception is partially responsible for how waste has been handled and discarded in the past without due consideration of its socio-economic, health, and environmental impacts. However, the perceived scarcity of resources, coupled with the exponential increase in the global population, has changed our impression of what actually constitutes waste. This is based on the mind-set that what constitutes waste in a certain area may constitute a valuable resource (for instance feedstock) in another economic context. As a result, the landscape of waste and waste industry are changing:

"It's just looking at waste more as resource now rather than just say you pick it up chuck it on the ground; I mean that is industry now we try to move away from calling it waste. Now we see ourselves as resource industry rather than waste industry. So actually what we are trying to do to keep these materials in economy useful as long as possible" (Par_015).

In other words, waste is no longer waste, but is now being regarded as a resource that could find its way back into production lines or manufacturing processes:

"One of the key achievements is something like reduction in disposal of biodegradable waste by given alternatives to other technologies like composting, AD stops co-disposal of solid and liquid waste going into the landfill thereby activation some other waste streams like spread the waste on land as fertiliser. So it has activated that particular industry, so that industry has employed so many people in terms of land spreading of waste on agricultural land which is one of the recovery avenues" (Par_013).

Considering the global amount of waste generation at 1.3 billion tonnes per year [41], this may be catastrophic if people still perceive this valuable resource as waste. Rather than for people to throw away what they buy [4] or not have the habit of recycling [5], people should be aware that the throw away culture is a thing of the past.

"People know now that they can't just throw your rubbish away, stuff got to be recycled whenever possible" (Par_006).

As a result, there should be more clarification on the materials that are recyclable and schemes should be designed to facilitate the collection of these materials in order to enhance the global circular economy initiatives. This could provide an opportunity to develop structured and unstructured markets for materials that are perceived to be waste, including items being produced from recycled materials. When designing recycling information therefore, there should be more clarity about items that could be recycled. As a result, a sufficient knowledge of materials that can be recycled (and/or sold) would not only influence the subjective reasoning of consumers but could also reinforce the actions of conscious (ethical) consumers.

How Should We Frame Recycling?

Recycling which is the preferred "waste treatment" method based on the tenets of waste hierarchy has been reported (such as [42]) as a gateway to other pro-environmental behaviours. In other words, a social agent that develops a habit of recycling is more likely to adopt other pro-environmental behaviours such as energy conservation. What constitutes recycling is well defined in policy statements (such as [43]). However, many people are confused about the whole idea of recycling and may likely affect how they recycle:

"In the UK you'll find that a lot of people still don't recycle although nowadays you've got street bins that household waste to go in, is still to getting people to realise we've got to recycle, that will be hardest" (Par_001).

Whereas the definition of recycling practically clarifies what is and what is not a recycling activity, it suggests that recycling is beyond the capacity of consumers. For example:

"I think recycling can be considered as just throwing away your trash, throwing away any thrash recycling is now I think considered as splitting up the type of trash" (Par_008).

In this study therefore, recycling is conceived as a material reprocessing (through "production/manufacturing") involving physical, chemical, mechanical, and thermal processes to derive initial or other products. As a result, labelling or profiling consumers as recyclers or non-recyclers (such as [13,44]) is taxonomically misleading and could undermine the important roles of consumers in the waste-recycling cycle. Recycling information may therefore be ill-defined and not properly structured in a way that could influence behaviour. The major challenge is to identify what constitutes the roles of individuals (consumers) along the chain of recycling in order to focus recycling information on those specific roles and responsibilities. Considering that *"there is a growing emphasis on recycling now and recovery of waste"* (Par_013); technically, the responsibility of individuals within the cycle is limited to the preparation (such as sorting, washing, drying, and possibly transportation) of materials for recycling.

From these perspectives, it is conceivable to suggest that the waste-recycling process should not be represented as a single activity but as multiple activities, with each preceding activity leading to the next activity in a cycle. As a result, waste recycling can be conceived as a technical process that is more than tossing a used material or an item into a designated bin in terms of the available collection scheme-commingle or source separation. In other words, commingling and/or source separation of waste are not recycling in their own rights, rather they are methods involved in preparing materials (waste) for collection. Recycling requires more efforts where putting (un)used items into collection bins (either through commingle or source separation) is only a first stage of recycling cycle while the responsibility of individuals (or consumers) should be termed "preparing for recycling" instead of recycling itself.

What (Materials) Should We "Prepare" for Recycling?

This aspect of waste-recycling information differentiates what to prepare for recycling from what is to dispose through incineration and/or landfills. In other words, it explains the materials that can be deposited in dedicated recycling bins and those that are going into trash/disposal bins. According to the research participants:

> *"I feel maybe people don't know what materials they can recycle, so there's a bit of confusion about can you recycle this, can you recycle that"* (Par_008).

The understanding of these materials will facilitate ease of recycling and also enhance the quality of materials being collected for recycling. This might differ from one location to another based on local facilities and capacities to handle (or process) the available materials:

> *"There is always a different system and different councils have different steps as well-some collect glass, some have to separate glass and some the collections (times) are different as well; some you've to walk around the corner to put your materials right there"* (Par_002).

What constitutes recyclables may also be influenced by the legal and economic requirements of a particular country or region. In the UK for example:

> *"The main piece of legislation is Environmental Protection Act 1990 that put in place something like recycling targets, local authority recycling plans, it made landfill more regulated and try to bring in landfill tax. That was a big drive in terms of the change in industry"* (Par_015).

Nevertheless, different materials are produced and prepared for recycling by the participants:

> *"Paper, cardboard, cereal boxes, newspapers, magazines—we don't have many magazines, but tins, cans, aerosols, foil trays and what else yeah plastic bottles and glass we get"* (Par_004).

This was supported by Austin et al. [21] who identified plastics, metals, glass, paper, wood, and bio-waste (mainly organic waste in nature) as major recyclables. In order to reduce any misconception about the materials that can be recycled, the key recyclables are identified in the EU Waste Framework Directive (2008/98/EC) for instance. While paper (including newspapers and magazines) was reported by all the participants, clothes were the least mentioned materials. This was supported by Barr [4] and Perrin and Barton [17] who observed that paper was the most recycled material while textile was the least recycled material by householders in the UK. The identification of these materials as major recyclables is primarily influenced by their economic, social, and environmental importance.

The provision of adequate information about these materials is necessary to create an informed awareness so as to facilitate ease of recycling and to ensure that "preparing for recycling" is convenient for people to undertake. This is based on the available evidence (such as [15,16,21]) of a strong relationship between specific information concerning the key recyclables and waste-recycling behaviour. However, the contribution of environmental knowledge in enhancing pro-environmental behaviours is still contentious and findings on the effect of environmental knowledge are mixed in the literature. For example, while [45] observed positive effects of environmental knowledge, [46] concluded that environmental knowledge has little or no effects on behaviours. Nevertheless, studies (such as [14,17,47]) have shown that sufficient knowledge of materials that can be recycled enhances recycling behaviour. It is intuitive to conclude that the provision of specific information on what to recycle could increase recycling awareness and subsequently influence recycling behaviour. On the contrary, provision of information alone may not be sufficient in its own right to influence recycling participation [17,48]. Accordingly, the relevance of recycling information may be discounted by the introduction of policy-based instruments [13] or market-based instruments [17] that may facilitate ease of recycling. On the other hand, the contents, including the medium, of disseminating recycling information are both important in enhancing recycling behaviour [48].

Although the provision of recycling information alone may not be sufficient in enhancing recycling behaviour [17], the absence of specific information could serve as a barrier to waste-recycling behaviour [1,49]. Recycling information, when available, may practically differentiate perceived recyclers from non-recyclers [44,50] and this may inform policy or strategy that enhances recycling. For instance:

> "Even though bins are provided it helps to put up the sign and specify what one goes into which kind of thing" (Par_006).

As a result, provision of sufficient information on recyclables is likely to influence recycling behaviour and also the quality of materials for recycling.

What Facilities Are Available for Recycling?

Having provided information on what constitutes waste, including materials that can be collected for recycling, there is a necessity to provide adequate information about where the materials can be deposited for the purpose of recycling. Accessibility to appropriate recycling facilities is a crucial factor in enhancing recycling participation and serves as one of the success criteria for any recycling scheme. According to the participants' accounts for instance:

> "If it (the facility) is easily accessible, it's feasible then I think a lot of more people will recycle" (Par_011).

This may either facilitate the ease of recycling when accessible or makes recycling more challenging to perform when not accessible:

> "At home at the moment we have general waste bin, we've got food waste as well that's on street service and the recycling I'll go around to (a supermarket's name) just down there and recycle glass, paper, cardboard, tins, cans, plastic bottle and I've also got a drink cutting recycling bin for tetra packs" (Par_004).

Accordingly, access to facilities and recycling behaviour are observed to be positively correlated [4,10,19,47].

In addition, the place of residence as well as the type of accommodation and the nature of community can influence the provision and accessibility of appropriate recycling facilities:

> "A lot of recycling is down to the area where you live" (Par_006).

"I think it's very much depends on the area you live in whether you recycle or not, whether you're wealthy or you live in a sort of less wealthy area" (Par_008).

"We stay in apartment at home which is got communal bins, we recycle as best as we can, papers are; we collect papers and put them in the recycling bins, everything else no I don't recycle at all" (Par_001).

These may be associated with economies of scale on the part of government and convenience on the part of resident:

"Where I live on my street, there aren't really any recycling bins either, we have one black general waste bin and is collected every second Tuesday; and many of my neighbours put their recycling in that bin and they all have cars however they don't drive down which is five minutes-drive down to a sort of recycling centre" (Par_011).

For example, rural dwellers may not produce enough waste to attract investments in recycling facilities and may result in a lack of appropriate facilities:

"If there's not enough containers to service an area you going to have a problem. I think that's where the real cost is—is the infrastructure on the ground where are not seeing a right investment or investment in right places both in real process and capabilities" (Par_002).

Whereas people, especially those in multi-family apartment, may consider recycling to be inconvenient due to a lack of storage space:

"We just don't have the bins, we don't have storage facilities; if we do have glass bottles or plastic bottles we don't have that facilities to store them, you know flats are like that, you don't space for storage so you just have to put them in the general bin" (Par_001).

Although the available options may be different in rural areas compared to urban (and suburban) communities, recycling facilities can range from simple caddies, desk bins, and outdoor kerbside containers to various recycling centers or drop-off points. Many of these dedicated sites, including supermarkets and other places (such as Household Waste Recycling Centers), have provisions for containers where specific materials can be deposited for recycling. For instance:

"So what I do is that we have a balcony so anything we need to recycle we actually put on the balcony and then when we decide to make a trip to either (supermarket's names) we take the recycling and put in the recycling centre" (Par_011).

In the UK, for example, besides some designated recycling locations (and drop-off points), the household waste recycling centers (HWRCs) previously known as Civic Amenity sites allow householders to transport their bulky materials for recycling. It is worth noting that HWRCs only receive bulky (household items) materials such as mattresses and furniture from householders. The intention of these drop-off points is to make recycling facilities available and as accessible as possible for residents and also to prevent fly tipping or open dumping. Nonetheless, a lack of adequate facilities was observed [51,52] as one of the barriers for participating in recycling schemes. If recycling facilities are really important in waste collection for recycling, then how adequate and convenient are these facilities for their users? What sort of information, and in what format, is available to the public about their usage? These are pertinent questions that should be addressed in order to make recycling facilities accessible. Inability to address these questions may reduce the likelihood of recycling participation and render these facilities inadequate when preparing materials for recycling. For example, [50] observed that recycling logistics have a positive influence on recycling behaviours while [13] reported that the appearance and size of recycling bins are some of the barriers to community waste recycling. While the existence of sufficient facilities enhances recycling behaviour [53] information about the facilities [10] could make the facilities more accessible. As a result, a well-designed scheme with provision of structured recycling information is more effective compared to provision of monetary incentives [54].

4.2.2. "Where" of Waste Recycling Information

The discussion has focused on the "what" of recycling information, this section therefore focused on the "where" of recycling information in terms of its sources and/or medium dissemination. For instance, one of the interviewees reckoned that:

"It has become more apparently feasible in the media in recent years" (Par_011).

Different sources of recycling information are identified in the literature on recycling behaviour (such as [10,16]). These sources included print (such as leaflets, local newspapers, government environmental newspapers, and posters) and broadcast or online media (for instance radio, television and the internet, including social media and intranet). For instance:

"I guess it kind of ties with the council given you specific bins to do this and I think you start to think more about it. And everywhere you look through the papers, media there's always about do you do your bit for the environment be it recycling, do you turn the lights off and that kind of stuff? So I think the advertising campaigns are effective" (Par_005).

In addition, anecdotal evidence suggests that personal contact and public consultation have been well adopted by some local councils, especially in the UK. Accordingly, access to recycling information and how that information is structured not only has a significant influence on waste-recycling behaviour but is also observed to mediate the influence of other factors [18,55].

The importance of recycling information sources cannot be underestimated as this may determine the degree of acceptance as well as participation in waste-recycling schemes. The credibility and reliability of recycling information may be influenced by the sources, which could consequently affect peoples' behaviour and participation in recycling schemes. For example, local council environmental newspapers [12] and leaflets [16] were observed to be more effective in influencing recycling behaviour compared to local newspapers. For example:

"I think publicity, social media—all these kind of things—are far more prominent and has been for the last 10 years or something like that. The awareness comes from social media, council publications I guess they influence us" (Par_005).

Compared to local or marketing newspapers, local councils' environmental newspapers or leaflets contain specific recycling information and feedback such as recycling performances, recycling targets, location of the local recycling facilities and description of recyclables that local newspapers would not necessarily report. In other words, local councils' communication strategies contain both declarative and procedural information that influence waste recycling participation as well as behaviour. In addition, [47] observed that door-stepping techniques (personal contact) are more effective in encouraging as well as increasing recycling rates. The views of the research participants were in support of these techniques in enhancing recycling:

"Obviously, the recycling office is going to visit schools and especially the primary kids, you get them involved, you get them enthusiastic and you know they will say oh the kids will go home and tell the parents what to do and what not to do and things" (Par_004).

"The kids are getting education at school, learning about recycling. They come to me then sometimes and ask me questions about can this go, which bin does this go" (Par_002).

Apart from creating awareness about the on-going schemes in any locality, an education campaign may also improve existing knowledge of what, where, and how to recycle. This was corroborated by Nixon and Saphores [12] who observed a relationship between publicity (and promotion) and recycling behaviour. This suggests that recycling publicity and promotion or a well-designed communication strategy can be an effective tool to engage or enhance recycling behaviour [16]. However, the rate of household waste recycling was observed to reduce soon after the campaign ended, which shows that the recycling behaviour could not be sustained beyond the period of information [52]. One possible reason for this could be that recycling barriers and/or other situational factors were not identified and addressed [1] prior to and during the campaign. Some of these barriers may include lack of facilities, proximity of the facilities, and recycling scheme design. In addition, recycling promotion or publicity adverts should be regularly updated in order to be more effective and engaging with respect to current waste-recycling issues. Multiple sources of recycling information may be considered when disseminating recycling information; this may be more helpful in increasing recycling knowledge than a single source of information [44]. However, recycling information through a reliable or credible channel with a certain degree of authority such as local councils may enhance and sustain recycling behaviour. As a result, multiple sources of recycling information may be considered when disseminating recycling information; this may be more reliable for the increase of recycling knowledge than a single source of information [12]. While an information campaign is important to enhance waste recycling, its relevance could diminish by the introduction of policy instruments and other external factors that may enhance simplicity of recycling [13,49].

4.2.3. "When" of Recycling Information

Recycling information may not be effective without a clear understanding of the stage at which the information should be provided. Previous projects have adopted and introduced specific recycling information, in particular prompts before and after the introduction of recycling schemes. While information was used prior to the scheme to create scheme awareness [10,56], performance feedback [47] was used after the scheme implementation. In addition, performance feedback has been reported to enhance the quality of materials for recycling by reducing the amount of contaminants [48]. Prior intervention information and performance feedback may therefore serve as motivations or incentives for participation in the scheme. Nonetheless, the provision of feedback on recycling performance was observed to be less effective compared to financial or monetary incentives [48]. Although feedback may be less effective compared to other incentives, recycling behaviour is significantly influenced by prior-scheme information and performance feedback [16,47]. While the effects of the timing of the introduction of recycling information are still ambiguous, there is a likelihood that prior-scheme recycling information would be more effective, especially in recruiting new participants. For instance:

> "When that blue bin turned up with the green waste caddy, I don't know anything about that as a householder; no leaflets through your door, no information about it; pull out the green caddy bin what the hell is this for, do I put my bag in that, there's a little mesh thing sitting where does this go? Does it go in my utility room, do I fill what? I don't even know how to use the system so the education we got from that was slightly that lustre, I think that's the key thing as well as you know you got to get that education before you rolling out make sure everybody is aware of what they are going to do and then sustain it as well" (Par_002).

On the other hand, recycling information being provided (in terms of performance feedback) after introducing the scheme could be more effective in enhancing participation rates. In other words, performance feedback could be adopted to encourage those who are already participating in the recycling scheme. As a result, the stage at which recycling information is introduced will influence the nature of information and could also affect how people engage and participate in the scheme. For example, provision of waste-recycling information prior to a recycling scheme may offer sufficient time for people to reflect on it so as to clearly understand the scheme, including their participation. This may enhance the participants' knowledge of different facilities, including the scheme's requirements relating to frequency of collection and materials to be collected for recycling. Like any other interventions, the target behaviour may return to the baseline when the information or prompts are discontinued. The reliability and effectiveness of recycling information (or prompts) may therefore diminish over a long period of time [25]. The major challenge is how to sustain the effectiveness of this aspect of recycling information without diminishing the values of the recycling scheme. One possibility is to continuously update the recycling information at regular intervals when the performance (in terms of the quantity and quality of materials being collected for recycling) is observed to be deteriorating. This means there needs to be a system for the continuous monitoring and improvement of the scheme throughout its entire lifecycle. This is necessary due to the fact that waste management (recycling) requirements are dynamic based on the increasing changes in waste management legislation and/or regulation including the complexity of human behaviours.

In this study, therefore, it is suggested that coherent waste recycling information that explains the where, what, and when (and how) of waste recycling should be made available at least three months prior to the introduction of a scheme. Whilst there is no theoretical or philosophical basis for the selection of this time frame, it may be socially and technically desirable to allow sufficient time for behavioural adjustment that could influence people's level of preparedness. If the major reasons for not recycling are inconvenience, distance, and other waste-handling issues [47], giving the participants some valuable information prior to the commencement of a scheme would provide an opportunity to assess their status against the scheme's demands. Participants would also be made aware of the facilities such as recycling centers and types of recycling schemes (such as commingle) existing in their local jurisdiction. For instance, participants may not be conversant with (or aware of) the recycling facilities in their area despite the proximity of such facilities (see [44,56] for a review). From this review, it is argued that adequate and specific recycling information should be provided before the implementation of a particular scheme in order to create scheme awareness. In order to enhance participation, performance feedback should also be provided at regular intervals during and after the scheme introduction in order to sustain the relevance of recycling information.

5. Conclusion

In this research, the need for coherent and all-encompassing waste-recycling information was discussed and presented. Using the research participants' views and findings from the literature, we argued that coherent information creates scheme awareness and also provides opportunities for planners to design effective schemes that may be more attractive in encouraging participation. In addition, this study deliberately re-positioned waste recycling responsibilities based on the acceptable and legal definitions of waste and recycling. According to this study, the definition of recycling absolves individuals from any responsibilities of recycling. Instead, preparation of materials for collection was argued in this research to be the major responsibility of individuals or householders. Recycling is more than tossing materials in dedicated bins: it requires additional efforts such as chemical, thermal, or mechanical processes. On this basis, this study presents recycling as a technical activity that could only be performed by specialised waste management (or recycling) firms with appropriate facilities or capacities for recycling. Further, this research extends the on-going discussion on the legality of the existing definition of waste and identifies key recyclables that householders or individuals could prepare for recycling.

In support of these arguments, waste recycling information was sub-classified into three distinct segments—what, where, and when. The rationale was to provide scheme designers, policy-makers, and participants with an opportunity to address and understand how the materials that should be prepared for recycling could be enhanced. While previous studies have identified different factors that may influence recycling behaviours, this paper offered support to studies that have demonstrated that recycling information or communication is an effective strategy in influencing participation. However, the effects of information can be diminished with time and context, there should be more clarification concerning the what, when, and where of waste recycling information. It was argued that policies, strategies, and waste management schemes are more effective when holistic information is incorporated into the scheme design.

However, the description of waste recycling information is not complete without further knowledge about why and how waste recycling should be carried out. The "why" would provide necessary information concerning the reasons or the necessities for recycling while "how" explains the process or the procedure in recycling. As a result, procedural information would facilitate ease of preparing materials for recycling and may influence scheme participations. All these attributes of recycling information should be taken into consideration and be applied in tandem when designing or developing waste management strategies or policies that could enhance public involvement. It is therefore anticipated that an understanding of the whole network of recycling process (or interlinks) as discussed in this research would result in the full knowledge of materials that should be prepared for recycling which may consequently enhance recycling practices.

6. Recommendations

As a result of these findings, we would like to make the following recommendations especially for policy makers as well as practitioners:

1. Information aiming at enhancing recycling participation should be more explicit in terms of what, when, and where including how to recycle

2. When designing and disseminating recycling information, information recipients should be made aware of the importance of recycling and why they should recycle in the first place. As a result, recycling information should be both prescriptive and procedural in terms of recycling (including the items to recycle) and participation

3. There is a need to constantly updating recycling information so as to keep up with dynamics of people's behaviour in terms of waste generation and also to reflect seasonal patterns considering the effect of time and contexts on recycling information

4. In order to incentivize and to enhance recycling behaviour, there should be a mechanism for feedback on recycling performance

5. The provision of recycling information (and/or communication) should facilitate ease and accessibility of recycling schemes and should target perceived recyclers and non-recyclers.

Acknowledgments: The authors are grateful for the contributions of Seonaidh McDonald (Aberdeen Business School, Robert Gordon University), Sarah Pedersen (Department of Communication, Robert Gordon University), and Evagelos Korobilis-Magas (Marketing Department, Aberdeen Business School, Robert Gordon University). In addition, we express our gratitude to the independent reviewers for their contributions and suggestions. The author(s) received no financial support for the research, authorship, and/or publication of this article.

Author Contributions: Before her death, Joanneke was instrumental to this manuscript and also proofread the initial draft. Adekunle designed and conducted the interviews, transcribed and analysed the data, and wrote the entire paper.

References

1. Oke, A. Workplace waste recycling behaviour: A meta-analytical review. *Sustainability* **2015**, *7*, 7175–7194. [CrossRef]

2. Stern, P. Toward a coherent theory of environmentally significant behavior. *J. Soc. Issues* **2000**, *56*, 407–424. [CrossRef]

3. Ekström, K.M. *Waste Management and Sustainable Consumption: Reflections on Consumer Waste*; Routledge Taylor and Francis Group: New York, NY, USA, 2015.

4. Barr, S. What we buy, what we throw away and how we use our voice: Sustainable household waste management in the UK. *Sustain. Dev.* **2004**, *12*, 32–44. [CrossRef]

5. Knussen, C.; Yule, F. "I'm not in the habit of recycling" The role of habitual behavior in the disposal of household waste. *Environ. Behav.* **2008**, *40*, 683–702. [CrossRef]

6. Moore, D. Thirty Percent of Residents "Confused" about What Can Be Recycled. Chartered Institution of Wastes Management Journal, 2015. Available online: http://www.ciwm-journal.co.uk/thirty-percent-of-residents-confused-about-what-can-be-recycled/ (accessed on 5 July 2015).

7. Kollmuss, A.; Agyeman, J. Mind the gap: Why do people act environmentally and what are the barriers to pro-environmental behavior? *Environ. Educ. Res.* **2002**, *8*, 239–260. [CrossRef]

8. McDonald, S.; Oates, C.J.; Alevizou, P.J. No through road: A critical examination of researcher assumptions and approaches to researching sustainability. *Rev. Mark. Res.* **2016**, *13*, 139–168.

9.   Garcés, C.; Lafuente, A.; Pedraja, M.; Rivera, P. Urban waste recycling behaviour: Antecedents of participation in a selective collection programme. *Environ. Manag.* **2002**, *30*, 378–390. [CrossRef] [PubMed]

10. Vicente, P.; Reis, E. Factors influencing households' participation in recycling. *Waste Manag. Res.* **2008**, *26*, 140–146. [CrossRef]

11. Berglund, C. The assessment of households' recycling costs: The role of personal motives. *Ecol. Econ.* **2006**, *56*, 560–569. [CrossRef]

12. Nixon, H.; Saphores, J.D.M. Information and the decision to recycle: results from a survey of us households. *J. Environ. Plan. Manag.* **2009**, *52*, 257–277. [CrossRef]

13. McDonald, S.; Oates, C. Reasons for Non-Participation in a Kerbside Recycling Scheme. *Resour. Conserv. Recycl.* **2003**, *39*, 369–385. [CrossRef]

14. Thøgersen, J. Monetary Incentives and Recycling: Behavioural and psychological reactions to a performance-dependent garbage fee. *J. Consum. Policy* **2003**, *26*, 197–228. [CrossRef]

15. Kaplowitz, M.D.; Yeboah, F.K.; Thorp, L.; Wilson, A.M. Garnering input for recycling communication strategies at a big ten university. *Resour. Conserv. Recycl.* **2009**, *53*, 612–623. [CrossRef]

16. Mee, N.; Clewes, P.S.; Phillips, P.S.; Read, A.D. Effective implementation of a marketing communications strategy for kerbside recycling: A case study from Rushcliffe (UK). *Resour. Conserv. Recycl.* **2004**, *41*, 1–26. [CrossRef]

17. Perrin, D.; Barton, J. Issues associated with transforming household attitudes and opinions into materials recovery: A review of two kerbside recycling schemes. *Resour. Conserv. Recycl.* **2001**, *3*, 61–74. [CrossRef]

18. Evison, T.; Read, A.D. Local Authority Recycling and Waste-Awareness Publicity and Promotion. *Resour. Conserv. Recycl.* **2001**, *32*, 275–292. [CrossRef]

19. Grodzinska-Jurczak, M.; Tomal, P.; Tarabuła-fiertak, M.; Nieszporek, K.; Read, A.D. Effects of an Educational Campaign on Public Environmental Attitudes and Behaviour in Poland. *Resour. Conserv. Recycl.* **2005**, *46*, 182–197. [CrossRef]

20. McDonald, S. Green Behaviour: Differences in Recycling Behaviour between the Home and the Workplace. In *Going Green: The Psychology of Sustainability in the Workplace*; Bartlett, D., Ed.; The British Psychological Society: Leicester, UK, 2011; pp. 59–64.

21. Austin, J.; Hatfield, D.; Grindie, A.; Bailey, J. Increasing recycling in office environments: The Effects of Specific Informative Cues. *J. Appl. Behav. Anal.* **1993**, *26*, 247–253. [CrossRef] [PubMed]

22. Brothers, K.J.; Krantz, P.J.; McClannahan, L.E. Office paper recycling: A function of container proximity. *J. Appl. Behav. Anal.* **1994**, *1*, 153–160. [CrossRef]

23. Kelly, T.C.; Mason, I.G.; Leiss, M.W.; Ganesh, S. University community responses to on-campus resource recycling. *Resour. Conserv. Recycl.* **2006**, *47*, 42–55. [CrossRef]

24. Humphrey, C.R.; Bord, R.J.; Hammond, M.M.; Mann, S.H. Attitudes and Conditions for Cooperation in a Paper Recycling Program. *Environ. Behav.* **1977**, *9*, 107–124. [CrossRef]

25. De Young, R. Changing behavior and making it stick: The conceptualization and management of conservation behavior. *Environ. Behav.* **1993**, *25*, 485–505. [CrossRef]

26. Bryman, A. *Social Research Methods*, 4th ed.; Oxford University Press: Oxford, UK, 2012.

27. Creswell, J.W. *A Concise Introduction to Mixed Methods Research*; Sage Publications: Thousand Oaks, CA, USA, 2014.

28. Tashakkori, A.; Teddlie, C. (Eds.) *Sage Handbook of Mixed Methods in Social & Behavioral Research*; Sage Publications: Thousand Oaks, CA, USA, 2010.

29. Bowen, G.A. Naturalistic inquiry and the saturation concept: a research note. *Qual. Res.* **2008**, *1*, 137–152. [CrossRef]

30. Miles, M.B.; Huberman, A.M. *Qualitative Data Analysis: An Expanded Sourcebook*; Sage Publications: Thousand Oaks, CA, USA, 1994.

31. Bryman, A. *Social Research Methods*, 5th ed.; Oxford University Press: Oxford, UK, 2016.

32. Strauss, A.; Corbin, J. *Basics of Qualitative Research: Techniques and Procedures for Developing Grounded Theory*; Sage Publications: Thousand Oaks, CA, USA, 1998.

33. Charmaz, K. *Constructing Grounded Theory: A Practical Guide through Qualitative Analysis*; Sage Publications: Thousand Oaks, CA, USA, 2006.

34. Lee, D.T.F.; Woo, J.; Mackenzie, A.E. The cultural context of adjusting to nursing home life: Chinese elders' perspectives. *Gerontologist* **2002**, *5*, 667–675. [CrossRef]

35. Creswell, J. *Qualitative Inquiry and Research Design: Choosing among Five Traditions*; Sage Publications: Thousand Oaks, CA, USA, 1998.

36. Guest, G.; Bunce, A.; Johnson, L. How many interviews are enough? An experiment with data saturation and variability. *Field Methods* **2006**, *1*, 59–82. [CrossRef]

37. Ritchie, J.; Lewis, J.; Nicholls, C.M.; Ormston, R. (Eds.) *Qualitative Research Practice: A Guide for Social Science Students and Researchers*; Sage Publications: Thousand Oaks, CA, USA, 2013.

38. Pitkow, J.E.; Kehoe, C.M. Emerging trends in the WWW user population. *Commun. ACM* **1996**, *6*, 106–108. [CrossRef]

39. Smith, M.A.; Leigh, B. Virtual subjects: Using the Internet as an alternative source of subjects and research environment. *Behav. Res. Methods Instrum. Comput.* **1997**, *4*, 496–505. [CrossRef]

40. Oates, C. J.; McDonald, S. Recycling and the Domestic Division of Labour: Is Green Pink or Blue? *Sociology* **2006**, *40*, 417–433. [CrossRef]

41. The World Bank. What a Waste A Global Review of Solid Waste Management. Urban Development Series Knowledge Papers. 2012. Available online: http://www-wds.worldbank.org/external/default/WDSContentServer/WDSP/IB/2014/09/17/000442464_20140917123945/Rendered/PDF/681350REVISED00t0a0Waste020120Final.pdf (accessed on 4 January 2014).

42. Berger, I.E. The demographics of recycling and the structure of environmental behavior. *Environ. Behav.* **1997**, *29*, 515–531. [CrossRef]

43. The European Parliament and the Council of the European Union. Directive 2008/98/EC of the European Parliament and of the Council. 2008. Available online: http://eur-lex.europa.eu/LexUriServ/LexUriServ.do?uri=OJ:L:2008:312:0003:0030:EN:PDF (accessed on 22 May 2013).

44. De Young, R. Exploring the difference between recyclers and non-recyclers: The role of information. *J. Environ. Syst.* **1988**, *18*, 341–351. [CrossRef]

45. Mainieri, T.; Barnett, E.G.; Valdero, T.R.; Unipan, J.B.; Oskamp, S. Green buying: The influence of environmental concern on consumer behavior. *J. Soc. Psychol.* **1997**, *137*, 189–204. [CrossRef]

46. Grob, A. A structural model of environmental attitudes and behaviour. *J. Environ. Psychol.* **1995**, *15*, 209–220. [CrossRef]

47. Timlett, R.E.; Williams, I.D. Public participation and recycling performance in England: A comparison of tools for behaviour change. *Resour. Conserv. Recycl.* **2008**, *52*, 622–634. [CrossRef]

48. Chung, S.S.; Poon, C.S. Hong Kong citizens' attitude towards waste recycling and waste minimization measures. *Resour. Conserv. Recycl.* **1994**, *10*, 377–400. [CrossRef]

49. Hage, O.; Söderholm, P.; Berglund, C. Norms and Economic Motivation in Household Recycling: Empirical Evidence from Sweden. *Resour. Conserv. Recycl.* **2008**, *53*, 155–165. [CrossRef]

50. Ebreo, A.; Vining, J. Motives as predictors of the public's attitudes toward solid waste issues. *Environ. Manag.* **2000**, *25*, 153–168. [CrossRef] [PubMed]

51. Barr, S. Factors influencing environmental attitudes and behaviors: A UK case study of household waste management. *Environ. Behav.* **2007**, *39*, 435–473. [CrossRef]

52. Knussen, C.; Yule, F.; Mackenzie, J.; Wells, M. An Analysis of Intentions to Recycle Household Waste: The roles of past behaviour, perceived habit, and perceived lack of facilities. *J. Environ. Psychol.* **2004**, *24*, 237–246. [CrossRef]

53. Luyben, P.; Cummings, S. Motivating beverage container recycling on a college campus. *J. Environ. Syst.* **1981**, *11*, 235–245. [CrossRef]

54. Iyer, E.S.; Kashyap, R.K. Consumer recycling: Role of incentives, information, and social class. *J. Consum. Behav.* **2007**, *6*, 32–47. [CrossRef]

55. Kalsher, M.J.; Rodocker, A.J.; Racicot, B.M.; Wogalter, M.S. Promoting Recycling Behavior in Office Environments. In Proceedings of the Human Factors and Ergonomics Society Annual Meeting, Seattle, WA, USA, 11–15 October 1993; SAGE Publications: Thousand Oaks, CA, USA, 1993; pp. 484–488.

56. McDonald, S.; Ball, R. Public Participation in Plastics Recycling Schemes. *Resour. Conserv. Recycl.* **1998**, *22*, 123–141. [CrossRef]

10

Challenges in Automotive Fuel Cells Recycling

Rikka Wittstock [1,†], **Alexandra Pehlken** [1,*,†] **and Michael Wark** [2,†]

[1]   Cascade Use, Carl von Ossietzky University of Oldenburg, Ammerlaender Heerstrasse 114-118,
      26129 Oldenburg, Germany; rikka.wittstock@uni-oldenburg.de

[2]   Department of Chemistry, Carl von Ossietzky University of Oldenburg, Ammerlaender Heerstrasse 114-118,
      26129 Oldenburg, Germany; michael.wark@uni-oldenburg.de

*   Correspondence: alexandra.pehlken@uni-oldenburg.de

†   These authors contributed equally to this work.

Academic Editor: Michele Rosano

Abstract: Fuel cell driven cars belong to the 'zero emission' vehicles and should contribute to lower CO_2 emissions. However, they contain platinum, which is known as a critical material in the European Union. This study investigated the potential contribution of recycling fuel cell vehicles (FCV) to satisfy the platinum demand arising from a widespread deployment of fuel cell vehicles in Europe. Based on a qualitative examination of the four consecutive steps in the recycling chain (collection, dismantling, disassembly and pre-processing, material recovery) of fuel cell vehicles, two recycling scenarios were developed. Using dynamic material flow analysis, these two recycling scenarios were applied to two scenarios for the market penetration of fuel cell vehicles in nine European lead markets to deliver both the associated impact on platinum demand and the contribution of recycling for meeting this demand. The diffusion of FCV in Europe will not cause a depletion of platinum resources in the short term, as the calculated 537.06 t and 459.24 t in cumulative platinum requirements are far below the currently estimated global reserves. However, concerns regarding the future development of platinum supply and demand remain.

Keywords: fuel cell; end-of-life vehicle; electric vehicle; recycling; material flow assessment; STAN

1. Introduction

In 2011, the European Council reconfirmed its determination to meet the 2 °C-target for global warming through European Union(EU)-wide reductions of greenhouse gas (GHG) emissions of 80% to 95% (compared to 1990 levels) by 2050 [1]. As road transport alone makes up one-fifth of the European Union's total emissions of carbon dioxide (CO_2) [2] in order to meet the overall reduction targets by 2050, transport-related GHG emissions must be cut by up to 67% [1]. Despite ongoing improvements to the emission standards of internal combustion engine vehicles, under the current European vehicle portfolio, it will not be possible to meet these ambitious reduction targets without the integration of so-called 'zero-emission vehicles', which include battery electric and fuel cell vehicles [3]. Although battery electric vehicles have been at the forefront of public attention, due to their short range, limited by battery capacities, they are best-suited for urban driving. Fuel cell vehicles, on the other hand, combine the emission benefits of battery electric vehicles with almost the same ranges and refuelling times provided by conventional internal combustion engine vehicles. With these characteristics, they are ideally suited to personal transportation in medium and larger cars, as well as longer trips [4], a market segment which accounts for around half of all passenger cars in the EU and is responsible for 75% of transport-related CO_2 emissions [5].

However, potential trade-offs and unintended rebound effects resulting from the endorsement of novel powertrain technologies must be considered carefully. As highlighted by a number of publications, electromobility not only entails environmental benefits, but also has substantial impacts on the demand for non-renewable resources, in particular for so-called special or technology metals [6–12].

With regards to fuel cell vehicles, platinum is of particular concern, as it represents the main catalyst material of the proton exchange membrane fuel cell that is used in automotive applications. The extraction of platinum group metals from primary deposits involves high environmental impacts, including emissions of sulphur dioxide, of CO_2 equivalents in the range of 13,000 tons per ton of platinum group metals [13], excessive water and energy consumption [14], as well as habitat destruction, air and water pollution and generation of dust, particulate matter and solid waste [15]. In addition, platinum is considered a critical metal in the EU due to its geological scarcity, its use in a variety of technologies and its highly concentrated supply base [16]. Because of the limited annual supply and unfavourable mining conditions, platinum prices are expected to remain high, which has been identified as a major barrier to the diffusion of fuel cell vehicles by raising the cost of production [17].

While various publications discuss the diffusion of fuel cell vehicles and the resulting platinum demand [6,7,9,12,17–20] few examine the role of recycling in satisfying the industry's platinum requirements. This is especially astonishing considering the fact that recycling rates vary significantly between applications. While industrially used catalysts are recycled with rates close to 90%, passenger cars represent one application with relatively low platinum recycling rates, as the recycling quotas for exhaust gas catalysts reach a mere 50% to 60% [21]. This already makes the automotive industry the largest net consumer of platinum today, even when growth of vehicle sales is ignored [17]. With fuel cell vehicles containing more platinum than combustion engine vehicles by a factor of >10, a widespread deployment could severely aggravate the temporal and structural scarcity already perceptible in the global platinum market. It is therefore in the interests of both fuel cell and car manufacturers to close the existing knowledge gaps and focus on the contribution of recycling for meeting their future platinum demand.

Hence, this study aims to investigate to what extent the recycling of platinum from fuel cell vehicles may contribute to satisfying this industry's future platinum demand. In order to provide an enhanced understanding of the current proton exchange membrane fuel cell technology and its material compositions, the paper sets out with a review of the state-of-the-art. The subsequent part then forms the qualitative basis for the later quantitative assessment of the recycling potential of fuel cell vehicles. Based on a review of technical and non-technical literature, this part examines the current status of fuel cell vehicle recycling as well as options for the design of a future recycling system. At the core of this study is a dynamic material flow analysis, modelling the flows of platinum throughout a fuel cell vehicle's life cycle, in order to determine the potential for meeting platinum demand through the recycling of end-of-life fuel cell vehicles.

2. Theoretical Context

2.1. Proton Exchange Membrane Fuel Cells

From a structural perspective, fuel cell vehicles can be considered a type of hybrid vehicle, in which the fuel cell replaces the internal combustion engine (Figure 1) [22]. Using atmospheric oxygen and compressed gaseous hydrogen supplied from the onboard tank, the fuel cell generates electricity, which powers the vehicle's electric motor. Due to their favourable attributes, such as low operating temperature, fast start-up and fast response to varying loads, fuel cells installed in current automotive applications are of the Polymer Electrolyte Membrane fuel cell (PEMFC) type.

PEMFC consist of a membrane electrode assembly (MEA) framed by bipolar flow field plates (Figure 2). The MEA comprises a compound structure of a polymer electrolyte membrane, a gas diffusion layer (GDL) and a catalyst layer. Serving as the fuel cell's electrolyte, the gas-proof but proton-conductive membrane consists of a 50 to 150 μm thick foil, which is known mainly by its registered trade name Nafion®. The membrane separates the cell's electrodes, which are formed of a gas diffusion layer and a catalyst layer that is usually composed of carbon-supported platinum or platinum-alloys [23].

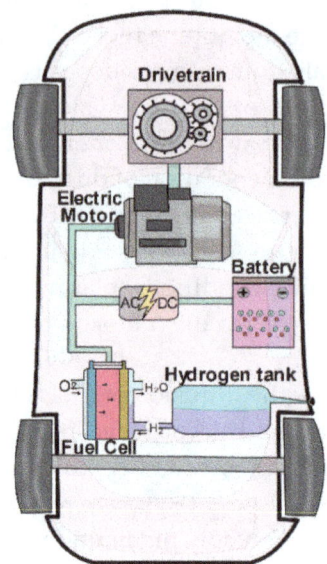

Figure 1. Simplified schematic view of a fuel cell driven vehicle (own figure).

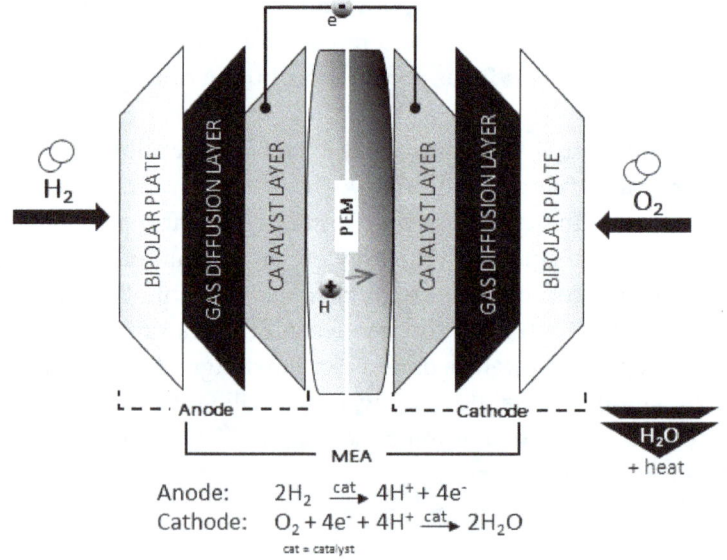

Figure 2. Simplified schematic view of Polymer Electrolyte Membrane fuel cell (PEMFC) (own figure).

At present, the most widely used membrane material continues to be Nafion®, which is made of a hydrophobic polytetrafluorethylene (PTFE, also known as Teflon®) backbone with hydrophilic side groups containing ether units and sulphonic acid end groups. Besides these fluorocarbon-containing membranes, intense research is also done on polybenzimidazole (PBI) membranes which are soaked with phosphorous acid. Such membranes do not suffer as much from swelling and water loss at temperatures above 100 °C, and thus allow operation temperatures of up to 200 °C. At temperatures above about 130 °C, the adsorption of carbon monoxide (CO) on the platinum surface decreases, and thus the catalyst poisoning effect by CO is strongly reduced. CO is an unwanted by-component in hydrogen fuel, if H_2 is obtained from methane conversion It was reported for PBI/phosphorous acid fuel cells that they can tolerate 3 vol % CO in H_2 at 200 °C while at 125 °C only 0.1 vol % are acceptable for reaching the same current density [24]. However, this type of high-temperature fuel cell is still in the research phase and not available on the commercial market.

In PEMFC, the GDL commonly consists of carbon cloth or carbon paper, hydrophobically treated with PTFE in order to prevent water from accumulating in the pores. An additional micro-porous layer consisting of carbon or graphite particles mixed with PTFE is typically adhered to the GDL to improve both electrical contact and water transport [25].

A platinum-based catalyst is generally used for both the oxidation reaction taking place at the anode and the reduction reaction occurring at the cathode. Particles of the catalyst (10 to 100 nm in size) are finely dispersed on a porous substrate, which usually consists of high surface area carbon powders, e.g., carbon black. Commercial production today typically employs procedures similar to those of the printing industry, in which the supported or unsupported catalyst material is mixed with solvents, binder (perfluorosulfonic acid or other ionomers in protonated form) and other additives to form a 'catalyst ink' and then applied in wet form. The resulting catalyst layer is between 10 and 30 μm thick [25–27].

2.2. Development of the Platinum Content

Platinum offers superior catalytic properties with regard to activation of the desired reactions, stability in the fuel cell's harsh chemical environments and long-term durability. However, its high costs in particular, have sparked research aimed at reducing or replacing platinum loadings while maintaining or even improving PEMFC performance. Dedicated research programs have resulted in significant technology advances, allowing a considerable reduction of PEMFC platinum loadings without compromising performance in recent years [28]. While research aimed at eliminating the use of platinum group metals as catalyst materials altogether and replacing it with alternatives is ongoing, it is unlikely to produce any viable options for the coming decade at least [17,29]. Hence, a further reduction of platinum load remains a vital prerequisite for commercialisation of FCV.

Table 1 shows the platinum content target for PEMFC electrocatalysts in 2020, as applied throughout the US Department of Energy's combined research programs on fuel cell technology [30]. Note that the data presented in Table 1 refers to so-called nanostructured thin-film catalysts, which are based on platinum alloyed with, for example, nickel, cobalt, manganese, iron or titanium and which currently remain at a laboratory stage.

Table 1. Technical targets for electrocatalysts in 80 kW automotive fuel cell systems (adapted from [31] (p. 8)).

Characteristics	Unit	Status 2013	Target 2020
Total platinum content	g/kW_{rated}	0.14	0.125
Total platinum loading	mg/cm^2 electrode area	0.15	0.125

Assuming a fuel cell stack of 80 kW, the platinum loading reported for 2013 in Table 1 translates into a total platinum requirement of around 11.2 g per FCV. However, as these are fuel cell stacks produced and tested under laboratory conditions only, the platinum load of fuel cell stacks installed in fuel cell vehicles currently 'on the road' is likely to be considerably higher. Estimates given by expert assessments range from 30 to 42 g of platinum per fuel cell vehicle (depending on the assumed stack power), though no official figures have been published by vehicle manufacturers [17,32,33].

3. State-Of-The-Art in the Recycling Chain of Fuel Cell Vehicles

With up to 42 g of platinum per vehicle, fuel cell vehicles could hence constitute a significant secondary resource in the case of further market penetration. The concurrent implementation of an effective recycling concept would not only serve to counteract the aggravation of scarcities, but also secure resource access and reduce raw material costs for the automotive industry. The co-development of efficient recycling systems is hence considered a prerequisite for successful market integration of fuel cell vehicles.

In order to determine the state-of-the-art in the recycling of fuel cell vehicles as well as anticipating challenges that could lead to suboptimal recycling rates, we performed a systematic literature search of scholarly databases, patent databases as well as online repositories of the US Department of Energy. To supplement scientific literature with information on practice-related projects, a further exploration using a combination of relevant search terms was performed using an online search engine. As the focus of our research lies on the impact of fuel cell vehicle diffusion on the European market, all legal and regulatory aspects refer to conditions within the European Union. The presentation of the literature search results follows a common partitioning of the recycling chain into consecutive steps.

3.1. Collection

In the European Union, end-of-life vehicle recycling is governed primarily by Directive 2000/53/EC, which will presumably also apply to the various powertrain technologies classed as electric vehicles. Under the Directive, automobile producers are liable for take-back, treatment and recycling of end-of-life vehicles (ELV) and incur the costs involved with these requirements; the end-of-life treatment is thus free of charge to consumers [33]. While the precise organisation of collection and dismantling systems differs between the EU's member states and typically involves a number of different business entities, all states certify 'Authorised Treatment Facilities' that issue a 'Certificate of Destruction' to consumers when their deregistered vehicle enters the ELV process [34,35]. In many member states, however, the number of ELV registered this way represents less than 50% of the deregistered vehicles, making it clear that by no means all ELV in Europe are disposed of as prescribed by the ELV Directive [34]. While the issue of ELV's unknown whereabouts is discussed on a European level, it is not the focus of this paper and will only be considered in so far as implications for the future fate of FCV can be deduced.

While the majority of deregistered vehicles are exported to other (mainly Eastern European) EU countries and could therefore be available for recycling in the receiving countries after a second use phase, in practice a major share of these vehicles is resold to countries even further East or to developing countries, where a suitable recycling infrastructure is lacking [36]. Only a small number of vehicles are exported directly to non-EU countries; the majority of these vehicles, however, end up in countries lacking the infrastructure required for environmentally sound recycling. Issues such as illegal exports of ELV falsely classified as used vehicles, lack of legal rigorousness, unlicensed scrap operators, abandoned and garaged ELV represent further significant barriers to maximising collection rates [35].

While collection deficits represent the single largest cause for platinum-group metal losses in the recycling chain of automotive catalysts [36] and the issues described above persist despite ongoing research on possible causes and counter-measures, it is disputable in how far such issues will also prove problematic to a future FCV fleet. Not only is the efficiency of ELV collection schemes likely to improve in future years as a result of corrective actions initiated today, but more importantly, the use of pure hydrogen as a fuel will provide a significant barrier for exports to non-European countries. Requiring a suitable hydrogen supply infrastructure, FCV will be unattractive in countries where such an infrastructure does not exist, which at the time of significant market diffusion in Europe is likely to still be the case in the poorer non-EU countries that provide the typical final destination of European ELV today. Exports of deregistered FCV to countries without a suitable recycling infrastructure should therefore only be a minor issue in the coming decades, meaning a large proportion of the platinum in use by this application will remain in Europe [20].

Nonetheless, the need for highly efficient logistics chains and economic incentives is postulated throughout the literature. Concepts discussed in current research projects include producer-led schemes, such as a reclaimable deposit or lease program for fuel cells, a platinum lease program, in which ownership of the contained platinum remains with the (possibly governmental) lessor throughout the fuel cell vehicle's life cycle, and vertical integration of collection, dismantling and recovery businesses, which would lead to an increase in recycling efficiency by reducing the number of actors involved [37–41].

3.2. Dismantling

As is the case with battery electric vehicles, fuel cell vehicles will presumably follow the same end-of-life processing routes as conventional ELV, with the provision that the fuel cell system must be removed prior to shredding [42]. Instructions for the salvaging and dismantling of FCV have been devised in the United States, which not only emphasize the safety hazards involved with inadvertently damaging high-voltage lines, fuel cell stacks, batteries and hydrogen tanks but also provide advice on how to distinguish FCV from conventional vehicles and on where to locate such high-risk components [43]. These components generally necessitate manual removal, discharge or other treatment before the vehicle can be further dismantled, which is likely to add significant costs.

The extent to which incorrect removal of fuel cell stacks may lead to platinum group metal (PGM) losses in future recycling activities will depend on a number of factors, including the degree of automation in dismantling businesses, the variety of FCV brands and models handled by a dismantler and the design and location of the fuel cell stack in different FCV models. However, as fuel cells are generally concentrated in a relatively compact stack (cf. Figure 1) only one component needs to be located and removed, rather than a number of small units dispersed throughout the entire vehicle, as is the case with, for example, electrical equipment or certain types of magnets.

As mentioned above, failure to remove the fuel cell stack and other components manually prior to further dismantling involves serious safety hazards. Due to the stack's high platinum concentration and the associated monetary value, manual removal is not only economically feasible but also less labour-intensive than the disassembly of exhaust gas catalysts, of which 4% fail to be removed prior to shredding [37]. Since the stack can be removed and transported intact, dust losses as described, for example, for exhaust gas catalysts are also considered insignificant [41]. It is, however, uncertain in how far fuel cell stacks of future FCV will be accessible and removable by independent workshops and ELV dismantlers. It may be assumed that for some time following the wider market penetration of FCV, only few, branded workshops will have the specialist knowledge, personnel and equipment required for removal of the fuel cell stacks, as is the case with battery electric vehicles [42].

3.3. Disassembly and Pre-Processing

While according to the ELV Directive, the reuse of components is to be favoured over recycling and material recovery, simply reusing a fuel cell vehicle's MEAs at the end of its lifetime is not feasible, as fuel cell failure is often caused by degradation of the MEA and attempting to replace or repair any of its components would risk irrevocably damaging the others [25]. Recycling of automotive fuel cells is hence aimed at recovering the valuable raw materials. The required disassembly and pre-processing steps, as well as the fractions produced in this stage, hence depend on the objectives for material recovery and the applied recovery processes. Due to the high value of the contained platinum, it is likely that platinum recovery will constitute the major driver for fuel cell recycling, with other materials, such as the polymeric membrane, being of minor interest. However, during operation of the fuel cell, small platinum particles may diffuse into the membrane. Excluding the membrane and other parts of the fuel cell from the recycling process may thus limit recycling efficiency in addition to making the recycling process more complicated.

Similar to the value-oriented disassembly of waste electric and electronic goods, the recycling of fuel cells requires a manual disassembly procedure. Although automated processing of small fuel cells with a low power range—such as those used as battery replacement—may be conceivable in the future, such processes require high volumes of end-of-life fuel cells and have hence not been developed to date [44].

Schiemann et al [44] provide exemplary information on the manual disassembly procedure of a PEM fuel cell stack into its respective fractions, an overview of which is given in Table 2. Note that the exact composition of the fractions produced will depend on the respective PEM technology and may differ between brand and age of the stack.

Table 2. Disassembly process for PEM fuel cells and fractions produced (adapted from [44] (pp. 28–31)).

Disassembly Step	Action Taken	Fraction Produced
Step 1	Removal of tie rods and casing	Connecting elements: high-grade steel
Step 2	Removal of end plates	Connecting elements: high-grade steel;
		Insulation: plastics;
		End plates: high-grade steel or aluminium alloy
Step 3	Withdrawal of the stack's individual layers, each consisting of bipolar plates, seal, gas diffusion layer and membrane electrode assembly	Seal: plastics
		Bipolar plates: graphite, polymeric binders, additives, separating agents
		Gas Diffusion Layer (GDL): non-woven fabric, carbon paper
		Membrane Electrode Assembly (MEA): polymeric membrane coated with graphite and platinum

Nonetheless, the procedure described above produces fractions that are well-known from the disassembly of electric and electronic equipment and thus facilitates the further processing and valuing in the secondary materials market; however, economic feasibility was not considered in this approach [44]. It must be noted that this procedure assumes that the GDL is entirely separate from the MEA or can at least be separated with minor effort.

While this may be true for certain brands, age of technology or application of the fuel cell, in practice, the GDL is usually hot-pressed and laminated to the catalyst coated membrane in order to form the MEA, making separation of the GDL and catalyst coated membrane difficult. Attempting to strip the GDL off such a 5-layer MEA in order to expose the catalyst-coated membrane is not only labour-intensive but could also result in platinum losses and should thus be avoided [45].

In addition, the fuel cell stacks of some manufacturers are moulded rather than screwed shut and therefore cannot be disassembled by simply removing the casing and individual layers [46]. In order to separate the MEA from other components, the end adapters need to be cut off and the casing removed. The stack can then be broken into individual units of MEA and bipolar plate, from which the catalyst coated membrane may be separated. It is conceivable that minimal dust losses occur when the stack is forcefully broken into individual units. When applied to high volumes of end-of-life fuel cells, the platinum losses resulting from such processes could be significant.

Once the precious metal-loaded MEA has been separated from other fuel cell components, the need for further processing prior to the material recovery step depends on the ensuing recovery methods as well as the targeted materials. Of the nine platinum recovery processes analysed in the following chapter, the majority require a further comminution of the MEA following its separation from other fuel cell stack materials, in order to disintegrate the MEA's laminate structure and expose the catalyst layer.

Shore [45] gives a more detailed description of the applied procedures and also mentions potential drawbacks of the employed technologies, such as matting together of MEA fragments and subsequent obstruction of milling equipment, overheating of equipment and failure to successfully comminute MEAs due to the elastic properties of the membrane. In order to counteract these problems, Shore reports two possible approaches: cryogenic milling and granulation at room temperature followed by milling [45]. If cryogenic milling forms part of the pre-processing stage, special attention must be paid to prevent loss of platinum-rich dusts, as the finely dispersed platinum may be carried off by the nitrogen gas liberated during gasification of liquid nitrogen.

Further difficulties in disassembly and pre-processing arise when it is assumed that in the future, specialised recycling facilities will handle a range of different types of fuel cells from different fields of application, as proposed by [41,44,47].

In a detailed analysis of the different types and brands of fuel cells available, Schiemann et al stress that it will likely be tremendously difficult for disposal companies to recognise the exact type of fuel cell system they are dealing with [44]. With a view to recycling efficiency, but especially health and safety issues and environmental protection, the authors emphasise that each type of fuel cell requires unique disassembly and processing steps and propose the use of flow charts, photographic guides and manufacturer catalogues to avoid misjudgement.

3.4. Material Recovery

As the presence of fluorine compounds in all three layers is customary according to the state-of-the-art, it is especially the MEA materials that present established recycling practices with considerable challenges. Given higher volumes, the feed of end-of-life PEMFC to established pyrometallurgical recycling processes for the recovery of platinum is not feasible, although lower volumes may be tolerated. Incineration of the fluoropolymers contained in Nafion®, the hydrophobically treated gas diffusion layer and the catalyst layer's binding agent leads to the formation of hydrogen fluoride (HF), a toxic compound that is both harmful to health and highly corrosive [46,48]. The application of established pyrometallurgical procedures would thus require the costly and time-consuming refurbishment of existing plants with special protective linings for the furnaces, as well as the installation of elaborate filter and scrubbing systems to eliminate HF from the off-gas [49]. Despite such investments, any fluorine constituents remaining in the PGM-containing slag could hamper the later separation [50]. In addition, through incineration the valuable membrane material Nafion® is essentially destroyed in the process and cannot be recycled [44]. From a recycling viewpoint, the absence of fluorine in the membranes would thus be a major advantage of alternative fuel cell systems, such as the PBI/phosphorous acid fuel cells.

Jha et al. [51] give an extensive overview on the hydrometallurgy recovery and recycling of platinum by leaching of spent catalysts. Despite the standard route of using aqua regia as a leachant, they are focusing on platinum leaching using acidic and alkaline solutions in the presence of oxidizing agents such as nitric acid and hydrogen peroxide, sodium cyanide and iodide solutions. In addition, they also highlight the environmental concerns. Since most publications deal with platinum recovery in general and do not address the PEMFC specifications, few papers are dealing directly with the membrane characteristics in fuel cells [52–55]. Most researchers already identify the potential of not only recovering the platinum from fuel cells; they focus on recovering the whole membrane for reusing it in new applications. However, this is still on laboratory scale and not available on the market today. From the environmental point of view, it would be the best way.

The intention of this paper is not to add another literature review to the community. Instead we performed a patent search in order to determine to what extent the aforementioned issues with regards to the formation of HF have been acknowledged by relevant actors and resulted in the development of alternative procedures. This search included patents registered under the European Patent Office, the US Patent & Trademark Office and the World Intellectual Property Organization (WIPO) in English language since the year 2000, which focus on the recovery of platinum group metals from PEMFC while avoiding HF emissions. Note that with a view to the research focus of this paper, the search focused solely on the recovery of PGM (as well as the more broadly phrased recovery of 'noble metals' or 'precious metals', which may include additional elements, such as gold or silver) from PEM fuel cells. Procedures aimed exclusively at recycling another type of fuel cell, different constituents, such as the membrane material, or other kinds of fluorine-containing devices, were neglected. The results of the patent search including a short description of the respective procedures are summarised in Table 3.

Table 3. List of patents for recovery of Platin Group Metals (PGM) from PEMFC avoiding hydrogen fluoride (HF) emissions.

Patent Number	Inventor/s	Proprietor	Materials Recovered (and Rate in %)	Procedure	Source and Technology
WO/2004/102711	Hagelüken, Kayser, Romero-Odeja, Kleinwächter	Unicore AG & Co. KG	PGM (95%)	Binding of HF through inorganic additive	[56], thermal pretreatment followed by hydro-metallurgy
WO/2006/024507	Koehler, Zuber, Binder, Baenisch, Lopez	Unicore AG & Co. KG	PGM and/or fluorine-containing constituents (no rate provided)	Separation of membrane and catalyst layer by treatment with medium in supercritical state	[50], thermal pretreatment (supercritical status) followed by hydro-metallurgy
WO/2006/115684					
WO/2007/149904	Shore, Robertson, Shulman, Fall, Matlin, Heinz	BASF Catalysts LLC	PGM and/or ionomer (up to 99%)	Leaching using aqua regia	[57–61], hydrometallurgy
WO/2009/029463					
WO/2009/149241					
WO/2010/132156					
EP2700726A1	Romero, Meyer, Voss	Heraeus Precious Metals GmbH	PGM (no rate provided)	HF-resistant furnace	[62], thermal treatment only
WO/2015/010793	Paepke	Daimler AG	PGM and/or ionomer (>98%)	Separation of membrane and catalyst layer through ultrasound and solvent addition	[63], pretreatment by ultrasonic bath followed by pyrometallurgy

As shown in Table 3, a number of recovery procedures are available that allow the recovery of platinum from end-of-life MEA incurring only minimal losses. However, while avoiding HF emissions, all procedures described above continue to entail certain drawbacks.

Mixing the MEA with an inorganic additive, such as calcium carbonate, has shown to be insufficient in binding HF, especially for materials which release hydrogen fluoride at low temperatures [62]. Using water in its supercritical state, the process proposed by [50] relies on a high-pressure operating environment, thus posing concerns for safety as well as untenable equipment costs. Similar reservations may be voiced for the HF-proof furnace and scrubbing process invented by [62], as its cost advantage compared to refurbishment of existing precious metal refining facilities is questionable. Finally, while the procedure proposed by [45] avoids HF emissions, it uses hydrochloric acid and nitric acid, both of which are harmful if skin contact occurs, and generates nitric oxides as well as solid MEA residues, which then require further treatment to avoid environmental impacts.

Insufficient yields of platinum may be problematic in those cases, where removal of the GDL is required in the pre-processing stage or the recovery process is aimed at separating membrane and catalyst layer. This is due to the fact that over a fuel cell's lifetime, platinum particles tend to migrate from the catalyst layer into the membrane and/or GDL, leading to the formation of nanocrystallites. The platinum nanoparticles have to be recovered by dissolution in acidic solutions forming Pt^{2+} ions. In a subsequent step the platinum ions are electrochemically reduced [57]. In addition, some loss of theoretically recoverable platinum in waste streams, such as the process water and solid waste from leaching, appears to be unavoidable [45]. Several of the above mentioned patents also fail to elaborate on any further refining needs in order to obtain platinum of a purity level that is suitable for commercial use. As [45] describes, the conventional combustion of MEA produces a platinum-rich ash, which must be further refined to remove impurities.

Finally, it should be noted that as of 2016, all processes have been implemented and tested at a laboratory level only. Although [45] provides a cost analysis and comes to the conclusion that large-scale operations based on the invented process as described above would be feasible from a financial perspective, no such information is available for any of the other recovery methods. It is therefore not possible to determine whether industrial-scale implementation is possible at all, nor how any of the procedures mentioned above compare to a HF-proof refurbishment of existing precious metal recovery facilities with special protective linings for the furnaces and off-gas treatment in terms of costs. Further research on recovery methods focussing on process efficiency and costs is likely to only take place once sufficient volumes of end-of-life fuel cells arise.

4. Material Flow Analysis for Fuel Cell Vehicles

Building on the insights gained in the qualitative investigation of the recycling chain of fuel cell vehicles, we use dynamic material flow analysis (MFA) to quantitatively assess the impact of a widespread diffusion of fuel cell vehicles in Europe on the global demand for platinum as well as the potential of recycling for meeting this demand.

4.1. Research Design and Calculation Approach

The dynamic MFA for this study is performed using the software tool STAN (short for SubSTance flow ANalysis), a freeware available from the Technical University Vienna, Austria (http://www.stan2web.net/). The calculations performed as a preliminary basis to create the model of platinum flows arising from the market penetration of fuel cell vehicles are derived from a top-down approach as summarised in Figure 3. An explanation of the parameters selected that govern both flow values and transfer coefficients within the MFA diagram is given below.

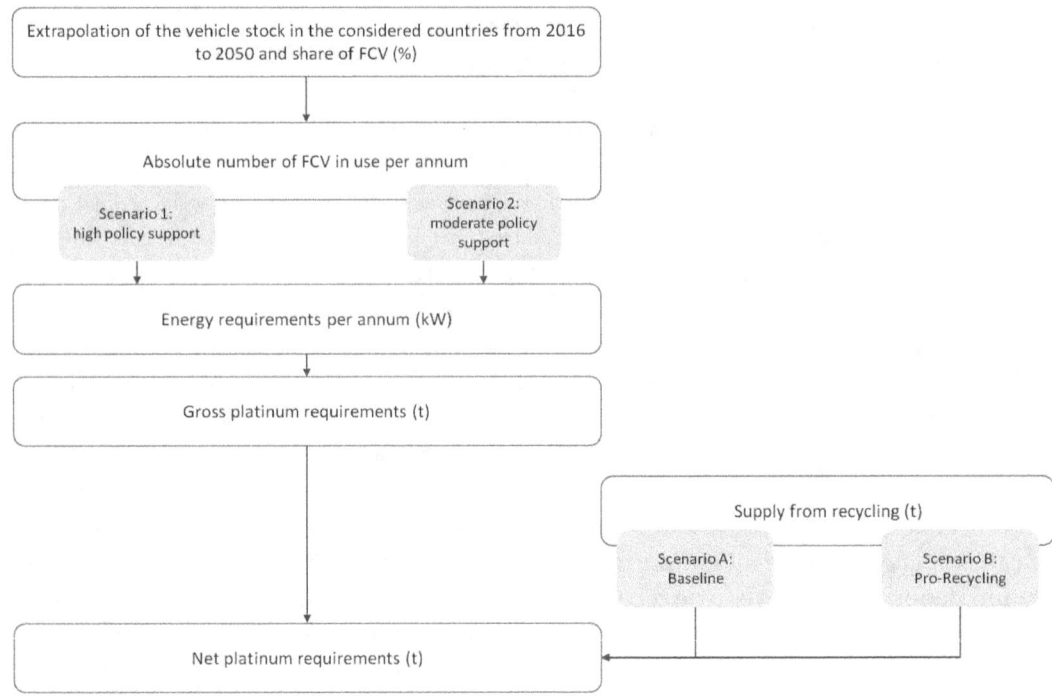

Figure 3. Calculation approach in preparation for the Material Flow Assessment.

4.2. System Boundaries and Assumptions

In order to narrow down the scope of the MFA, a number of system boundaries and parameters were set.

Several studies are available projecting possible scenarios of FCV market diffusion, some of which also include predictions of associated platinum requirements [6,7,9–11,17–20]. Aiming to evaluate possible recycling pathways for FCV for a closed European system, it was especially important to select a study focusing on Europe specifically. This requirement already precluded a number of publications, as the majority take on a global perspective, with some also considering specific lead markets.

From the remaining publications, the HyWays study of the European Commission [64] was chosen because it covers the most far-reaching time horizon (2010 to 2050) and for the equally important reason that the contained data is easily accessible. Based on sociological, technological and economic analyses as well as extensive interaction between science and industry experts from a range of member states, the study produced market penetration scenarios for various hydrogen-based technologies, including four on FCV. In doing so, the study focuses on the following ten European countries, which hence also form the geographical focus of the MFA presented in this paper: Finland, France, Germany, Greece, Italy, the Netherland, Norway, Poland, Spain and the United Kingdom. Of these, Greece was later excluded in this study due to a lack of data on vehicle ownership.

In view of technological developments since the publication of the study, two scenarios assuming moderate technological progress were selected, which in turn differentiate between high policy support (Scenario 1) and moderate policy support (Scenario 2). Scenario 1 is hence based on the assumption that market barriers, such as high investment costs, can be counteracted due to a political endorsement of particularly FCV thus pushing the market penetration of FCV. In scenario 2, on the other hand, political endorsement applies to sustainable mobility in general rather than to a specific powertrain technology, i.e., it puts no focus on FCV.

Applying the market penetration ratios given in [64] to an extrapolation of the vehicle stock in the nine lead markets delivers a projected development of the total vehicle stock as well as the FCV market shares as illustrated in Figure 4. The four FCV market penetration scenarios provided by [64] include two rather optimistic scenarios that are characterised by high to very high policy support and fast technological progress and assume that mass production of FCV will begin by 2013. In contrast, the two more conservative scenarios based on modest rates of technical learning delay the earliest mass-market rollout to 2016 and are characterised as follows:

> "*the hypothetic start of mass production has been shifted to 2016 and the number of first movers reduced to 4 which will ramp-up their plant utilization rate from 5% to 90% within a five year time frame (maximum production capacity of each of the four plants 100,000 units per year). After reaching full utilization of the production capacities of the first movers [...] after 5 years [...], it was assumed that followers are entering the market in a similar way and the first movers are doubling their production capacity*". [64] (p. 16)

Given the current state of the FCV market, the assumptions underlying the two more conservative scenarios appear closer to reality. Hence, only the two conservative scenarios, will be considered for this MFA.

With a view to the geographical focus of this thesis, the mean value of the growth rates suggested below will be applied, with the latest publication on passenger cars in use in the EU (Eurostat 2014) used as the baseline amount. It must be noted that several studies forecast the level of vehicle ownership to peak at some point in the future and then decline. While the MFA would undoubtedly benefit from an accurate calculation of the vehicle fleet's growth per year, for simplicity reasons only an average annual growth rate of 0.3% will be used. Based on the vehicle stock reported by Eurostat (2014) and the above mentioned average annual growth rate, the vehicle fleet in the nine analysed EU member states will develop as illustrated in Figure 4. In 2050, the market penetration rates given by the two scenarios are equivalent to 87,521,198 fuel cell vehicles (=7,001,695,862 kW) in Scenario 1 and 78,769,078 fuel cell vehicles (=6,301,526,276 kW) in Scenario 2, respectively.

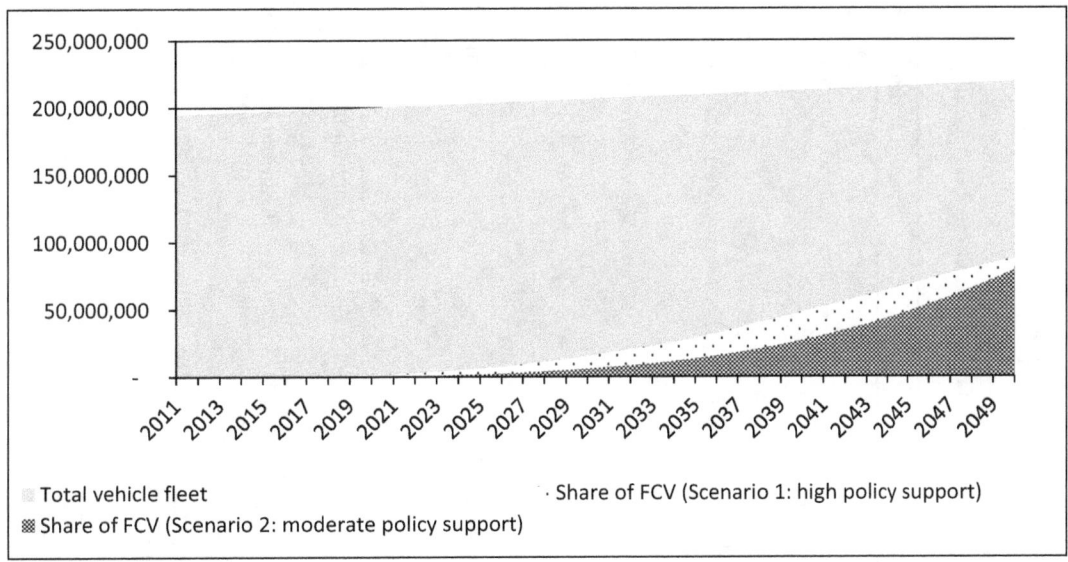

Figure 4. Development of the vehicle stock from 2011 to 2050 and share of fuel cell vehicles.

In addition, various other parameters were gained from a review of the relevant literature as well as expert interviews and are summarised in Table 4.

Table 4. Parameters and assumptions for material flow analysis.

Parameter	System Boundaries/Assumptions	Sources
Geographic boundary	Finland, France, Germany, Italy, the Netherlands, Norway, Poland, Spain, the United Kingdom	[64]
Time scale	2016 to 2050	Authors' choice based on data availability
Market penetration scenarios	High policy support, modest learning rate (Scenario 1); modest policy support, modest learning rate (Scenario 2)	[64]
Average annual growth rate of vehicle fleet	0.03%	Mean value of [5,65–67]
Recycling scenarios	Baseline (Scenario A), Pro-Recycling (Scenario B)	Authors' choice
Type of vehicle	Light-duty passenger vehicle	Authors' choice based on vehicle segmentation and GHG emissions
Average vehicle lifetime	10 years	[25]
Average vehicle power	80 kW	[30]
Average vehicle weight	1447 kg	[25]
Average weight of fuel cell stack	40.8 kg	[25]
Platinum load per kW	Continuously declining from 2020: 0.125 g, 2030: 0.07 g, 2050: 0.07 g	[31,68]
Loss rate of catalyst ink in manufacturing	Continuously declining from 2020: 15%; 2030: 5.2%; 2050: 1.1%	Authors' estimation based on [69–74] and average annual improvement rate of ~0.8%
Use phase loss	0.67%	[41,75]

As evident from the preceding chapter, exact figures for platinum losses arising from deficits in the respective steps of the recycling chain cannot be given before real-world experience with a recycling system for fuel cells is gained. In order to capture a range of possible developments, two recycling scenarios with diverging assumptions, as listed in Table 5, were developed using information gained during the preceding qualitative analysis of the recycling chain. These are to be interpreted as a "best case" and "worst case" scenario, with the latter based on recycling rates of platinum-rich exhaust gas catalysts, as reported by [37], but supplemented with the limited factual information on the recycling of fuel cells that is available. The Baseline-Scenario (A) therefore represents the actual recycling possibilities, including the whole recycling chain, as collection, processing, etc. The Pro-Recycling Scenario (B), on the other hand, applies the most optimistic estimates for each parameter as derived from the preceding literature review and includes very little losses and very few so-called "unknown whereabouts" in car recycling. The efficiency of the entire recycling chain depends on the efficiency of each step. For example, in the Baseline Scenario (A) total recycling chain efficiency can be calculated as follows: 33% × 96% × 96% × 95% = 28.9%. Despite efficiencies of above 95% for all remaining steps, the low rate of ELV collection thus delivers a low overall recycling efficiency, stressing the fact that careful management of each step in the recycling chain is necessary. In reality, a realistic number might lie between these two scenarios.

Table 5. Parameters of the two recycling scenarios (based on expert knowledge and own assumptions).

Parameter	Baseline Scenario (A)	Pro-Recycling Scenario (B)
Share of End-of-Life Vehicles collected by recycling facilities	33%	85%
Share of fuel cell stacks recovered in dismantling step	96%	99%
Share of platinum-containing components recovered in pre-processing step	96%	98%
Share of platinum recovered in material recovery step	95%	98%
Total recycling chain efficiency	28.9%	80.8%

All parameters as well as results of the preliminary calculations were entered into the STAN software to deliver annual and cumulative gross platinum requirements from 2016 to 2050, as well as the share of demand met by recycling of fuel cell vehicles. The material flow diagram can be viewed in the Supplementary Materials section.

5. Results and Discussion

Based on the parameters described above, the material flow analysis delivers the following results with regards to the development of platinum demand and recycling potential arising from a deployment of fuel cell vehicles in Europe.

The results of Table 6 indicate that the diffusion of FCV in Europe will not cause a depletion of platinum resources in the short term, as the calculated 537.06 t and 459.24 t in cumulative platinum requirements are far below the currently estimated global reserves of 66,000 t of platinum group metals [76], even though any contribution from the recycling of FCV is not yet included in this extrapolation.

In contrast, a comparison of the maximum annual gross demand established by the two scenarios, as illustrated in Figures 5 and 6, with the current global supply shows that the future platinum requirements for European FCV production only could place a significant strain on the global platinum market, although US and Asian vehicle markets are not even considered in this extrapolation. With 23.42 t and 39.66 t, respectively, by 2050 the European FCV industry alone would require a maximum of 14.5% and 24.5% of the 161.74 t produced in 2014 [77]. This could not only lead to significant price increases, but also raises the possibility of supply shortages as well as the dependency of a number of industries on a critical resource and emphasises the importance of recycling.

Table 6. Cumulative platinum demand and recycling potentials.

Task	Scenario 1 (High Policy Support)	Scenario 2 (Moderate Policy Support)
Cumulative platinum demand for Fuel Cell Vehicle (FCV) production	537.06 t	459.24 t
Cumulative platinum content of end-of-life FCV	293.51 t	155.23 t
Cumulative amount of platinum recovered (Scenario A: Baseline)	84.8 t (208.71 t lost)	44.85 t (110.38 t lost)
Cumulative amount of platinum recovered (Scenario B: Pro-Recycling)	230.02 t (63.49 t lost)	119.17 t (36.06 t lost)

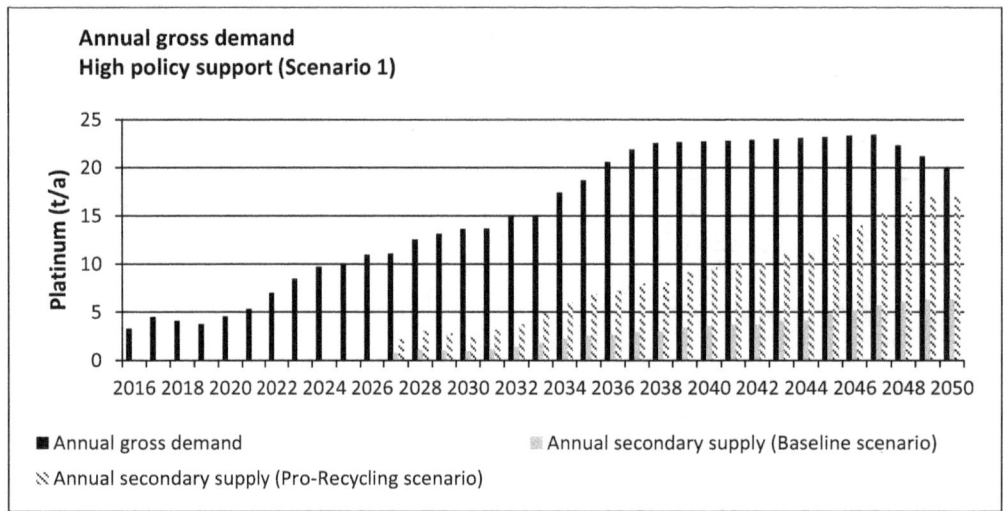

Figure 5. Scenario 1—Annual gross demand for platinum and contribution of secondary supply.

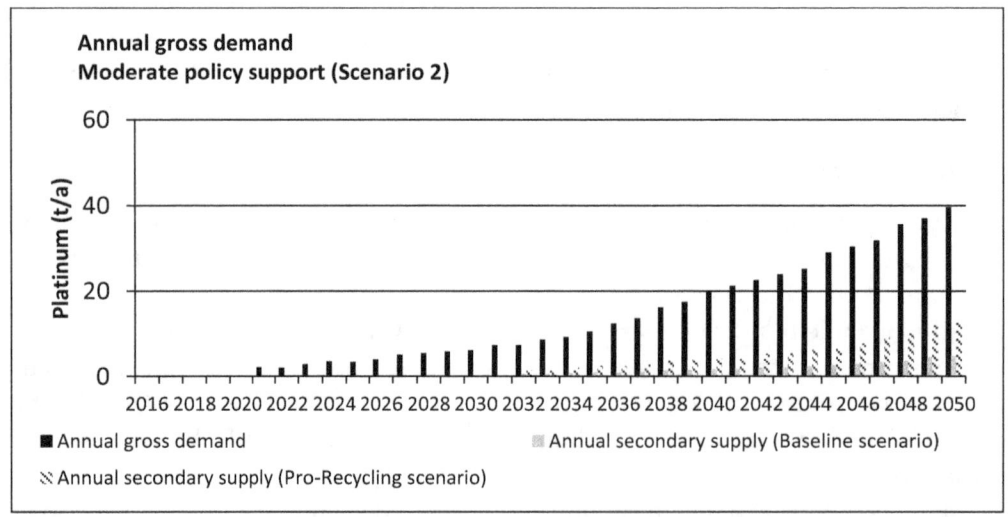

Figure 6. Scenario 2—Annual gross demand for platinum and contribution of secondary supply.

With regards to the role of recycling in meeting the platinum requirements of a growing FCV fleet, Figures 7 and 8 show a significant recycling potential, the exploitation of which could greatly reduce this industry's dependence on the volatile platinum market. Up to 85.6% (17.14 t) of the platinum required for European FCV production could be obtained from recycling of end-of-life FCV by 2050 (Scenario 1; Pro-Recycling), while the other three scenario combinations deliver much more conservative results of 31.5% (6.32 t) (Scenario 1; Baseline) (cf. Figure 7), 32.6% (26.74 t) (Scenario 2; Pro-Recycling) and 12.3% (4.86 t) (Scenario 2; Baseline) (cf. Figure 8). Although the latter three estimates appear less promising in terms of meeting the industry's platinum requirements, even the lowest of these values still represents 4.86 t of platinum (worth €179,820,000 at today's market price in 2015 [78]) that would otherwise have to be produced from primary sources.

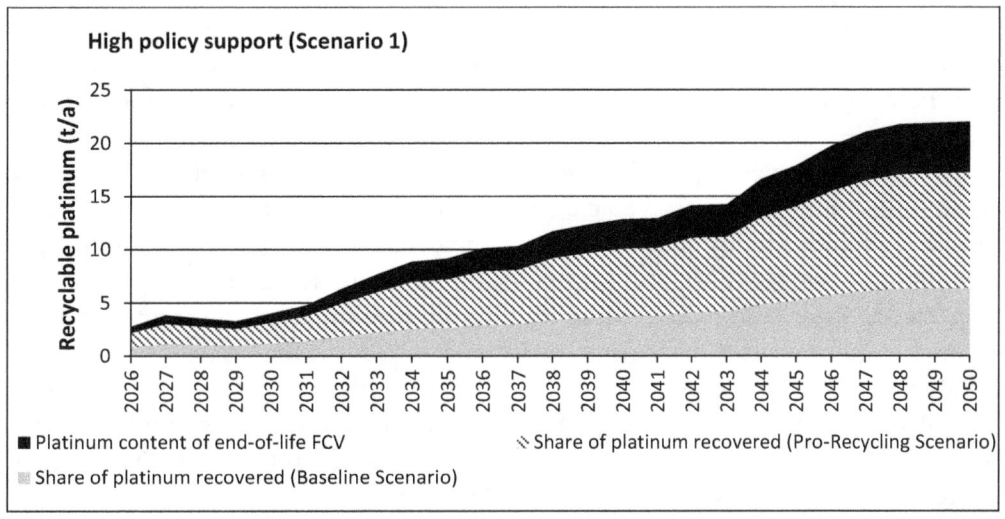

Figure 7. Scenario 1—Recycling potential and share of platinum recovered.

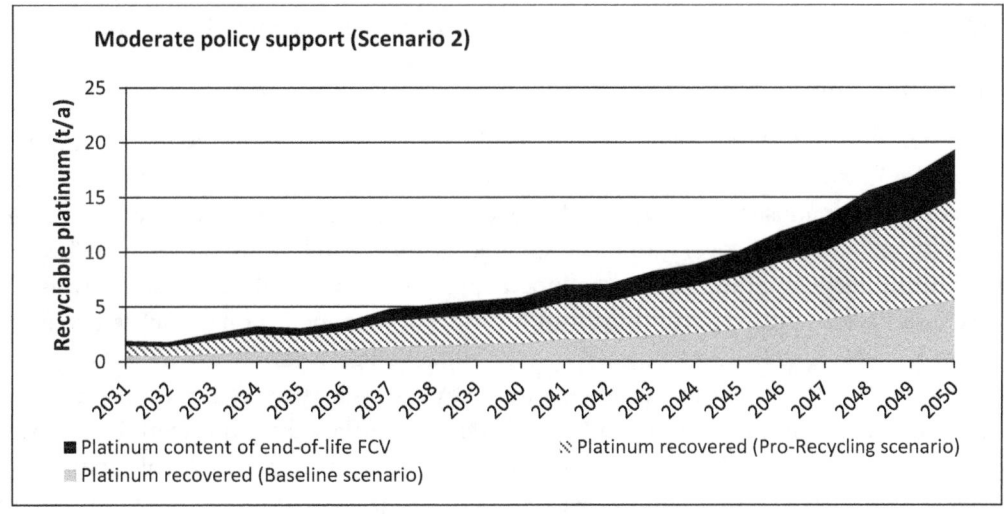

Figure 8. Scenario 2—Recycling potential and share of platinum recovered.

Nonetheless, in both Scenario (1) and (2), the discrepancy between the two recycling scenarios becomes apparent. This is underlined in Figures 7 and 8, which plot the recycling potential (in terms of platinum content available in end-of-life vehicles) of Scenario (1) and (2) against the realised quantities of secondary supply under the assumptions of the Baseline and Pro-Recycling cases. As the results show, with a maximum of 4.76 t and 4.48 t, respectively, significant quantities of platinum are lost every year even in the Pro-Recycling scenario. These amounts are, however, far below the maximum annual losses of 15.64 t (Scenario 1) and 13.72 t (Scenario 2), respectively, that fail to be recovered from end-of-life FCV under the Baseline scenario.

Considering the entire time span from 2016 to 2050, Table 6 lists the cumulative platinum contents of end-of-life FCV as well as the shares recovered under the four scenario combinations. While the platinum losses are substantial even in the Pro-Recycling scenarios, the losses documented as part of the Baseline scenario are excessively high and thus cast doubts on the endorsement of fuel cell vehicles for sustainable development.

Their presumed high recycling potential is one of the reasons that a number of studies portray FCV and the impacts of their market diffusion on platinum demand in a rather positive way. However, when comparing the results of both the qualitative and quantitative assessment of this study with previous publications, one may conclude that several studies tend to overestimate the platinum supply achievable from FCV recycling. For example, Råde, Tiax, Saurat and Bringezu [6,19,20] assume recycling rates of 90%, 95% and 75%, respectively. This paper, on the other hand, upon examining each step in the recycling chain, proposes much more conservative recycling rates between 28.9% and 80.8%.

As analysed in Chapter 3, highly efficient recycling chains as the ones cited by Råde, Tiax, Saurat and Bringezu [6,19,20] may be conceivable in theory in the future, but are incommensurable to the quotas realised for vehicles today and exceed those achieved in the majority of platinum applications by far. Such assumptions, however, also impact on the platinum demand arising from a diffusion of FCV as suggested in these studies, and hence involve the potential for significantly underestimating the net platinum requirements. Considering the fact that 98% is the highest recovery rate reported by any of the patents analysed in Chapter 3 for the final step of material recovery, all the preceding stages in the recycling chain would need to collect the respective material handled with 99% efficiency in order to achieve an overall recycling rate of 95% [19]. Without major changes to the current collection and reporting system, such rates appear unattainable.

As shown in Chapter 3, the recycling of platinum from fuel cell vehicles is still in its very early stages, largely due to the fact that significant volumes of end-of-life fuel cells do not exist yet. Values stated for possible recycling chain efficiencies, fractions recovered and amounts lost must therefore be understood as an indication only and the underlying assumptions considered with due circumspection.

6. Conclusions and Implications for Further Research

This study investigated the potential contribution of recycling to satisfy the platinum demand arising from a widespread deployment of fuel cell vehicles in Europe. Based on a qualitative examination of the four consecutive steps in the recycling chain (collection, dismantling, disassembly and pre-processing, material recovery) of fuel cell vehicles, two recycling scenarios with "best case" and "worst case" future recycling rates were developed. Using dynamic material flow analysis, these two recycling scenarios were applied to two scenarios for the market penetration of fuel cell vehicles in nine European lead markets to deliver both the associated impact on platinum demand and the contribution of recycling for meeting this demand. As demonstrated by the highly diverging results, the quantitative assessment of future platinum flows from FCV recycling is afflicted with uncertainties and may be considered highly speculative. Nonetheless, the results of this study's MFA imply that well-devised recycling systems could contribute quantities of secondary platinum in an order of magnitude that could mitigate existing raw material dependencies and possible future supply shortages considerably.

It appears as though stakeholders of the precious metal industry that are already involved in the recycling of platinum-containing applications have acknowledged both the opportunities and challenges arising from a further diffusion of fuel cells and fuel cell vehicles in particular. A number of procedures for recovering platinum from end-of-life fuel cells with high efficiencies and under the premise of avoiding HF emissions exist, yet it must be stressed that none have been implemented at an industrial scale. Future research should thus not only investigate the large-scale applicability of such procedures, but also environmental impacts and profitability in comparison to established pyrometallurgical recovery processes. Hydrometallurgical recovery processes seems to show promising results, especially relating to the recovery rate of more than 90% but further research is needed to achieve the best environmental performance. A big advantage in some hydrometallurgical processes is the fact that the membrane could be recovered as a whole membrane and might be available for reuse purposes, whereas the pyrometallurgical process recovers only materials and not components. From a recycling perspective, further research on PBI-based fuel cell systems that lower the presence of fluorine in the membrane also appears to be promising, although their efficiency and stability remain unsatisfying at the current development status. In the gas diffusion layer (GDL), some polytetrafluoroethylene (PTFE) will always be necessary to ensure the necessary hydrophobicity. However, compared to Nafion® membranes, the total fluorine content will be reduced by about 80%.

Even more problematic is the fact that less is certain about the preceding stages in the recycling chain, which appear to have received little attention by producers, recycling businesses, researchers and other stakeholders. This is especially concerning when considering that deficits in the initial stages of the recycling chain (above all collection) account for a major share of material loss in the recycling chains of other platinum-rich consumer goods and could potentially do so for fuel cell vehicles as well. Integrated research efforts that include all stages and actors of a potential fuel cell recycling chain and are aimed at devising practice-oriented strategies would therefore be highly beneficial in order to achieve a closed European system and minimise losses of a critical resource.

Supplementary Materials: The following are available online at http://www.mdpi.com/2313-4321/1/3/343/s1, Figure S1: STAN material flow diagram including flow codes, Figure S2: STAN use phase sub-system of the material flow diagram including flow codes.

Acknowledgments: Part of this research was made possible through founding the research group "Cascade Use" at Oldenburg University, funded by the German Federal Ministry of Education and Research (No. 01LN1310A). We also appreciate the useful information given by our colleagues from NEXT ENERGY EWE Research Centre for Energy Technology in Oldenburg.

Author Contributions: All authors have contributed equally to this work. However, Rikka Wittstock performed all Material Flow Assessments with the STAN software with data provided by Alexandra Pehlken and Michael Wark.

Abbreviations

The following abbreviations are used in this manuscript:

CO	Carbon Monoxide
CO_2	Carbon Dioxide
ELV	End-of-Life Vehicle
EU	European Union
FCV	Fuel Cell Vehicle
GDL	Gas Diffusion Layer
GHG	Greenhouse Gas
H	Hydrogen
HF	Hydrogen Fluoride
MEA	Membrane Electrode Assembly
MFA	Material Flow Analysis
PBI	Polybenzimidazole
PEM	Polymer Electrolyte Membrane
PEMFC	Polymer Electrolyte Membrane Fuel Cell
PGM	Platinum Group Metals
PTFE	Polytetrafluorethylene
US	United States of America
WIPO	World Intellectual Property Organization

References

1. European Commission. *Communication from the Commission: A Roadmap for Moving to a Competitive Low Carbon Economy in 2050. COM (2011) 112 Final*; European Commission: Brussels, Belgium, 2011.
2. European Commission. Reducing Emissions from Transport. Available online: http://ec.europa.eu/clima/policies/transport/index_en.htm (accessed on 27 October 2015).

3.   Mohrdieck, C.; Venturi, M.; Breitrück, K.; Schulze, H. Mobile Anwendungen. In *Wasserstoff und Brennstoffzelle.*
     Technologien und Marktperspektiven; Töpler, J., Lehmann, J., Eds.; Springer: Wiesbaden, Germany, 2014;
     pp. 59–111.

4.   Bundesministerium für Verkehr, Bau und Stadtentwicklung. *Elektromobilität—Deutschland als Leitmarkt und*
     Leitanbieter; Bundesministerium für Verkehr, Bau und Stadtentwicklung: Berlin, Germany, 2011.

5.   McKinsey & Company. A Portfolio of Power-Trains for Europe: A Fact-Based Analysis. The Role of
     Battery-Electric Vehicles, Plug-In Hybrids and Fuel Cell Electric Vehicles. Available online: http://www.fch.
     europa.eu/node/786 (accessed on 7 September 2016).

6.   Råde, I. Requirement and Availability of Scarce Metals for Fuel-Cell and Battery Electric Vehicles.
     Ph.D. Thesis, Chalmers University of Technology and Gothenburg University, Gothenburg, Sweden, 2001.

7.   Gordon, R.B.; Bertram, M.; Graedl, T.E. Metal stocks and sustainability. *Proc. Natl. Acad. Sci. USA* **2006**, *103*,
     1209–1214. [CrossRef] [PubMed]

8.   Elshkaki, A. Systems Analysis of Stock Buffering. Development of a Dynamic Substance Flow-Stock Model
     for the Identification and Estimation of Future Resources, Waste Streams and Emissions. Ph.D. Thesis,
     Leiden University, Leiden, The Netherlands, 6 September 2007.

9.   Angerer, G.; Marscheider-Weidemann, F.; Lüllmann, A.; Erdmann, L.; Scharp, M.; Handke, V.; Marwede, M.
     Rohstoffe für Zukunftstechnologien; Fraunhofer IRB Verlag: Stuttgart, Germany, 2009.

10. Buchert, M.; Schüler, D.; Bleher, D. *Critical Metals for Future Sustainable Technologies and Their Recycling*
     Potential; Öko-Institut e.V. for United Nations Environment Program: Freiburg, Germany, 2009.

11. Yang, C. An impending platinum crisis and its implications for the future of the automobile. *Energy Policy*
     2009, *37*, 1805–1808. [CrossRef]

12. Buchert, M.; Jenseit, W.; Dittrich, S.; Hacker, F.; Schüler-Hainsch, E.; Ruhland, K.; Knöfel, S.; Goldmann, D.;
     Rasenack, K.; Treffer, F. *Ressourceneffizienz und ressourcenpolitische Aspekte des Systems Elektromobilität*;
     Öko-Institut e.V.: Darmstadt, Germany, 2011.

13. Saurat, M.; Bringezu, S. Platinum Group Metal Flows on Europe, Part I. *J. Ind. Ecol.* **2008**, *12*, 754–767.
     [CrossRef]

14. Mudd, G.M. Platinum group metals: A unique case study into the sustainability of mineral resources.
     In Proceedings of the 4th International Platinum Conference: Platinum in Transition 'Boom or Bust',
     Johannesburg, South Africa, 11–14 October 2010; The South African Institute of Mining and Metallurgy:
     Johannesburg, South Africa, 2010; pp. 113–120.

15. Cairncross, E. Health and Environmental Impacts of Platinum Mining: Report from South Africa. 2014.
     Available online: http://www.thejournalist.org.za/wp-content/uploads/2014/09/Environmental-health-
     impacts-of-platinum-mining1.pdf (accessed on 29 August 2015).

16. European Commission. *Report on Critical Raw Materials for the EU*; European Commission: Brussels,
     Belgium, 2014.

17. Bernhart, W.; Riederle, S.; Yoon, M. *Fuel Cells—A Realistic Alternative for Zero Emission?* Roland Berger
     Strategy Consultants: Stuttgart, Germany, 2014.

18. Carlson, E.J.; Anderson, A.; Clearly, B. Precious Metal Availability and Cost Analysis for PEMFC
     Commercialization. Hydrogen, Fuel Cells and Infrastructure Technologies FY 2003 Progress Report.
     Available online: https://www.eecbg.energy.gov/hydrogenandfuelcells/pdfs/iva4_carlson.pdf (accessed
     on 11 July 2015).

19. Tiax LLC. *Platinum Availability and Economics for PEMFC Commercialisation. Report to US DOE*; Tiax LLC:
     Cambridge, MA, USA, 2003.

20. Saurat, M.; Bringezu, S. Platinum Group Metal Flows on Europe, Part II. *J. Ind. Ecol.* **2009**, *13*, 406–421.
     [CrossRef]

21. Hagelüken, C. Recycling the PGM—A European Perspective. *Platin. Met. Rev.* **2012**, *56*, 29–35. [CrossRef]

22. Chan, C.C.; Bouscayrol, A.; Chen, K. Electric, Hybrid and Fuel-Cell Vehicles: Architectures and Modeling.
     IEEE Trans. Veh. Technol. **2010**, *59*, 589–598. [CrossRef]

23. Kurzweil, P. *Brennstoffzellentechnik. Grundlagen, Komponenten, Systeme, Anwendungen*, 2nd ed.; Springer:
     Wiesbaden, Germany, 2013.

24. Li, Q.; He, R.; Gao, J.A.; Jensen, J.O.; Bjerrum, N.J. The CO Poisoning Effect in PEMFCs Operational at
     Temperatures up to 200 °C. *J. Electrochem. Soc.* **2003**, *150*, A1599–A1605. [CrossRef]

25. Simons, A.; Bauer, C. A life-cycle perspective on automotive fuel cells. *Appl. Energy* **2015**, *157*, 884–896. [CrossRef]

26. Kaz, T. Herstellung und Charakterisierung von Membran-Elektroden-Einheiten für Niedertemperatur Brennstoffzellen. Ph.D. Thesis, University of Stuttgart, Stuttgart, Germany, 22 August 2008.

27. Koraishy, B.; Meyers, J.M.; Wood, K.L. Manufacturing of Membrane Electrode Assemblies for Fuel Cells. 2009. Available online: http://www.sutd.edu.sg/cmsresource/idc/papers/2009-_Manufacturing_of_membrane_electrode_assemblies_for_fuel_cells.pdf (accessed on 4 September 2015).

28. Spendelow, J.; Marcinkoski, J. Fuel Cell System Cost 2013. DOE Fuel Cell Technologies Office Record No. 14012. Available online: http://www.hydrogen.energy.gov/pdfs/14012_fuel_cell_system_cost_2013.pdf (accessed on 30 June 2015).

29. Holton, O.T.; Stevenson, J.W. The Role of Platinum in Proton Exchange Membrane Fuel Cells. *Platin. Met. Rev.* **2013**, *57*, 259–271. [CrossRef]

30. US Department of Energy. Hydrogen and Fuel Cells Program: Library. Available online: http://www.hydrogen.energy.gov/library.html (accessed on 30 June 2015).

31. US Drive FCTT. Fuel Cell Technical Team Roadmap. Available online: http://energy.gov/eere/vehicles/downloads/us-drive-fuel-cell-technical-team-roadmap (accessed on 23 June 2015).

32. Lehmann, J.; Luschtinetz, T. *Wasserstoff und Brennstoffzellen. Unterwegs mit dem saubersten Kraftstoff*; Springer: Wiesbaden, Germany, 2014.

33. Sui, P.-C. University of Victoria, Victoria, BC, Canada. Personal Communication, 2015.

34. Heiskanen, J.; Kaila, J.; Vanhanen, H.; Pynnönen, H.; Silvennoinen, A. A look at the European Union's End-of-Life Vehicle Directive—Challenges of treatment and disposal in Finland. In Proceedings of the 2nd International Conference on Final Sinks: Sinks—A Vital Element of Modern Waste Management, Espoo, Finland, 16–18 May 2013.

35. Fergusson, M. *End of Life Vehicles (ELV) Directive—An Assessment of the Current State of Implementation by Member States*; Policy Department Economy and Science of the European Parliament: Brussels, Belgium, 2007.

36. Wilts, H.; Bleischwitz, R. Combating Material Leakage: A Proposal for an International Metal Covenant. *Surv. Perspect. Integr. Environ. Soc.* **2011**, *4*, 1–9.

37. Hagelüken, C.; Buchert, M.; Stahl, H. *Stoffströme der Platingruppenmetalle. Systemanalyse und Maßnahmen für eine Nachhaltige Optimierung der Stoffströme der Platingruppenmetalle*; GDMB Medienverlag: Clausthal-Zellerfeld, Germany, 2005.

38. Axion Recycling. UK Closed-Loop Fuel Cell Component Recycling Is a Step Closer. Available online: http://www.axionpolymers.com/tag/fuel-cell/ (accessed on 30 August 2015).

39. Schittl, G. Recyclingpotenzial von Kritischen Rohstoffen in Technologien zur Energieumwandlung. Master's Thesis, University of Natural Resources and Life Sciences, Vienna, Austria, 7 March 2012.

40. Cerri, I.; Lefebvre-Joud, F.; Holtappels, P.; Honegger, K.; Stubos, T.; Millet, P. *Scientific Assessment in Support of the Materials Roadmap Enabling Low Carbon Energy Technologies: Hydrogen and Fuel Cells*; Joint Research Centre, Institute for Energy and Transport: Petten, The Netherlands, 2012.

41. Kromer, M.A.; Joseck, F.; Rhodes, T.; Guernsey, M.; Marcinkoski, J. Evaluation of a platinum leasing program for fuel cell vehicles. *Int. J. Hydrogen Energy* **2009**, *34*, 8276–8288. [CrossRef]

42. Kwade, A.; Bärwaldt, G. *LithoRec—Recycling von Lithium-Ionen-Batterien. Abschlussbericht des Verbundprojekts*; Technische Universität Braunschweig: Brunswick, Germany, 2012.

43. National Alternative Fuels Training Consortium. Safety Booklet Automotive Recycling. Available online: http://naftc.wvu.edu/cleancitieslearningprogram/firstrespondersafetytraining/towing-recovery (accessed on 2 July 2015).

44. Schiemann, J.; Kerßenboom, A.; Prause, H.J.; Peil, S. *Handbuch Verwertung von Brennstoffzellen und deren Peripherie-Systeme*; Institut für Energie- und Umwelttechnik e.V.: Duisburg, Germany, 2007.

45. Shore, L. Platinum Group Metal Recycling Technology Development. Final report to BASF Catalysts LLC and US DOE Office of Hydrogen, Fuel Cells, and Infrastructure Technologies Program. Available online: http://www.osti.gov/scitech/biblio/962699 (accessed on 29 July 2015).

46. Wegner, R.; Fokkens, E.; Holdt, H.; Bukowsky, H.; Trautmann, M.; Nettesheim, S.; Jakubith, S.; Scholz, P.; Mollenhauer, T.; Theuring, S.; et al. React—Rückgewinnung und Wiedereinsatz von Edelmetallen aus Brennstoffzellen. In *Recycling und Rohstoffe*; Thomé-Kozmiensky, K.J., Goldmann, D., Eds.; Vivis TK-Verlag: Nietwerder, Germany, 2012; Volume 5, pp. 429–441.

47. Handley, C.; Brandon, N.P.; van der Vort, R. Impact of the European Union vehicle waste directive on end-of-life options for polymer electrolyte fuel cells. *J. Power Sources* **2002**, *106*, 344–352. [CrossRef]

48. US Environmental Protection Agency. Hydrogen Fluoride. Available online: http://www.epa.gov/ttnatw01/hlthef/hydrogen.html (accessed on 03 July 2015).

49. Stolten, D. *Hydrogen and Fuel Cells: Fundamentals, Technologies and Applications*; John Wiley & Sons Ltd.: Bognor Regis, UK, 2010.

50. Koehler, J.; Zuber, R.; Binder, M.; Baenisch, V.; Lopez, M. Process for Recycling Fuel Cell Components Containing Precious Metals. WIPO Patent No. WO/2006/024507, 9 March 2006.

51. Jha, M.K.; Lee, J.; Kim, M.; Jeong, J.; Kim, B.-S.; Kumar, V. Hydrometallurgical recovery/recycling of platinum by the leaching of spent catalysts: A review. *Hydrometallurgy* **2013**, *133*, 23–32. [CrossRef]

52. Duclos, L.; Svecova, L.; Laforest, V.; Mandil, G.; Thivel, P.-X. Process development and optimization for platinum recovery from PEM fuel cell catalyst. *Hydrometallurgy* **2016**, *160*, 79–89. [CrossRef]

53. Xu, F.; Mu, S.; Pan, M. Recycling of membrane electrode assembly of PEMFC by acid processing. *Int. J. Hydrogen Energy* **2010**, *35*, 2976–2979. [CrossRef]

54. Zhao, J.; He, X.; Tian, J.; Wan, C.; Jiang, C. Reclaim/recycle of Pt/C catalysts for PEMFC. *Energy Convers. Manag.* **2007**, *48*, 450–453. [CrossRef]

55. Oki, T.; Katsumata, T.; Hashimoto, K.; Kobayashi, M. Recovery of Platinum Catalyst and Polymer Electrolyte from Used Small Fuel Cells by Particle Separation Technology. *Mater. Trans.* **2009**, *50*, 1864–1870. [CrossRef]

56. Hagelüken, C.; Kayser, B.; Romero-Ojeda, J.; Kleinwächter, I. Process for the Concentration of Noble Metals from Fluorine-Containing Fuel Cell Components. WIPO Patent No. WO/2004/102711, 25 November 2004.

57. Shore, L.; Matlin, R. Simplified Process for Leaching Precious Metals from Fuel Cell Membrane Electrode Assemblies. WIPO Patent No. WO/2009/029463, 5 March 2009.

58. Shore, L.; Matlin, R.; Heinz, R. Method and Apparatus for Recovering Catalytic Elements from Fuel Cell Membrane Electrode Assemblies. WIPO Patent No. WO/2009/149241, 10 December 2009.

59. Shore, L.; Matlin, R.; Heinz, R. Method for Recovering Catalytic Elements from Fuel Cell Membrane Electrode Assemblies. WIPO Patent No. WO/2010/132156, 18 November 2010.

60. Shore, L.; Robertson, A.B.; Shulman, H.S.; Fall, M.L. Process for Recycling Components of a PEM Fuel Cell Membrane Electrode Assembly. WIPO Patent No. WO/2006/115684, 2 November 2006.

61. Shore, L. Process for Recycling Components of a PEM Fuel Cell Membrane Electrode Assembly. WIPO Patent No. WO/2007/149904, 27 December 2007.

62. Romero, J.; Meyer, H.; Voss, S. Device and Method for the Thermal Treatment of products Containing Fluorine and Precious Metals. European Patent No. EP2700726A1, 26 February 2014.

63. Paepke, M. Method for Recycling Membrane/Electrode Units of a Fuel Cell. WIPO Patent No. WO/2015/010793, 29 January 2015.

64. European Commission. *HyWays—The European Hydrogen Roadmap. Project Report EUR 23123*; Office for official publications of the European Communities: Luxembourg, Luxembourg, 2008.

65. Bundesministerium für Verkehr, Bau und Stadtentwicklung. *Szenarien der Mobilitätsentwicklung unter Berücksichtigung von Siedlungsstrukturen bis 2050*; Bundesministerium für Verkehr, Bau und Stadtentwicklung, Traffic and Mobility Planning GmbH, Deutsches Institut für Urbanistik, Institut für Wirtschaftsforschung Halle: Magdeburg, Germany, 2006.

66. Shell Germany. Shell PKW Szenarien bis 2040. Available online: http://www.shell.de/aboutshell/media-centre/annual-reports-and-publications/shell-pkwszenarien.html#vanity-aHR0cDovL3d3dy5zaGVsbC5kZS9wa3dzdHVkaWU (accessed on 9 October 2015).

67. World Energy Council. Global Transport Scenarios 2050. Available online: https://www.worldenergy.org/publications/2011/global-transport-scenarios-2050/ (accessed on 9 October 2015).

68. Adamson, K. The Fuel Cell and Hydrogen Annual Review. 2015. Available online: http://www.4thenergywave.co.uk/annual-review/ (accessed on 9 October 2015).

69. Ultrasonic Systems Inc. Fuel Cell. Available online: http://www.ultraspray.com/markets/fuel-cell (accessed on 6 October 2015).

70. Sonaer Ultrasonics. Fuel Cell Coatings. Available online: http://sonozap.com/Fuel_Cell_Coating.html (accessed on 6 October 2015).

71. Sonotek. Ultrasonic Nozzle Overview. Available online: http://www.sono-tek.com/ultrasonic-nozzle-overview/ (accessed on 6 October 2015).

72. Ehrenberger, S.; Deutsches Zentrum für Luft- und Raumfahrt, Brunswick, Germany. Personal Communication, 2015.

73. Wagner, P.; Next Energy, Oldenburg, Germany. Personal Communication, 2015.

74. Betrieblicher Umweltschutz Baden-Württemberg. Nasslackieren (Spritzlackieren). Available online: http://www.bubw.de/?lvl=465 (accessed on 9 October 2015).

75. Dyck, A.; Next Energy, Oldenburg, Germany. Personal Communication, 2015.

76. USGS. Mineral Commodities Summary 2011. Available online: http://minerals.usgs.gov/minerals/pubs/mcs/ (accessed on 8 July 2015).

77. World Platinum Investment Council. Platinum Quarterly: Q4 2014. Available online: http://www.platinum investment.com/files/WPIC_Platinum_Quarterly_Q4_2014.pdf (accessed on 11 July 2015).

78. Johnson Matthey. Price Charts. Available online: http://www.platinum.matthey.com/prices/price-charts (accessed on 11 June 2015).

11

Cathode Ray Tube Recycling in South Africa

Pontsho Ledwaba [†] **and Ndabenhle Sosibo** *

Mintek, Small Scale Mining and Beneficiation, 200 Malibongwe Drive, Randburg 2125, South Africa; pontsho.ledwaba@wits.ac.za

* Correspondence: ndabenhles@mintek.co.za

† Current address: Centre for Sustainability in Mining & Industry, Wits University, 1 Jan Smuts Avenue, Braamfontein 2000, South Africa

Academic Editor: Michele Rosano

Abstract: Households and businesses produce high levels of electrical and electronic waste (e-waste), fueled by modernization and rapid obsolescence. While the challenges imposed by e-waste are similar everywhere in the world, disparities in progress to deal with it exist, with developing nations lagging. The increase in e-waste generation highlights the need to develop ways to manage it. This paper reviews global and South African e-waste management practices with a specific case study on Cathode Ray Tube (CRT) waste. CRTs present the biggest problem for recyclers and policy makers because they contain lead and antimony. Common disposal practices have been either landfilling or incineration. Research into South African CRT waste management practices showed there is still more to do to manage this waste stream effectively. However, recent developments have placed e-waste into a priority waste stream, which should lead to intensified efforts in dealing with it. Overall, these efforts aim to increase diversion from landfill and create value-adding opportunities, leading to social and environmental benefits.

Keywords: electrical and electronic waste; recycling; legislative frameworks; environmental management; landfilling

1. Introduction

The modernization of the 21st century, buoyed by rapid urbanization, population growth, and the once-booming economy, has led to high levels of end-of-life electronic waste (e-waste) [1–5]. The rapid obsolescence of consumer electronics and accessories such as cellular phones and computers is also adding disproportionately to the e-waste stream. The world production of e-waste sits between 20 and 50 million tons yearly, with current recycling rates ranging between 15 and 20 percent worldwide [6,7]. In 2014, the total e-waste produced on the African continent was 1.9 million tons. Egypt (0.37 million tons), South Africa (0.35 million tons), and Nigeria (0.22 million tons) were the leading producers [6]. There are few current reports available on e-waste in the African continent, which further blunts efforts towards managing this waste stream. Pérez-Belis et al. captured the existing body of knowledge spanning from 1992 to August 2014 [8]. In developing countries, Africa included, besides the increase in e-waste produced locally, shipping of e-waste from countries such as the USA is a problem [9]. Developing nations encouraged the practice since waste senders paid the receivers. Besides the willingness of the receivers, the lack of government regulations against such practices worsened the problem. Such diffusion of e-waste further diluted the accurate tracking of growth of this waste, especially in developing countries. Globally, volumes of e-waste are increasing by nearly 3 to 5 percent yearly since the 1990s, and e-waste continues to grow three times faster than municipal solid waste [10].

E-waste growth presents equally a problem, due to the toxicity of its parts, and an opportunity, as valuable minerals contained in the waste are recoverable [11]. Several studies have shown that the purity of metal parts in e-waste can easily be superior to that of rich-content minerals [12–14]. Waste printed circuit boards (WPCBs) are an example of a waste stream that holds high economic value [15]. Cucchiella et al. showed that WPCBs from information technology and telecommunications equipment have NPVs equal to 19,966,000 € and 6,606,000 € for mobile plants of 240 tons/year and 576 tons/year. The NPV for consumer equipment was a positive 1,050,000 €. Interest in electronic waste (e-waste) recovery and recycling has therefore been on the rise over the past two decades. The rise in recycling is attributable to other causes, including depleting mineral deposits, declining metal recoveries, and grades. Deepening mineral deposits, a concentration of strategic minerals in politically unstable regions, and general risks associated with primary mining are some of the causes [16].

The increase in waste produced from electronic and electrical equipment (EEE) has increased the need to create interventions to manage this waste. The level of interventions differs from country to country, with other countries having progressed. Like most other developing countries, South Africa's e-waste management industry is still in the early stages. Until recently, South Africa did not recognize e-waste as a threatening waste stream. According to the Department of Environmental Affairs (2015), e-waste currently makes up between 5% and 8% of the municipal solid waste in South Africa. There is an expectation that this will grow steadily over the coming years. According to the United Nations Environment Program (UNEP), obsolete computers in both South Africa and China will increase by 500% in 2020 compared with 2007 levels [17]. This has prompted the government and other supporting institutions to develop e-waste management methods whose focus is not only on waste disposal but also waste reduction, waste reuse, recycling, and metal recovery.

By definition, e-waste is a term used to refer to all EEE that has reached the end of its useful life. There are 10 different categories of e-waste according to the European Waste Electric and Electronics Equipment (WEEE) Directives 2002/96/ European Commission and 2012/19/ European Union. These include all products driven by electricity, and include small and large household products, IT equipment, electrical and electronic tools, etc. [18].

There are three main components of electronic equipment: metals (ferrous and nonferrous), plastics, and glass. For example, a typical computer contains about 32% ferrous metals, 23% plastic, and 15% glass [19]. Recovering metals from e-waste has been the focus for most recyclers, and extensive research is advancing to develop environmentally sound recovery methods to salvage valuable metals from e-waste. Early efforts included crude methods such as acid-washing and open incineration, which led to serious environmental problems [20–22].

2. Global E-Waste State

South Africa, Brazil, Russia, China and India form part of the BRICS consortium (BRICS—acronym for the five member countries). BRICS is a group of countries characterized by their strong economic growths and a need for a stronger political voice in world governance [23]. Although diverse, the BRICS group of countries still accounted for 17.7% of extra-EU exports in 2014, making it the most important EU export destination [24]. This is a sevenfold increase compared to 1999. BRICS countries contributed one-fifth of the world's economy, producing 27% of the world Gross Domestic Product, and accounts for 43% of the world population [25]. A Goldman Sachs report highlighted the importance of BRICS, arguing that the bloc will eclipse the current G7 richest countries of the world by 2050 [26].

This analysis makes it clear that the BRICS region deserves a lot more attention in any technological developmental studies. There are big differences between the BRICS countries including social, economic, and political features according to their history and region. Studies have shown that this group of countries is diverse and its economic features must be studied country by country [27]. The above observations called for the current study to focus solely on South Africa in contrast to global practices. The following sections present several countries and their approach to e-waste management and relevant legislation.

2.1. United States of America (USA)

The USA has one of the fastest-growing e-waste streams in the world because of citizens' high buying power and frequent upgrading/discarding of electronic products. According to UNU-IAS SCYCLE (2015), the USA produced 7072 metric kilotons of e-waste in 2014 [28]. This equates to 22.1 kg per inhabitant. Although e-waste accounts only for 10% of the total municipal solid waste produced in the USA, it is growing at a rate 2–3 times faster than any other waste produced [29]. Usually, the USA has had two alternatives of dealing with this waste: (1) disposal in U.S. landfills or (2) export of the waste from the USA. Between 2003 and 2005, about 80%–85% of all e-waste produced in the USA ended in U.S. landfills [30]. Further studies showed that lead exceeded the U.S. Environmental Protection Agency's (EPA) Toxicity Characteristics Leaching Procedure federal limits [31]. Realizing the environmental and human impact of landfill disposal of e-waste led to an exploration of reusing and recycling of e-waste and discouraging e-waste landfilling.

The second approach to deal with e-waste has been export to developing countries, which has received much attention worldwide. The practice of exporting e-waste to developing nations is sometimes in contravention of the Basel Convention Agreement. However, for a long time there were no U.S. national laws banning this practice. A notice to the receiving country was seen as compliance with the Basel Convention Agreement [32].

Different U.S. states developed various ways to deal with e-waste in the form of incentives or penalties. One of these ways was passing an Electronic Waste Recycling Fee by the State of California, imposed on certain video display devices (laptops, televisions, monitors, etc.). Covered Electronic Devices (CEDs) was the term referring to products subject to that fee [33]. Shown in Table 1 is the fee schedule [34]. Fees were to be paid upon purchase of the items. The e-waste fees income funded safe and affordable collection and recycling of CEDs that contain dangerous materials.

Table 1. California's electronic waste recycling fees.

Electronic Waste Recycling Fee				
Categories	2005–2008	2009–2010	2011–2012	2013–Present
>4 and <15 inch	$6	$8	$6	$3
=15 and <35 inch	$8	$16	$8	$4
+35 inch	$10	$25	$10	$5

The EPA drafted the Plug-In to eCycling Guidelines for Materials Management in 2004. This guided eCycling partners on handling and discarding end-of-life electronics [35]. The Hazardous Waste Management and Cathode Ray Tubes Final Rule 40, CFR Parts 9, 260, 261, 271 law followed in 2006. This focused on reuse and recycling as well as exporting obsolete CRT parts. The U.S. Department of the Interior drafted the Responsible Recycling (R2) Practices for Use in Accredited Certification Programs for Electronics Recyclers, Best Practice Document in 2008 [36]. The purpose of the document was to come up with an accepted set of practices for the electronics recycling industry. At the same time several U.S. states developed and passed bills on electronic waste handling and treatment. These included Arkansas, California, Connecticut, Georgia, Hawaii, Illinois, Indiana, Kentucky, Louisiana, Massachusetts, Maine, Maryland, Michigan, Minnesota, Mississippi, Missouri, Nebraska, Nevada, and New Hampshire. Others were New Jersey, New Mexico, North Carolina, Oklahoma, Oregon, Pennsylvania, Rhode Island, South Carolina, Texas, Virginia, Washington, West Virginia, and Wisconsin.

All the different bills passed in the USA centered on several collection methods, such as curbside, special drop-off events, permanent drop-off, take back, and point of purchase [37]. Although these alternatives presented the best-case scenarios in the world, there is still room to improve, since most big U.S. companies export their electronic products throughout the world. Fragmentation between many different U.S. states complicates this further.

The efforts by the USA in dealing with e-waste provide a good lesson for developing countries such as South Africa. The number and ages of bills passed in the USA in the last decade shows the scale of the challenge and the time needed to achieve success in e-waste management.

2.2. European Union

E-waste in the European Union is also the fastest-growing waste stream, with an expected growth to more than 10 million tons by year 2021 [38]. The EU member states produced an estimated 9523 metric kilotons of e-waste in 2014, equivalent to 18.7 kg per inhabitant [38]. The USA levels dwarf the 9523 metric kilotons, but the per-person rate closely matches.

The EU passed the WEEE Directive 2002/96/EC and 2012/19/EU limiting the use of certain hazardous substances in EEE (RoHS Directive 2002/95/EC). Directive 2002/96/EC came into force in 2003, providing for e-waste collection schemes whereby consumers returned their end-of-life electronics for free as well as setting up collection points. The directives centered on five areas, namely:

- electronic product design
- e-waste collection
- e-waste recovery
- e-waste treatment and financing and
- electronics user awareness.

Directive 2012/19/EU came into force in 2012, reflecting the growing need for an all-inclusive regulation of e-waste. RoHS Directive 2002/95/EC came into force in 2003, aimed at regulating heavy metals such as lead, mercury, cadmium, and chromium (VI), and flame retardants such as polybrominated biphenyls (PBB) or polybrominated diphenyl ethers (PBDE), to be substituted with safer alternatives. Further revision took place and became effective in 2013 (RoHS Directive 2011/65/EU), further including a normative document named Logistics of Waste Electrical and Electronic Equipment, WEEELABEX Standard Version 9.0 dealing with e-waste handling before treatment. Following this, the Collection, Logistics & Treatment Requirements for End-of-Life Household Appliances Containing Volatile Fluorocarbons or Volatile Hydrocarbons, Standard EN 50574-1, 2012 and EN 50574-2, 2014 came into force in 2012 and 2013. These documents gave normative requirements for assessing depollution for treatment of household e-waste containing volatile fluorocarbons or volatile hydrocarbons.

The EU, Switzerland, Norway, Belgium, Sweden, and the Netherlands have all exceeded the minimum requirements for collection and recycling targets. These countries have systems that perform above the rest in scope and compliance levels [38,39].

2.3. Japan

Japan enforced e-waste regulations in 2001. These laws require that manufacturers and importers take back end-of-life electronics for recycling [40]. The "Home Appliance Recycling Law", enacted in 1998 and enforceable from 2001, required importers and producers to recycle televisions, refrigerators, washing machines, and air conditioners. The end users paid the recycling fee, the retail outlet provided the collection points, and the producing company did the recycling. Collection was extended to local government authorities, who also provided collection services of goods for disposal. In cases where a producer was not available or it was impracticable to route goods back, an alternative, appointed company assumed that role. Recycling fees ranged between ¥1800 and ¥5000 depending on the e-waste item for recycling. An assessment of the Japanese methods shows that 74% of e-waste reached a recycler compared to the U.S. average of 12.5% [41].

Japan produced 2200 metric kilotons of waste in 2014, an equivalent of 17.3/kg per inhabitant [42].

2.4. China

China is one of the largest producers and consumers of e-waste in the world [43]. According to UNU-IAS SCYCLE (2015), China produced 6033 metric kilotons of e-waste in 2014 alone [44]. This is equal to 4.4 kg per inhabitant. Estimates showed that 90% of the 70%–80% of the global e-waste exported to Asia ends up in China [45].

China has signed the Basel Convention and has several regulations on e-waste. These are:

- Pollution Control Management Method for Electronic Information Products
- Administrative Measures for the Prevention and Control of Environmental Pollution by Electronic Waste and Regulation on the Administration of the Recovery and Disposal of Waste Electrical and Electronic Products.

The content of these regulations include:

- the control and restraining the use of hazardous and toxic substances in electronic and electric products,
- registration of dismantlers, recyclers, and disposers with the local Environmental Protection Bureau (EPB) and
- e-waste recycler permits and the establishment of a fund for e-waste recycling and disposal. Producers pay the recycling fund.

There are about 105 enterprises dedicated to dismantling e-waste in China at fixed places. However, a trade-in incentive was discontinued in 2011 for these enterprises and this led to most of the recycling going to small, informal traders [46]. These informal recyclers recycle for income, and endanger their lives and the environment because of a lack of knowledge. This has resulted in a complex informal network that works outside the laws and makes legislation useless. A lack of proper penalties for noncompliance with the applicable laws made enforcement nearly impossible [47].

2.5. India

India faces an increasing problem of e-waste mainly because of the growth rate of the country's Information Technology (IT) and electronics industry. These have increased India's consumption of electronic items and IT hardware, which afterwards leads to high obsolescence rates and thus high e-waste production [48]. In 2014, India produced 1641 metric kilotons of e-waste (1.3 kg per inhabitant) [49]. For an emerging economy like that of India, these levels of e-waste offer an alternative livelihood for the urban and rural poor through recycling, but it comes at a cost. There are significant risks to health and the environment as these recyclers often use primitive methods.

An unorganized, informal sector collects, transports, and dismantles most of India's e-waste. According to a Greenpeace report, only 3% of the e-waste produced in India is collected by the institutions approved to recycle it [50].

India signed the Management of e-Waste, Guidelines in 2008. The objective of the guideline was to provide guidance for the identification of various sources of e-waste and prescribe procedures for handling it in an environmentally sound manner. These guidelines provided guidance and a broad outline, and pit the onus on the recyclers to determine specific treatment methods according to the risk potential. In 2013 India drafted a law, Electronic Waste (Handling and Disposal), which specified the responsibilities of producers. The responsibilities were:

- collecting e-waste produced during manufacture of equipment
- collection of e-waste produced from end-of-life of the producers' product
- setting up collection centers
- financing and organizing a method to meet costs involved in e-waste management
- providing contact details of approved collection centers to consumers and

- creating awareness though publications, advertisement, posters, etc.

The Ministry of Environment, Forest and Climate Change of India issued E-Waste (Management) Rules in March 2013.

2.6. South Africa

South Africa is another important developing country that has significant e-waste. According to UNU-IAS SCYCLE, South Africa produced 346 metric kilotons in 2014, a rate of 6.6 kg per inhabitant [51]. The rate per inhabitant is in the range of that of China, with that of India far lower (1.3 kg per inhabitant). No lessons can be drawn from the state of Chinese e-waste recycling. Instead, several deficiencies are common to both countries. First, e-waste collection is a complex combination of formal and informal methods for both countries. In South Africa, usually all the collected materials end with government-linked recycling centers, where processing takes place. There are few buy-back schemes in South Africa.

South Africa does not have a dedicated legislation for this waste stream compared with other developed countries. The next sections present some existing legislation that covers this waste stream in general.

3. Case Study: Cathode Ray Tubes

CRTs have multiple recyclable parts and so present good opportunities as compared to different e-waste streams. Recycling of CRTs is not usually efficient because of the dangerous nature of its various parts. CRT glass is the biggest problem because it contains lead, strontium, antimony, and barium, among other dangerous chemicals. These elements have the potential to harm the environment and human health if not handled properly. Currently, CRT glass waste handling is mainly through incineration or dumping in landfills, resulting in soil and groundwater pollution. With limited landfill capacities, there is an increasing need to find applications for waste CRT glass to minimize the volume going to landfills and create business opportunities from the waste material.

The authors based their review on this background for South Africa. The study assessed current research done on CRT recovery and recycling, aiming to provide an understanding of the current gaps and opportunities that exist within the South Africa. This study serves as a background to further experimental work by the authors. This work includes pyrometallurgical recovery of metals and recovery of plastics from CRTs.

4. CRT Characteristics and Composition

Cathode ray tubes have been used for more than 70 years as critical parts for televisions and Personal Computers (PCs). The development of the first CRT was during the 1890s as an oscilloscope to view and measure electrical signals [52]. Since then, CRTs have been a popular medium used to send and receive images electronically in television units. Although CRT technology has undergone many revisions, the core design and role have mainly remained the same over the last few decades [53–60].

Estimates are that CRTs make up between 60 and 70 percent of the weight of a computer monitor or television [30]. The remaining material consists of plastic, a printed circuit board (PCB), cabling, and wires [31]. There are two types of CRTs, namely monochrome (black and white) and color. The monochrome and color are structurally similar with only a few technical differences. The major difference between the two CRTs is the chemical composition of the glass contained in the CRT. Figure 1 shows the structure of a CRT.

Panel, funnel, and neck glass are the three main parts of CRTs [61–63]. The panel glass makes up about 65 percent of the total weight. The funnel glass accounts for about 30 percent of the weight and the funnel glass occupies only 5 percent [64].

The chemical composition of CRTs varies from manufacturer to manufacturer, version, and time of production [65]. In addition, the different types of glass parts have different chemical compositions. Table 2 provides a typical chemical composition of both the monochrome and the color CRTs [66].

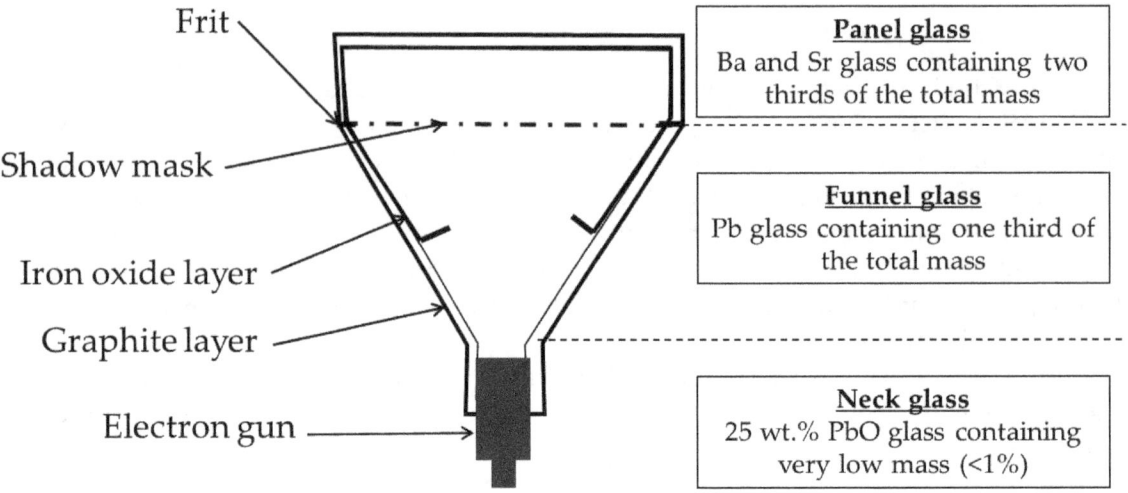

Figure 1. Schematic diagram showing CRT components: the panel glass, funnel glass, and neck glass, with brief information on each section.

Table 2. Typical chemical composition of monochromatic and color CRT [34].

Oxide	Monochrome CRT (Weight %)			Color CRT (Weight %)		
	Panel	Funnel	Neck	Panel	Funnel	Neck
SiO_2	66.05	65.49	56.50	61.23	56.72	50.00
Al_2O_3	4.36	4.38	1.00	2.56	3.42	1.00
K_2O	6.65	5.72	9.00	5.56	5.73	10.00
Na_2O	7.63	7.05	4.00	8.27	6.99	2.00
CaO	0.00	0.00	0.00	1.13	3.12	2.00
MgO	0.01	0.00	0.00	0.76	2.02	0.00
BaO	11.38	11.92	0.00	10.03	4.03	0.00
SrO	0.99	0.94	0.00	8.84	1.99	0.00
PbO	0.03	0.00	29.00	0.02	15.58	34.00

Based on the chemical composition, there are no big differences between the monochromatic and color CRTs. Silica (SiO_2) forms the highest percentage in both CRTs. In both types, the panel glass contains little lead (Pb). Panel glass manufacturing only involves a barium–strontium glass, which is free of lead. Lead in CRTs is found in the funnel and neck components, and concentrated in the neck glass [67]. The role of lead is absorbing UV and X-ray radiation produced by the CRT [68]. Given the overall composition of the CRT, the main concern is the presence of lead. This is because lead is toxic and has adverse health and environmental effects if not handled properly [69,70].

5. CRT Mass Flow

As previously noted, advances in e-waste recovery and recycling differ from country to country. Some countries such as the United States have progressed more than others, especially when compared to developing countries [11,22,70–73]. However, EEE production and the challenges it brings are a worldwide phenomenon. Globally, the need to recover and recycle e-waste is increasing as the landfilling the dangerous substances contained in the waste is no longer a good option. Most landfills are getting full. Also, environmental laws are stricter and e-waste is regarded as high-risk [74].

In South Africa, the e-waste industry is still developing and therefore there are still many barriers that exist. One of the main challenges slowing the development of e-waste management strategies is the lack of baseline data. An extensive body of knowledge exists for other waste streams, and there are enough government and private initiatives to deal with these forms of waste [75–78]. Limited qualitative and quantitative data are available on e-waste in South Africa. However, things have begun moving in the right direction, with South Africa having now recognized e-waste as one of six priority waste streams. The number of e-waste research studies is now increasing. Dominik Zumbuehl conducted one such study that was specific to CRT recycling [79]. The research was a mass flow assessment of CRT in the Cape Metropolitan Municipality, South Africa. The main aim of the research was providing the basis for development of CRT management systems. As part of the assessment, Zumbuehl mapped out the CRT value chain, showing all key role players. Figure 2 shows a typical mass flow of CRT monitors and televisions.

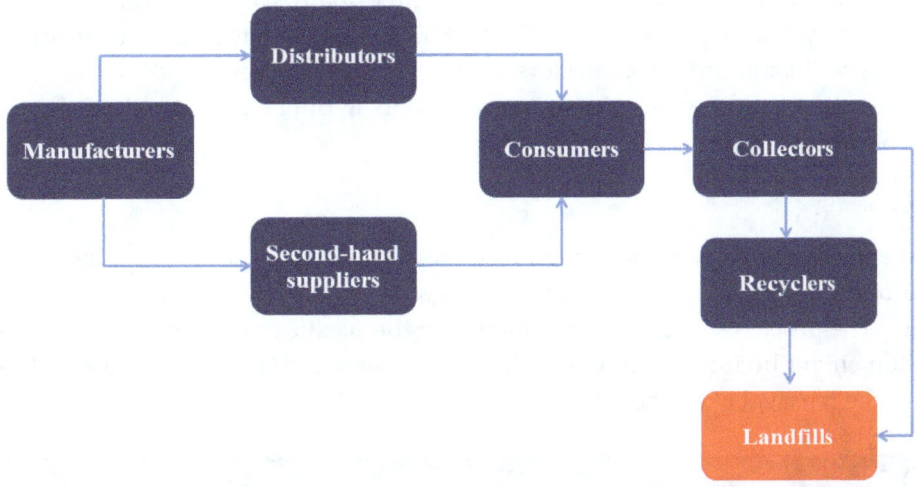

Figure 2. A typical mass flow of CRT monitors and televisions.

South Africa imports CRT monitors as there are no local producers in the country. The major brands manufacturing CRTs are Samsung, Proline, Sony, Mecer, etc. South African distributors sell CRT products to consumers. There is also a choice of buying from second-hand suppliers that import obsolete computers and refurbish them for the secondary market. Once the products have reached their end-of-life, they are regarded as waste. The recovery begins at this point. Several companies collect e-waste material to either sell the valuable material (that is, PCBs) to recyclers or to dispose of it in landfills. The collectors dismantle the monitors or televisions to separate plastics, PCBs, wires, metals, and glass. Selling or landfilling of the different parts then takes place.

As mentioned above, minimal data exist on the number CRTs entering landfills. Zumbuehl estimated landfilling of 7950 CRT monitors in Cape Metropolitan Municipality, yearly [79]. There are no up-to-date national figures per province, except for the annual estimate of 0.35 million tons quoted by Balde et al. [6]. A high percentage of CRTS still ends up in landfills or incinerators.

5.1. CRT Recycling Technologies

Figure 3 provides a CRT processing flow chart. E-waste recycling comprises four main steps: collection, pre-processing, sorting, separation, and processing. The first stage of recycling is collection from consumers once the EEE have reached their EOL.

Figure 3. A generic CRT processing flow sheet.

The next stage after collection is pre-processing. Sorting and separation follow. After separation, the different e-waste parts go through to the different streams of production. The following subsections discuss the different stages of handling CRT waste (and other e-waste).

5.1.1. Collection

As noted above, the main consumers are private, corporate, and government institutions. Collection of obsolete ICT equipment from these three main sources forms most of the waste. The collection is not yet widely accessible in South Africa. However, the main collection method used today is the door-to-door approach. Collectors or recyclers usually facilitate the collection. Most of the recyclers or collectors have now placed e-waste containers in public places such as shopping malls to try to entice consumers.

5.1.2. Pre-Processing

The next stage in the value chain is pre-processing. This step involves stripping and dismantling of CRT monitors and TVs. This process is manual, using hand tools such as screwdrivers, grippers, and hammers. The process starts with the removal of the plastic casing, followed by the cables and wires, and then circuit boards. Removal of the CRT glass is the last step. Figure 4 shows the CRT monitors after the removal of all the other parts.

Figure 4. Stripped CRT monitors.

5.1.3. Sorting and Separation

The method used during sorting and separation stages depends on the final application of the parts. Certain applications require crushing of CRTs into different size fractions and separation according to the different types of glass (i.e., leaded or non-leaded glass). SwissGlas, in Switzerland, does this on an industrial scale, processing nearly five tons of CRT cullets hourly [80]. Different glass types can be separated before crushing. Several techniques exist for separating the panel glass from the funnel glass. The next bullet points discuss these briefly.

- *Hotwire technique*: The hotwire technique involves the use of a heated electric wire, wound at the panel–funnel glass boundary. Heating is kept up for a certain time before cool air is blown onto the surface, leading to thermal shock. This is the preferred method for processors because it is cheap, easy to use, and efficient. The hotwire technology is suitable for small to medium processing plants. The current market price of the technology is between R900,000 and R2 million [65,81].

- *Laser cutting method*: Proventia Automation from Finland developed this method. The technology uses a carbon dioxide laser beam that cuts the CRT below the frit and separates the CRT into the funnel and panel glass. This method has distinct advantages as it does not use any chemicals or water [36]. However, this method is expensive, costing roughly 500,000 euros (R8.9 million). This device can separate up to 75 CRTs an hour [82].

- *Diamond cutting method*: This technology uses a diamond wire to separate the two glasses. A continuous loop of wire cuts into the glass as the CRT passes through the cutting plane. This method is slow and produces dust. This technology can process up to 70 CRTs an hour. It costs between R4 million and R5 million [81].

- *Acid melting*: Acid melting separates the funnel and the panel glass using nitric acid. During this method, the boundary dissolves in a hot acid bath. This method is not efficient—it produces large amounts of wastewater, resulting in high disposal costs [65].

- *Water jet technique*: This technology uses a high-pressure spray of water containing an abrasive, directed at the target surface. The water is focused through a single or double nozzle-spraying arrangement set at a specific distance. It is also efficient and only takes 30 s to separate a CRT monitor. China is piloting this technology [80].

- *Comparison of separation techniques*: Table 3 is a comparison of the different separation techniques. There are several technologies available on the market for CRT separation. The setup costs for these technologies are high.

Table 3. Comparison of separation techniques.

Indicators	Hot Wire	Laser Cutting	Diamond Cutting	Water Jet	Acid Melting
Investment costs	Low	High	High	High	Low
Variable costs	Low	Low	High	High	Low
Quality of glass	High	High	High	High	High
Wet process	No	No	No	Yes	Yes

Laser and hot wire technologies are the most attractive solutions in the market because of the low variable costs. Despite difference in the costs, all available technologies can provide a separation that leads to a high-quality product. However, health and safety and potential environmental issues should be taken into consideration when making the choice.

5.1.4. Processing

Normal disposal methods for CRTs have been landfilling and incineration because of their low recyclable value. Landfilling is no longer the cheapest alternative because not only do landfills occupy massive amounts of land, but there is a risk of soil contamination and groundwater pollution. Incineration was at one point a solution to reduce the space needed for disposal; however, it now poses serious threats of air pollution. Thus there is a need for new and efficient methods for CRT waste management. Subsequent sections discuss the different recycling strategies.

5.2. Overview of Recycling Strategies

Extensive research exists at an international level on the different CRT recycling approaches [63–66,78,79]. Proposals exist on how to deal with CRT waste. Two categories exist, namely closed-loop recycling (waste to new CRT) and open-loop recycling (waste CRT to new products) [18,67]. Figure 5 shows the two recycling routes with key end-products. The next section further discusses the two methods.

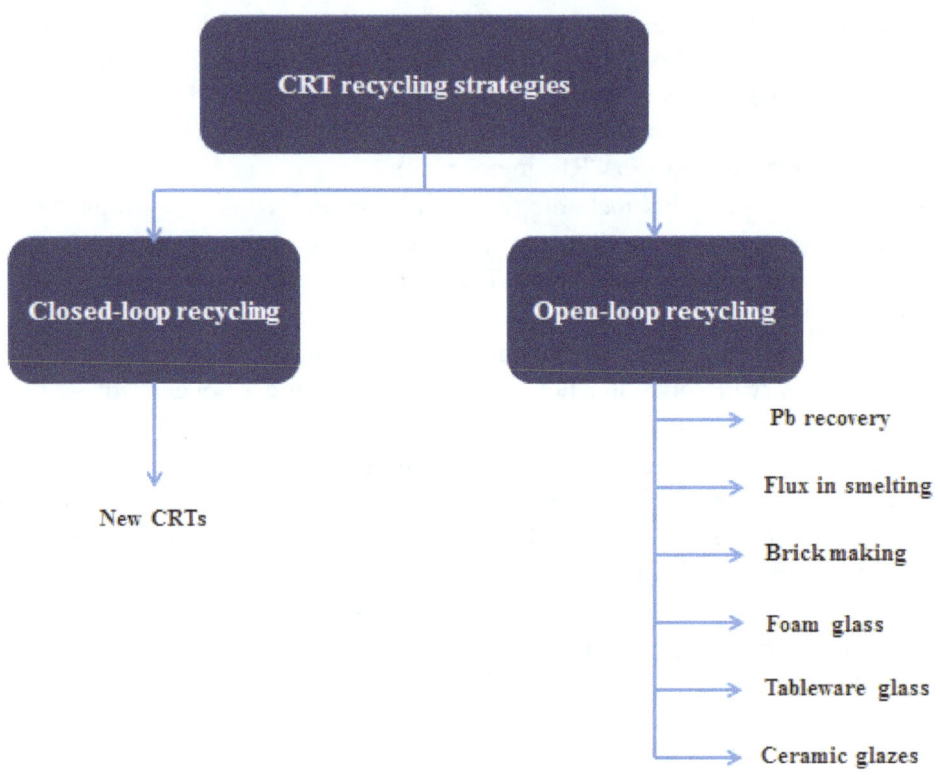

Figure 5. Potential recycling strategies for CRT glass.

5.2.1. Closed-Loop Recycling

Closed-loop recycling is the industrial-scale recycling of waste CRT to produce new CRTs [54]. During this process, the waste CRT glass is divided into the nonleaded panel glass and the lead-containing glass. Electric-wire or laser cutting or acid melting methods separate the panel and the funnel glass, as discussed above. According to Yu-Gong et al., this method has shown great progress in managing CRTs [36]. Compared with landfilling and incineration, this method is environmentally and economically sound. According to Zumbuehl, Samsung Corning in Germany and Thomson-Polcolor in Poland are manufacturing new CRT glass from waste material [80]. Added benefits from the use of waste CRT are energy and raw material savings. Introduction of LCD technologies in the market threaten the sustainability of this business. With advances in ICT, CRT demand is decreasing rapidly.

5.2.2. Open-Loop Recycling

The second proposed form of recycling is open-loop recycling. Open-loop recycling involves the use of waste CRTs to produce a wide variety of new products. Open-loop recycling is attractive because it offers both environmental and economic benefits [65]. A brief discussion of the various areas where waste CRT is used follows.

- *Smelting*: In smelting, CRT glass can replace sand as a flux. This works if the metals contained in the glass (lead) are compatible and recoverable. Metal recovery ensures the slag is free of toxic chemicals [79].

- *Brick manufacturing*: Various studies have investigated the use of CRT glass in brick manufacturing. This application is only suitable for panel glass since it contains no lead. Results have been promising so far, according to the first published studies by Staffordshire University in the United Kingdom. Tests showed that the physical properties of the product made it suitable for non-engineering applications, such as decorative bricks and cladding tiles [53].

- *Foam glass production*: Foam glass is an insulating material made from post-consumer waste glass. Production of foam glass from waste glass has reached a commercial scale. The use of waste CRT therefore stands to be successful in this application as manufacturers have showed no technical barriers [18,83,84].

- *Tableware glass*: Researchers in Murano, Italy used CRT glass in tableware production. The project test-work was in collaboration with a commercial glass factory. The project yielded positive results with good-quality glass products produced. The leaching tests conducted on the products also came out positive [18].

- *Insulating glass*: Recytube, one of the more popular CRT research projects, studied the use of CRT during manufacturing fiberglass. The quality of fiberglass from CRT glass was comparable to commercial fiberglass. Even though the laboratory test results proved positive, there is still some resistance from manufacturers [18].

- *Ceramic glazes*: Different raw materials compose glaze suspensions. Various studies have examined the use of CRT glass as one of the raw materials. Because CRT glass contains barium, strontium, and zirconium, it gives brightness, chemical resistance, and a matte effect to the glaze mixture [18].

A wide range of applications exist for dealing with waste CRT. Concerns about introducing hazardous substances into new products remain, though. Most current work involves lab testing of this.

6. CRT-Specific Legislation

As previously noted, the e-waste industry in South Africa is still at its infancy. South Africa only recognized e-waste as a priority waste stream recently. Currently, there is no specific legislation that regulates e-waste management in the country. South Africa has progressive legislation that recognizes principles of environmental protection and rehabilitation. The following few paragraphs provide a discussion from an international perspective.

The USA is one of the few countries that have CRT-specific legislation, namely the Hazardous Waste Management and Cathode Ray Tubes, Final Rule, 40 CFR Parts 9, 260,261,271 of 2006. This CRT Rule streamlines the management standards for used CRTs and encourages reuse and recycling instead of landfilling or incineration. This legislation proposes several CRT reuse and recycling methods.

The EPA has the following requirements:

- annual reports from exporters of used CRTs for recycling,
- notice when CRTs are exported for recycling,
- notice when CRTs are exported for reuse and,
- normal business records kept by exporters of used CRTs for reuse and translated into English on request.

In addition, the state of Iowa (Cathode Ray Tube Device Recycling Rules, IAC 567 Chapter 122, 2004) and the state of Virginia (Cathode Ray Tubes and Mercury Thermostat Recycling, Va. Code 10.1-1425.26, as Amended 2010) have passed their own state legislation dedicated to CRTs. Most other developed countries have included terms for CRTs within other e-waste legislations (discussed in Section 2).

China, as a developing nation, has legislation specifically for CRTs, namely the Requirements of Disassembly and Treatment for Waste CRT Televisions and Display Devices, Standard GB/T 31377-2015.

6.1. South African Legislation

The Constitution of the Republic of South Africa guides the industry, first and foremost, along with the National Environmental Management Act under the Waste Act 59 of 2008. Legislation that is applicable to e-waste includes: Environment and Conservation Act (No. 73 of 1989), Occupational Health and Safety Act (No. 85 of 1993), and the Hazardous Substances Act (No. 15 of 1973). The next section discusses these in detail.

6.1.1. The Constitution of the Republic of South Africa

The Constitution of the Republic of South Africa is the main legislation in the country and ensures that basic human rights are upheld. Environmental protection is of prime importance in South Africa, given the many environmental problems left behind by past apartheid policies and practices. Section 24 of the Constitution stipulates that: *"everyone has the right to an environment that is not harmful to their health or well-being; and the right to have the environment protected, for the benefit of the present and future generations, through reasonable legislative and other measures that prevent pollution and ecological degradation; promote conservation and secure ecologically sustainable development and use of natural resources while promoting justifiable economic and social development"* [85]. This gives all citizens the right to live in an environment that is not harmful to their health.

6.1.2. The National Environmental Management Act: Waste Act

The National Environmental Management Act (Waste Act) is founded on the principles of the Constitution. The Waste Act gives a framework for sound environmental management practices for all development practices. The Act, passed in 2008, regulates and controls management of all waste, including e-waste. Before this Act, there was no specific legislation to manage waste in the country. The Act refers to avoidance or minimization and remediation of pollution, including waste reduction, reuse, recycling, and proper waste disposal [86].

6.1.3. Environment Conservation Act (No. 73 of 1989)

The Environment Conservation Act (ECA) provides for protection and controlled use of the environment. Section 20 of the ECA makes specific reference to waste disposal and assigns the role of issuing permits for waste disposal to the Department of Water Affairs and Forestry.

6.1.4. Hazardous Substances Act (No. 15 of 1973)

E-waste contains many hazardous substances. The Hazardous Substances Act regulates hazardous substances and their disposal. The Act classifies different types of hazardous substances into four groups and imposes detailed requirements when handling, selling, using, and applying them.

6.1.5. The Occupational Health and Safety Act (No. 85 of 1993)

The main objective of Occupational Health and Safety Act (OHSA) is to ensure that people work in an environment that is not harmful to their wellbeing. Other rules in the Act deal with handling, use, exposure control, use of PPE, and storage or disposal of hazardous chemicals. These include:

- Lead Regulations,
- Hazardous Chemical Substances Regulations,
- Environmental Regulations for Workplaces and
- General Safety Regulations.

6.1.6. Other Applicable Legislation

Other applicable laws are:

- National Water Act (No. 36 of 1998)
- Atmospheric Pollution Prevention Act (No. 45 of 1965)
- Air Quality Act (No. 39 of 2004)
- National Health Act (No. 61 of 2003)
- Precious Metal Regulation
- Second-Hand Goods Legislation
- Consumer Protection Act (No. 68 of 2008)

6.1.7. The Path forward for South Africa: Waste Research Development and Innovation (RDI) Roadmap (2015—2025)

The Department of Science and Technology (DST) leads the RDI roadmap to stimulate waste R&D, waste innovation, and human capital development in the waste sector. The government recognizes the waste sector as an important platform contributing to job creation and economic development. The basis for the Waste RDI Program is:

- increasing diversion of waste from landfill,
- creating value-adding opportunities,
- optimization of value-adds,
- leading to significant economic, social and environmental benefits, and
- creating a sustainable regional secondary economy.

The Waste RDI roadmap provides strategic direction, to coordinate and manage the country's waste investment in the following areas:

- Strategic Planning
- Modeling and Analytics
- Technology Solutions
- Waste Logistics Performance
- Waste and Environment
- Waste and Society

E-waste is a priority waste stream in the roadmap. E-waste currently makes up between 5% and 8% of the municipal solid waste in South Africa. The volumes of e-waste are continually increasing. There is an urgent need to develop e-waste management systems that address the current challenges. The challenges are consumer awareness, collection, recycling, and waste disposal. The roadmap plans to divert 50 percent of e-waste from landfills by 2024 [87].

7. Conclusions

There has been massive interest in e-waste the world over, fueled by the continued increase in producing EEE and the volumes of waste EEE produced. The problems emanating from poor disposal and management of e-waste, and the possible economic opportunities, have pushed countries to reconsider old ways of managing e-waste. Research is being conducted at an international level to come up with efficient solutions to deal with e-waste.

CRT-related research is one of the fastest-growing areas and receives a good deal of attention. For some time, landfilling and incineration were the best solutions for managing CRT waste. This has changed since, largely due to the strict environmental laws and policies. There is a trend towards recovering and recycling waste CRTs because they pose a risk to the environment and human health.

In South Africa, the e-waste recycling industry is fairly young, and therefore limited baseline data are available at the country level. E-waste currently forms between 5% and 8% of municipal solid waste in South Africa and there is an expectation that these volumes will soon increase significantly. The per-inhabitant e-waste production rates for South Africa are similar to those of China. However, the Chinese e-waste management methods are not effective owing to lax enforcement of laws. Similarly to China, South Africa has ended up with a mixture of formal and informal collection methods that is a lot harder to regulate. However, several initiatives are under consideration to solve the e-waste problem using a customized approach that recognizes local complexities.

Acknowledgments: The authors thank Mintek (SA) for permission to publish the results and financial support.

Author Contributions: P.L. conceived the project; P.L. and N.S collected and analyzed literature data. P.L. drafted the first manuscript; N.S. supervised and edited the manuscript.

References

1.   Kiddee, P.; Naidu, R.; Wong, M.H. Electronic waste management approaches: An overview. *Waste Manag.* **2013**, *33*, 1237–1250. [CrossRef] [PubMed]

2.   Kiddee, P.; Naidu, R.; Wong, M.H. Metals and polybrominated diphenyl ethers leaching from electronic waste in simulated landfills. *J. Hazard Mater.* **2013**, *252*, 243–249. [CrossRef] [PubMed]

3.   Li, J.; Zeng, X.; Chen, M.; Ogunseitan, O.A.; Stevels, A. "Control-alt-delete": Rebooting solutions for the e-waste problem. *Environ. Sci. Technol.* **2015**, *49*, 7095–7108. [CrossRef] [PubMed]

4.   Babu, B.R.; Parande, A.K.; Basha, C.A. Electrical and electronic waste: A global environmental problem. *Waste Manag. Res.* **2007**, *25*, 307–318.

5.   Li, J.; Zeng, X.; Stevels, A. Ecodesign in consumer electronics: Past, present, and future. *Crit. Rev. Environ. Sci. Technol.* **2015**, *45*, 840–860. [CrossRef]

6.   The Global E-Waste Monitor 2014: Quantities Flows and Resources. Available online: http://i.unu.edu/media/ias.unu.edu-en/news/7916/Global-E-waste-Monitor-2014-small.pdf (accessed on 1 August 2016).

7.   Mundada, M.N.; Kumar, S.; Shekdar, A.V. E-waste: A new challenge for waste management in India. *Int. J. Environ. Stud.* **2004**, *61*, 265–279. [CrossRef]

8.   Pérez-Belis, V.; Bovea, M.D.; Ibáñez-Forés, V. An in-depth literature review of the waste electrical and electronic equipment context: Trends and evolution. *Waste Manag. Res.* **2015**, *33*, 3–29. [CrossRef] [PubMed]

9.   Osibanjo, O.; Nnorom, I.C. The challenge of electronic waste (e-waste) management in developing countries. *Waste Manag. Res.* **2007**, *25*, 489–501. [CrossRef] [PubMed]

10. E-waste Facts: Causes International. Available online: https://www.causesinternational.com/ewaste/e-waste-facts (accessed on 1 August 2016).

11. Sahin, M.A.A.; Erust, C.; Altynbek, S.; Gahan, C.S.; Tuncuk, A. A Potential Alternative for Precious Metal Recovery from E-waste: Iodine Leaching. *Sep. Sci. Technol.* **2015**, *50*, 2587–2595. [CrossRef]

12. Jujun, R.; Yiming, Q.; Zhenming, X. Environment-friendly technology for recovering nonferrous metals from e-waste: Eddy current separation. *Resour. Conserv. Recycl.* **2014**, *87*, 109–116. [CrossRef]

13. Laner, D.; Rechberger, H. Treatment of cooling appliances: Interrelations between environmental protection, resource conservation, and recovery rates. *Resour. Conserv. Recycl.* **2007**, *52*, 136–155. [CrossRef]

14. Ilgin, M.A.; Gupta, S.M. Environmentally conscious manufacturing and product recovery (ECMPRO): A review of the state of the art. *J. Environ. Manag.* **2010**, *91*, 563–591. [CrossRef] [PubMed]

15. Cucchiella, F.; D'Adamo, I.; Koh, S.L.; Rosa, P. A profitability assessment of European recycling processes treating printed circuit boards from waste electrical and electronic equipments. *Renew. Sust. Energ. Rev.* **2016**, *64*, 749–760. [CrossRef]

16. Nyanjowa, W.; James, Y. *Material Flow Analysis of Printed Circuit Boards and Lithium Ion Batteries in Gauteng, South Africa*; Internal report; Mintek: Randburg, South Africa, 2015.

17. National Consultative Conference on Electronic and Electrical Waste (E-Waste) Management in South Africa. Available online: https://www.environment.gov.za/speech/molewa_government_e-waste_conference (accessed on 1 August 2016).

18. Andreola, F.; Barbieri, L.; Corradi, A.; Lancellotti, I. CRT glass state of the art: A case study: Recycling in ceramic glazes. *J. Eur. Ceram. Soc.* **2007**, *27*, 1623–1629. [CrossRef]

19. Buekens, A.; Yang, J. Recycling of WEEE plastics: A review. *J. Mater. Cycles. Waste* **2014**, *16*, 415–434. [CrossRef]

20. Ruan, J.; Xu, Z. Approaches to improve separation efficiency of eddy current separation for recovering aluminum from waste toner cartridges. *Environ. Sci. Technol.* **2012**, *46*, 6214–6221. [CrossRef] [PubMed]

21. Ruan, J.; Xue, M.; Xu, Z. Risks in the physical recovery system of waste refrigerator cabinets and the controlling measure. *Environ. Sci. Technol.* **2012**, *46*, 13386–13392. [CrossRef] [PubMed]

22. Huo, X.; Peng, L.; Xu, X.; Zheng, L.; Qiu, B.; Qi, Z.; Zhang, B.; Han, D.; Piao, Z. Elevated blood lead levels of children in Guiyu, an electronic waste recycling town in China. *Environ. Health Perspect.* **2007**, *115*, 1113–1117. [CrossRef] [PubMed]

23. The Role of BRICS in the Developing World. Available online: http://www.ab.gov.tr/files/ardb/evt/1_avrupa_birligi/1_9_politikalar/1_9_8_dis_politika/The_role_of_BRICS_in_the_developing_world.pdf (accessed on 10 October 2016).

24. Fedoseeva, S.; Zeidan, R. A dead-end tunnel or the light at the end of it: The role of BRICs in European exports. *Econ. Mod.* **2016**, *59*, 237–248. [CrossRef]

25. Yao, X.; Watanabe, C.; Li, Y. Institutional structure of sustainable development in BRICs: Focusing on ICT utilization. *Technol. Soc.* **2009**, *31*, 9–28. [CrossRef]

26. Wilson, D.; Puruthaman, R. *Dreaming with BRICs: The Path to 2050*; Global Economics Paper No. 99; Goldman Sachs: New York, NY, USA, 2003.

27. Armijo, L.E. The BRICs countries (Brazil, Russia, India, and China) as analytical category: Mirage or insight? *Asian Perspect.* **2007**, *31*, 7–42.

28. United Nations University Sustainable Cycles (2015) Step E-Waste World Map. Database Available from STEP—Solving the E-waste Problem 2015. Available online: http://www.step-initiative.org/Overview_USA.html#Regulatory (accessed on 10 October 2016).

29. From the EPA: E-Waste News and Statistics. Available online: http://ewastecollective.org/from-the-epa-e-waste-facts-and-statistics/ (accessed on 10 October 2016).

30. Jang, Y.-C.; Townsend, T.G. Leaching of Lead from Computer Printed Wire Boards and Cathode Ray Tubes by Municipal Solid Waste Landfill Leachates. *Environ. Sci. Technol.* **2003**, *37*, 4778–4784. [CrossRef] [PubMed]

31. Basel Convention on the Control of Transboundary Movements of Hazardous Wastes and Their Disposal. Available online: http://www.basel.int/portals/4/basel%20convention/docs/text/baselconventiontext-e.pdf (accessed on 10 October 2016).

32. Electronic Waste Recycling Fee. Available online: http://www.boe.ca.gov/pdf/pub95.pdf (accessed on 10 October 2016).

33. Tax Rates—Special Taxes and Fees. Available online: http://www.boe.ca.gov/sptaxprog/tax_rates_stfd.htm (accessed on 10 October 2016).

34. Plug-In to eCycling Guidelines for Materials Management. Available online: http://www.epeat.net/documents/reference-docs/epa-plug-in-to-ecycling-guidelines.2004-05.pdf (accessed on 10 October 2016).

35. Responsible Recycling ("R2") for Use in Accredited Certification Programs for Electronics Recyclers. Available online: https://www.doi.gov/sites/doi.gov/files/migrated/greening/electronics/upload/R2-Document-2.pdf (accessed on 10 October 2016).

36. Kang, H.Y.; Schoenung, J.M. Electronic waste recycling: A review of US infrastructure and technology options. *Resour. Conserv. Recycl.* **2005**, *45*, 368–400. [CrossRef]

37. Study on Collection Rates of Waste Electrical and Electronic Equipment (WEEE). Available online: http://ec.europa.eu/environment/waste/weee/pdf/Final_Report_Art7_publication.pdf (accessed on 10 October 2016).

38. Sthiannopkao, S.; Wong, M.H. Handling e-waste in developed and developing countries: Initiatives, practices, and consequences. *Sci. Total Environ.* **2013**, *463*, 1147–1153. [CrossRef] [PubMed]

39. Widmer, R.; Oswald-Krapf, H.; Sinha-Khetriwal, D.; Schnellmann, M.; Böni, H. Global perspectives on e-waste. *Environ. Impact Assess. Rev.* **2005**, *25*, 436–458. [CrossRef]

40. Honda, S. Japan's experiences in environmentally sound management of e-waste. In Proceedings of the E-waste Workshop at IETC, Osaka, Japan, 6–9 July 2010.

41. United Nations University Sustainable Cycles (2015) Step E-Waste World Map. Database available from STEP—Solving the E-waste Problem 2015. Available online: http://www.step-initiative.org/Overview_Japan.html (accessed on 10 October 2016).

42. Chi, X.; Streicher-Porte, M.; Wang, M.Y.; Reuter, M.A. Informal electronic waste recycling: A sector review with special focus on China. *Waste Manag.* **2011**, *31*, 731–742. [CrossRef] [PubMed]

43. United Nations University Sustainable Cycles (2015) Step E-Waste World Map. Database Available from STEP—Solving the E-waste Problem 2015. Available online: http://www.step-initiative.org/Overview_China.html (accessed on 10 October 2016).

44. Ongondo, F.O.; Williams, I.D.; Cherrett, T.J. How are WEEE doing? A global review of the management of electrical and electronic wastes. *Waste Manag.* **2011**, *31*, 714–730. [CrossRef] [PubMed]

45. Wang, Z.; Guo, D.; Wang, X. Determinants of residents' e-waste recycling behavior intentions: Evidence from China. *J. Clean. Prod.* **2016**, *137*, 850–860. [CrossRef]

46. State Council. Regulations for the Administration of the Recovery and Disposal of Electric and Electronic Products. Order of the State Council of the People's Republic of China, No. 551; 2009. Available online: http://www.chinarohs.com/chinaweee-decree551.pdf (accessed on 10 October 2016).

47. Wath, S.B.; Dutt, P.S.; Chakrabarti, T. E-waste scenario in India, its management and implications. Environmental monitoring and assessment. *Environ. Monit. Assess.* **2011**, *172*, 249–262. [CrossRef] [PubMed]

48. United Nations University Sustainable Cycles (2015) Step E-Waste World Map. Database Available from STEP—Solving the E-waste Problem 2015. Available online: http://www.step-initiative.org/ (accessed on 10 October 2016).

49. Takeback Blues: An Assessment of E-waste Takeback in India. Available online: http://www.greenpeace.org/india/Global/india/report/2008/8/take-back-blues.pdf (accessed on 10 October 2016).

50. United Nations University Sustainable Cycles (2015) Step E-Waste World Map. Database available from STEP—Solving the E-waste Problem 2015. Available online: http://www.step-initiative.org/Overview_South_Africa.html (accessed on 10 October 2016).

51. Thomson, J.J. Cathode rays. *Philos. Mag.* **1897**, *44*, 293–316. [CrossRef]

52. Merlin, D. Cathode Ray Tube. U.S. Patent 2,053,268, 8 September 1936.

53. Branson, H. Cathode Ray Tube. U.S. Patent 2,274,586, 24 February 1942.

54. Gabor, D. Cathode Ray Tube. U.S. Patent 2,795,729, 11 June 1957.

55. Yoshiharu, K.; Senri, M.; Akio, O.; Susumu, Y. Cathode Ray Tube. U.S. Patent 3,448,316, 3 June 1969.

56. Nobuo, K.; Akio, O.; Takizo, S. Cathode Ray Tube. U.S. Patent 3,909,524, 30 September 1975.

57. Osakabe, K. Cathode Ray Tube. U.S. Patent 4,772,827, 20 September 1988.

58. Kawamura, H.; Kobara, K.; Kawamura, T.; Miura, K. Cathode-Ray Tube. U.S. Patent 5,291,097, 1 March 1994.

59. Kim, D.N.; Lee, B.W. Cathode Ray Tube. U.S. Patent 6,335,588, 1 January 2002.

60. Andreola, F.; Barbieri, L.; Corradi, A.; Lancellotti, I. Cathode ray tube glass recycling: An example of clean technology. *Waste Manag. Res.* **2005**, *23*, 314–321. [CrossRef] [PubMed]

61. Méar, F.; Yot, P.; Cambon, M.; Ribes, M. The characterization of waste cathode-ray tube glass. *Waste Manag.* **2006**, *26*, 1468–1476. [CrossRef] [PubMed]

62. Sua-iam, G.; Makul, N. Use of limestone powder during incorporation of Pb-containing cathode ray tube waste in self-compacting concrete. *J. Environ. Manag.* **2013**, *128*, 931–940. [CrossRef] [PubMed]

63. Mueller, J.R.; Boehm, M.W.; Drummond, C. Direction of CRT waste glass processing: Electronics recycling industry communication. *Waste Manag.* **2012**, *32*, 1560–1565. [CrossRef] [PubMed]

64. Tian, X.M.; Wu, Y.F. Recent development of recycling lead from scrap CRTs: A technological review. *Waste Manag.* **2015**, in press. [CrossRef]

65. Shi, X.; Li, G.; Xu, Q.; He, W.; Liang, H. Research Progress on Recycling Technology of End-of-life CRT Glass. *Mater. Rev.* **2011**, *11*, 1–29.

66. Tsydenova, O.; Bengtsson, M. Chemical hazards associated with treatment of waste electrical and electronic equipment. *Waste Manag.* **2011**, *31*, 45–58. [CrossRef] [PubMed]

67. Pant, D.; Singh, P. Chemical modification of waste glass from cathode ray tubes (CRTs) as low cost adsorbent. *J. Environ. Chem. Eng.* **2013**, *1*, 226–232. [CrossRef]

68. Pant, D.; Joshi, D.; Upreti, M.K.; Kotnala, R.K. Chemical and biological extraction of metals present in E waste: A hybrid technology. *Waste Manag.* **2012**, *32*, 979–990. [CrossRef] [PubMed]

69. Garlapati, V.K. E-waste in India and developed countries: Management, recycling, business and biotechnological initiatives. *Renew. Sust. Energ. Rev.* **2016**, *54*, 874–881. [CrossRef]

70. Iqbal, M.; Breivik, K.; Syed, J.H.; Malik, R.N.; Li, J.; Zhang, G.; Jones, K.C. Emerging issue of e-waste in Pakistan: A review of status, research needs and data gaps. *Environ. Pollut.* **2015**, *207*, 308–318. [CrossRef] [PubMed]

71. Milovantseva, N.; Fitzpatrick, C. Barriers to electronics reuse of transboundary e-waste shipment regulations: An evaluation based on industry experiences. *Resour. Conserv. Recycl.* **2015**, *102*, 170–177. [CrossRef]

72. Song, Q.; Li, J. Environmental effects of heavy metals derived from the e-waste recycling activities in China: A systematic review. *Waste Manag.* **2014**, *34*, 2587–2594. [CrossRef] [PubMed]

73. Cucchiella, F.; D'Adamo, I.; Koh, S.L.; Rosa, P. Recycling of WEEEs: An economic assessment of present and future e-waste streams. *Renew. Sust. Energ. Rev.* **2015**, *51*, 263–272. [CrossRef]

74. National Waste Information Baseline Report. Available online: http://sawic.environment.gov.za/documents/1880.pdf (accessed on 1 August 2016).

75. South African Waste Snapshot. UrbanEarth. Available online: http://www.urbanearth.co.za/sites/default/files/urban_earth_sa_waste_snapshot.pdf (accessed on 1 August 2016).

76. South African Waste Sector—2012: An Analysis of the Formal Private and Public Waste Sector in South Africa. Available online: http://www.wasteroadmap.co.za/download/waste_sector_survey_2012.pdf (accessed on 1 August 2016).

77. Environmental Protection and Infrastructure Programmes: 15 Years of Innovative Environmental Protection and Job Creation: 1999–2014. Available online: https://www.environment.gov.za/sites/default/files/reports/epip15years_review.pdf (accessed on 1 August 2016).

78. Guerrero, L.A.; Maas, G.; Hogland, W. Solid waste management challenges for cities in developing countries. *Waste Manag.* **2013**, *33*, 220–232. [CrossRef] [PubMed]

79. Zumbuehl, D. *Mass Flow Assessment (MFA) and Assessment of Recycling Strategies for Cathode Ray Tubes (CRTs) for the Cape Metropolian Area (CMA), South Africa*; Swiss Federal Institute of Technology Zurich: Zürich, St. Gallen, Switzerland, 2006.

80. MRT CRT Separator: Diamond Cutting Technology. Karlskrona: MRT System. Available online: http://www.mrtsystem.com/wp-content/uploads/2013/05/CRT-Separator-diamond-cutting.pdf (accessed on 1 August 2016).

81. Materials Recovery from Waste Cathode Ray Tubes (CRTs). Available online: http://ewasteguide.info/files/ICER_2004_WRAP.pdf (accessed on 1 August 2016).

82. Menad, N. Cathode ray tube recycling. *Resour. Conserv. Recycl.* **1999**, *26*, 143–154. [CrossRef]

83. König, J.; Petersen, R.R.; Yue, Y. Fabrication of highly insulating foam glass made from CRT panel glass. *Ceram. Int.* **2015**, *41*, 9793–9800. [CrossRef]

84. Guo, H.W.; Gong, Y.X.; Gao, S.Y. Preparation of high strength foam glass–Ceramics from waste cathode ray tube. *Mater. Lett.* **2010**, *64*, 997–999. [CrossRef]

85. Republic of South Africa. *Constitution of the Republic of South Africa, Act 108 of 1996*; Government Gazette: London, UK, 1996; Volume 378.

86. Republic of South Africa. *National Environmental Management: Waste Act*; Government Gazette: London, UK, 2009; Volume 525.

87. A Waste Research, Development and Innovation (RDI) Roadmap for South Africa (2015–2025): Towards a Secondary Economy. Available online: https://www.environment.gov.za/sites/default/files/docs/roadmappresentation.pdf (accessed on 1 August 2016).

12

Using Pro-Environmental Information to Modify Conservation Behavior: Paper Recycling and Reuse

Bruno Wichmann *, Martin Luckert, Katrina Bissonnette, Alyssa Cumberland, Claire Doll, Tanishka Gupta and Yuan Shi

Department of Resource Economics & Environmental Sociology, University of Alberta, 515 General Services Building, Edmonton, AB T6G-2H1, Canada; mluckert@ualberta.ca (M.L.); kbissonn@ualberta.ca (K.B.); cumberla@ualberta.ca (A.C.); cadoll@ualberta.ca (C.D.); tanishka@ualberta.ca (T.G.); yuan.shi@ualberta.ca (Y.S.)
* Correspondence: bwichmann@ualberta.ca

Academic Editor: Michele Rosano

Abstract: In cases where market policy instruments (e.g., taxes and quotas) are impractical tools to induce conservation behavior, information campaigns may be a valuable option. We use a difference-in-differences strategy to estimate the effectiveness a signage campaign for inducing paper recycling and reuse behavior in computer labs. We find that the implementation of signage with pro-environment appeals increases the probability of conservation behavior (i.e., recycling or reuse) by approximately 13%, despite the fact that pre-treatment levels of paper recycling and reuse were already at approximately 85%. Our results suggest that pro-environment campaigns can be an effective conservation tool and may be an important policy instrument for policy makers to consider.

Keywords: paper recycling; paper reuse; social norms; pro-environment appeals; computer labs

---

1. Introduction

Taxes, subsidies, and quotas are policy instruments sometimes used by policymakers to encourage socially desired behavior. However, there are some economic environments where the implementation of such market-based instruments is challenging. Taxes and quotas are not always politically feasible or practical, and subsidies may add significant burdens to limited government budgets [1]. Moreover, there may be instances, such as with recycling and reuse, where costs of monitoring behavior would be prohibitive. In such cases, policymakers are increasingly interested in mechanisms to influence individual decision-making that are not based on typical manipulations of market prices and quantities.

Policy instruments to address problems in waste management may face particular challenges compared to, for example, conservation programs that discourage consumption. While consumption may be taxed or subsidized at places of sales, it can be impractical to monitor individual behavior toward some types of recycling and waste disposal. For example, in some situations (e.g., schools, universities, and corporate offices) it could be difficult to tax paper disposal that could be recycled, or conversely subsidize paper that is reused or recycled. Accordingly, while energy and water conservation programs may focus on consumption reduction associated with monetary incentives through lower utility bills, recycling and reuse programs may need to rely on pro-environmental appeals.

Such an approach may be based on the "supply of environmentalism" related to psychological and economic concepts [2]. The provision of information may alter behavior by providing "psychic taxes" (with negative information) or "psychic subsidies" (with positive information). The information provided to influence behavior may appeal to social norms, which Scott and Marshall [3] define as a set of rules that determine appropriate and inappropriate values, beliefs, attitudes and behaviors. The authors explain that failure to follow the rules of social norms can result in punishments, including social exclusion. Following the theoretical model of Ferraro and Price [4], we assume that moral values are a function of social norms and may affect the conservation decisions of rational economic agents. The provision of information in these contexts may appeal to social norms that change the benefits and costs to individuals, and thereby influence behavior.

There have been numerous information-based initiatives that have attempted to change behaviors that can generate negative social spillovers, such as excessive alcohol consumption, drug use, gambling, and littering [5]. Other studies have shown that information campaigns can be effective in promoting environmental conservation behavior, such as reduced water use [5–8], reduced bottled water consumption [9], increased re-use of hotel towels [10], reduced electric energy consumption [1,11], and increased sustainable transportation behavior [12].

There is also research that suggests that information about recycling can increase recycling behavior [13–15]. The success of such information campaigns could arise because the information provided appeals to, and/or reinforces, accepted social norms and/or attitudes (e.g., [16]). For example, Abbott et al. [17] find evidence that social norms influence recycling in England. Thomas and Sharp [18] observe that a number of studies have found that over time, an increasing number of individuals are recycling regularly to the point where it can be deemed a common behavior. They conclude that recycling is becoming socially normalized in the United Kingdom. Nolan [19] provides a discussion about the normativeness of recycling.

Hornik et al. [20] offer a literature review of the determinants of recycling behavior, highlighting the role of incentives and facilitators. They discuss the limitations of economic incentives in promoting long-term sustainable changes in recycling behavior. They also argue that awareness of the importance of recycling and knowledge about recycling programs is an important facilitator of recycling behavior. Hornik et al. [20] conclude that strategies that increase recycling education and improve social image may induce more consumers to begin recycling.

Another challenge in the literature around information induced behavior arises because information is only one of many types of determinants that can influence behavior, cross-sectionally, and over time. To address this identification issue, it can be useful to have observations before and after a treatment, and controls that do not receive the current information treatments, thereby allowing a difference-in-differences approach. Very few studies have evaluated the impact of information on environmental behavior in such an experimental framework (e.g., [4]) and even fewer with respect to recycling behavior (e.g., [21]). Moreover, to our knowledge, this is the first study to investigate the effectiveness of an information campaign when recycling levels are already high.

The objective of this study is to evaluate the impact of a signage campaign. Our case study concerns the use of paper in university computer labs. We use survey data to obtain difference-in-differences estimates of the effectiveness of the norm-based strategies in promoting paper conservation behavior. By examining university computer labs, we explore the use of pro-environmental appeals in a setting where implementation of typical policy instruments (e.g., taxes, subsidies and quotas) is impractical; a situation common in working environments and offices that regularly consume large amounts of paper (Section 4.1 presents a discussion). Moreover, our sites' recycling and reuse baseline levels were approximately 84%. Our paper explores whether the implementation of signage with pro-environment appeals increases the probability of conservation behavior despite already high levels of recycling and reuse, thereby indicating whether such campaigns can significantly influence conservation behavior over the "last mile".

2. Methods

2.1. The Recycling and Reuse Program

We designed a program to encourage paper recycling and reuse. The program consisted of improved access to recycling bins and reuse trays. In addition, the program introduced signage regarding paper recycling and reuse. In order to isolate the impacts of the signage, these features were implemented differentially across labs. While we improved access to paper recycling and reuse in both labs, we implemented signage with pro-environmental appeals solely in the treatment lab. Therefore, our treatment effect refers to the impact of signage (in addition to improved infrastructure). Details are provided below.

Improved Access: Findings in the literature suggest that there are four key components that contribute to effectively inducing changes in recycling habits: specialized bins for paper recycling, lids on the bins to prevent disposal of unwanted material, bins being in close proximity to directional signs, and the placement of receptacles in all areas of consumption [22,23]. These requirements were satisfied in both labs. Moreover, the number and size of recycling bins and reuse trays reflected the capacity of the computer labs.

Signage: Signage was introduced in only one of the labs (i.e., the treatment lab). Our signage program consisted of three groups of posters mounted on three different walls next to printers. Each group of posters contained three components (see Figure 1). Component (a) was designed to prompt computer lab users to increase their usage of trays facilitating paper reuse. Component (b) was a "print green" poster that contained one of three pro-social appeals regarding consumption and resource use:

> *"Canadians are among the world's largest consumers of paper products, using 6 million tonnes of paper and paperboard annually"*

> *"24 trees are cut down for every tonne of paper and paperboard produced"*

> *"Approximately 324 litres of water are needed to produce one kilogram of paper"*

To improve paper reuse success, we designed a third poster, component (c), which provided instructions on how to reuse paper.

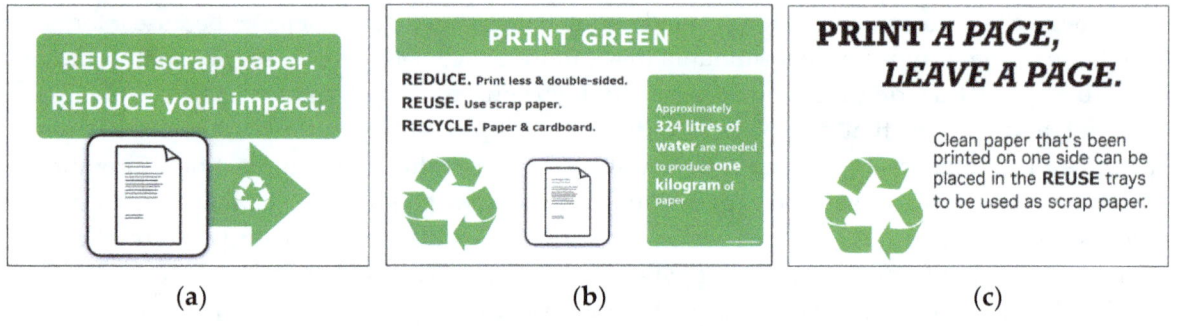

Figure 1. Signage implemented in the treatment lab; components (**a**), (**b**) and (**c**).

2.2. Experimental Design

Our experimental design is motivated by the desire to isolate the effects of the signage treatment from other factors that could influence recycling and/or reuse. There are two categories of factors that could confound this identification: factors that vary cross-sectionally, and factors that vary over time. A difference-in-differences approach can be used to attempt to control for such factors. Therefore, our experimental design includes observations before and after the treatment, and controls during the treatment that do not receive the information treatments.

We surveyed users of two computer labs: the control lab and the treatment lab. These are two large labs centrally located on campus with a distance of approximately 500 m between one another (i.e., a 6 min walk). Both labs may be used by any student on campus.

A baseline survey was implemented in the fall of 2012 in both labs. Next, the recycling and reuse program described in the previous section was implemented exclusively in the treatment lab in January 2014. In March 2014, we re-surveyed both labs using the same survey instrument that was implemented in the fall of 2012.

Interviewers randomly approached users at both computer labs over the course of a week, to fill out a short survey. Users who had already been surveyed were identified and not surveyed again to avoid double counting. Interviewers stated that they were collecting information to help the university office of building and grounds services. A key question was: "What do you most often do with used paper in this computer lab?" Respondents were instructed to choose one of four options: (a) recycle it; (b) use it for scrap paper; (c) throw it in the garbage; (d) other.

Table 1 summarizes our experimental design by showing the number of responses collected in both labs before and after the treatment.

Table 1. Number of observations per treatment cell.

Labs	Fall 2012 (Before Treatment)	Winter 2014 (After Treatment)	Total
Control	86	146	232
Treatment	46	124	170
Total	132	270	402

2.3. Program Evaluation

The experimental design facilitates the use a difference-in-differences regression model to estimate the effect of pro-environment signage on recycling and reuse behavior of computer lab users. This approach allows us to compare our treatment group with a control group, thereby accounting for non-treatment effects over time. We estimate the following probit model:

$$Prob(Y = 1) = G(\beta_0 + \beta_1 Post + \beta_2 Treatment + \beta_3(Post^*Treatment)), \qquad (1)$$

where Y is a binary indicator for the selected behavior (i.e., $Y = 1$ if paper is recycled, $Y = 0$ otherwise), G is the c.d.f. of the standard normal distribution, Post is an indicator for the Winter 2014 (post-treatment) survey, and Treatment is an indicator for the treatment lab.

In our case, the treatment effect is equal to the marginal effect of a change in *Post*Treatment* from zero to one, holding all other variables fixed. As a result, the estimate of the treatment effect on the probability of observing the selected behavior is:

$$G(\hat{\beta}_0 + \hat{\beta}_1 Post + \hat{\beta}_2 Treatment + \hat{\beta}_3(Post * Treatment)) - G(\hat{\beta}_0 + \hat{\beta}_1 Post + \hat{\beta}_2 Treatment), \qquad (2)$$

where $\hat{\beta}$ denotes probit parameter estimates, with average levels of Post and Treatment used to evaluate the above expression.

An important identifying assumption of difference-in-differences estimation of treatment effects is that behavior of control and treatment groups are identical, except for the effect of the treatment. We have no reason to believe that users of the control lab have a systematically different approach towards paper recycling or reuse than users of the treatment lab. Nevertheless, we use the Fall 2012 data (pre-treatment) to test if the proportions of selected behaviors (i.e., recycling, reuse and garbage) are statistically different between the two labs. We also note that a given piece of paper can only be associated with one behavior, two of which are conservation-oriented (i.e., recycling and reuse). Therefore, we also present proportions of the combined behaviors of recycling and reuse. Results are presented in Table 2. For all cases, differences between the two labs are not statistically significant. This is empirical evidence that behaviors in both labs are similar before treatment, providing support for the assumption of random treatment assignment.

Table 2. Tests of equality of proportions between the control and treatment labs prior to treatment (Fall, 2012).

Labs	Proportion of Recycling	Proportion of Reuse	Proportion of Recycling or Reuse	Proportion of Garbage
Control	0.430 (0.053)	0.430 (0.053)	0.861 (0.037)	0.128 (0.036)
Treatment	0.370 (0.071)	0.435 (0.073)	0.804 (0.058)	0.174 (0.056)
	H_0: Prop(Control) $-$ Prop(Treatment) $= 0$			
p-value	0.499	0.960	0.401	0.473

Note: Standard Errors are in parenthesis.

3. Results

Figure 2 shows the distribution of behaviors by lab, before and after the treatment. In the treatment lab, the proportions of respondents recycling and reusing paper increased from Fall 2012 to Winter 2014, while the proportion of respondents throwing paper in the garbage decreased. However, these changes may not have been caused by the information provided by our signs, as social norms as pro-environmental attitudes and behavior may have been increasing over time. Indeed, we see in the control group that recycling also increased over time, while reuse dropped and garbage did not change much.

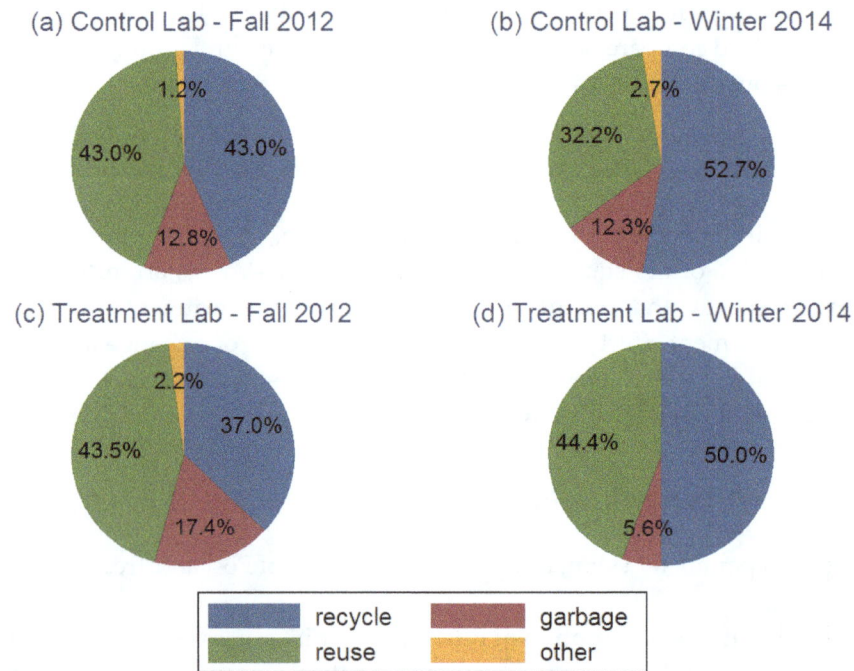

Figure 2. Behavior by lab, before and after treatment.

The difference-in-differences econometric model is able to provide an estimate of the treatment effect of our signage campaign on paper recycling and reuse behavior, while controlling for potential non-treatment changes over time. Table 3 shows the marginal effects of our explanatory variables on the probability of observing the corresponding behavior. Our interest is on the estimates of the treatment effect, i.e., the marginal effect of *Post*Treatment* on behavior.

We find that our campaign decreased the probability of throwing paper in the garbage by approximately 10% (fourth column of Table 3). This treatment effect is statistically significant at the 5% level. However, influences of the signage campaign on the individual behaviors of recycling and reuse, though positive, are not significant. We note, however, that all three behaviors are substitutes for one another, in that a given piece of paper may only end up in one place. Therefore, an alternative approach to examining behavior is to aggregate both conservation behaviors (i.e., paper recycling and reuse) in one model (third column of Table 3). In this model, we find that our signage campaign increased the probability of conservation behavior (i.e., recycling or reuse) by approximately 13%, with statistical significance at the 1% level. As a robustness check, we also estimate a linear probability model and find similar results. Specifically, for the Recycling or Reuse model, the estimated coefficient of *Post*Treatment* is equal to 0.150 (with $p < 0.05$); for the Garbage model, we estimate this coefficient to be -0.113 (with $p < 0.1$).

Table 3. Difference-in-Differences Probit Regressions.

Variable	Recycling	Reuse	Recycling or Reuse	Garbage
Post	0.097 (0.067)	−0.111 * (0.068)	−0.009 (0.040)	−0.004 (0.040)
Treatment	−0.063 (0.092)	0.004 (0.089)	−0.045 (0.055)	0.036 (0.052)
Post*Treatment	0.035 (0.111)	0.120 (0.109)	0.128 *** (0.048)	−0.097 ** (0.048)

Note: The table reports marginal effects. Regressions include a constant. Standard errors are in parenthesis. * $p < 0.1$, ** $p < 0.05$, *** $p < 0.01$.

4. Discussion

4.1. Interpretation of Results and Limitations

In cases where market instruments are impractical, information campaigns have been touted as a means to induce conservation-oriented behavior. Such situations may arise in office environments where it can be impractical to monitor individual behavior toward recycling and waste disposal. Moreover, whereas the use of economic instruments frequently requires significant changes in public policies, which may not be politically appealing, signage campaigns may proceed on an ad hoc basis, undertaken by individual firms without changes in public policy.

Few studies have shown links between such campaigns and changes in behaviors. Moreover, such investigations rarely have data that allow them to control for the host of intervening factors that could also contribute towards changed conservation-oriented behavior. In this study, we conduct a field experiment, with data collected temporally and cross-sectionally. However, there are limitations. The absence of information about characteristics of respondents prevents us from controlling for potential impacts of these differences. Nevertheless, given that both labs are available to all students and are centrally located with small distance from one another, the populations that these labs serve are likely the same. To the extent that we were able to capture a random sample of students in both labs, and that potential temporal changes equally influenced both labs, our difference-in-differences research design allows us to control for changes over time and isolate impacts of the campaign.

Our case study results show that, as a result of an information campaign, throwing paper in the garbage decreased by approximately 10%, and recycling and reusing increased by approximately 13%. Note that these behavioral changes occurred in an environment where almost 85% of users were already recycling or reusing, so there was not much room left for improvement. Recent research indicates that the changed behavior is lower when conservation is already high. For example, Ferraro and Price [4] show that the effect of social norms on reducing water consumption is 94% greater for high water users than that for low water users. In cases where recycling and reuse programs have not yet been implemented and paper conservation behavior is low, gains from information campaigns would likely be much higher than our results indicate.

Though such changes in two university computing labs may not seem substantial, implementing such programs on a wide scale could have large impacts on stocks and flows of paper. In 1980, The Economist suggested businesses should strive towards the "paperless office." But since then, global paper consumption has increased by 50% [24]. The average office worker uses approximately two pounds of paper per day, which corresponds to approximately 10,000 sheets of copy paper every year, or a total of 4 million tons of copy paper a year in the US [25]. Therefore, if information campaigns were able to reduce paper waste by approximately 13%, the result could increase paper conservation in American offices by 520,000 tons a year. Moreover, according to the American Forest & Paper Association, the US supply of printing-writing paper was 20.07 million tons in 2014, while, in the same year, 11.6 million tons of paper were recovered for recycling (a 58% recovery rate) [26]. If information campaigns were to increase this rate by 13%, printing-writing paper conservation would increase by 2.7 million tons a year.

There are, however, a number of limitations to our study that should be considered in interpreting our results. One potential issue is cross-contamination, which could occur if the treatment (i.e., the placement of signs) influences the control group. As both labs are located on the same campus, we cannot be certain that users interviewed in the control lab were not also users, and influenced by, the signs posted in the treatment lab. Note, however, that a contamination bias would work against the identification of a treatment effect. To the extent that contamination occurs, users of the control lab would increase conservation and the treatment effect found from using our difference-in-differences approach would be less. Accordingly, our estimates may represent lower bounds of the effect of the sign campaign on paper recycling and reuse behavior.

We also note that the findings of this study are drawn from survey data on stated behavior. There is a long standing literature that discusses the advantages and drawbacks of stated preference methods in comparison to revealed preference approaches. For example, Adamowics et al. [27] note that a revealed preference study could use, for instance, data on the weight of recycling bins before and after treatment. For our study, resources available did not allow us to perform a revealed preference investigation. Future research could examine possible differences between stated and revealed preferences estimates of the impact of information campaigns on conservation behavior.

Another potential area for further research arises from recent work suggesting that recycling can lead to increased consumption of waste papers [28,29]. To investigate this possibility, data on the whole quantity of paper used before and after the introduction of improved access and pro-environmental signage would be informative.

4.2. Pro-Environmental Information as Policy Instruments

Though pro-environmental information can be effective in promoting conservation behavior, a few notes of caution are in order. When economic instruments are used to alter behavior, a key part of the theory and process is the identification and quantification of a negative externality associated with the behavior in question. For example, paper reuse and recycling could be related to positive externalities associated with not harvesting trees, thereby maintaining sequestered carbon and/or providing for wildlife habitat. The values associated with the external costs or benefits are an integral part of setting the level of taxes or subsidies designed to alter behavior. Unlike economic approaches, which typically try to tie specific costs and benefits of changed behavior to specified levels of taxes and subsidies, approaches that provide information have no such basis. For example, in designing our campaign, we did not consider to what extent the conservation behavior that we sought to promote was beneficial.

So long as more of a desired behavior is better, then considering the level of desired behavior may not be a concern. This was the case at the University where our experiment was conducted. However, there may be increased costs and diminishing returns associated with pushing conservation behavior too far. For example, if a paper campaign attempts to divert the last 1% of paper going to garbage, then the paper going to reuse paper may have little white space remaining, and/or the paper going to recycling may be polluted with high levels of impurities (e.g., tapes, glues, and adhesives). Therefore, such approaches, if taken too far, could be putting paper through costly sorting processes, only to eventually end up in landfills. The basic problem here is that without accounting for benefits and costs (internal and external) we do not know what behavior is optimal. Information campaigns could, therefore, result in behavior that may seem morally correct, but could sacrifice efficiency and use more resources. Another potential weakness of trying to change behavior with information is that such approaches can only be effective if they appeal to and reinforce accepted social norms. Existing social norms are not always based on current knowledge and may be slow to evolve. Therefore, addressing arising problems where social norms have not yet been formed may have to be associated with intensive information campaigns if altered behavior is desired.

From a public policy perspective, trying to shape public opinion is not without its problems. When economists seek to design regulatory policies with economic instruments, they generally do so by accepting public values that exist, and attempt to internalize those public values that are not being correctly accounted for. The basis for this libertarian approach is that in a democratic society, values are generally not right or wrong, but the core upon which policy should be based. In contrast, information campaigns may be based on trying to change and mold public values, a concept that is typically more associated with authoritarian societies.

Nevertheless, information campaigns may provide a useful avenue for policy makers, either by themselves, or as complements to other conservation policies. For example, the southeast of Brazil is currently facing its worst drought in nearly a century. As a result, the largest reservoir system serving Brazil's largest city (Sao Paulo) is currently near depletion [30]. In this situation, conservation of water is clearly essential, but policy solutions have not been forthcoming, perhaps because of political procrastination in an election year (note that Brazil had a presidential election in 2014 [31]). With respect to household waste, Waite et al. [32] argue that England is unlikely to meet the EU target to reuse, recycle and compost 50% of its household waste by 2020. They identify the unpopularity of collection charges as one of the challenges towards meeting the goal. These are two cases where appeals to social norms with pro-environmental campaigns could potentially help, similar to Ferraro and Price's [4] findings that households that were exposed to social norms consumed approximately 4.8% less water than the average control group household.

5. Conclusions

In this study, we report results of a field experiment designed to evaluate the impact of a signage campaign on the use of paper in university computer labs. Baseline information on paper conservation behavior was collected from users of two similar computer labs at the University of Alberta, Canada. Next, a paper conservation campaign was implemented in one lab through the use of signage with social appeals. A subsequent survey collected post-campaign information on both control and treatment labs. These data allow us to obtain difference-in-differences estimates of the effectiveness of the norm-based strategies in promoting paper conservation behavior. We find that the implementation of signage with pro-environment appeals increases the probability of conservation behavior (i.e., recycling or reuse) by approximately 13%, despite already high levels of recycling and re-use.

Information campaigns may be an effective means of influencing behavior where market instruments are not viable, or as a complement to other policy alternatives. However, unless we have a clear, and agreed upon, idea of what that behavior should be, such approaches could lead us in directions that are not socially desirable. Fortunately, information campaigns may be implemented on small or large scales, by individual firms or as part of large government programs, so that the approach can be tailored and selectively applied to situations where desired conservation behavior has been identified and agreed upon.

Such was the case with our experiment where signage was added to recycling opportunities. Market instruments were not viable within the context of computing labs, and recycling and reuse were identified by the University as desirable outcomes. Within this context, the results of the information program were impressive in that, despite the fact recycling and re-use were already prevalent, significant increases were achieved for both recycling and re-use. The pervasive and enduring role of paper in offices suggests that these results are not only of statistical significance, but also, at large scales, significant in influencing substantial amounts of recycling and reuse.

Acknowledgments: Thanks to the University of Alberta Office of Sustainability for providing the pre-treatment data. Also, the University of Alberta Buildings and Grounds Services facilitated the implementation of signage, recycling bins, and paper reuse trays. Special thanks go to Jessie Kwasny who greatly contributed to the operationalization of the project.

Author Contributions: B.W. and M.L. conceived and designed the experiments; K.B., A.C., C.D., T.G., and Y.S. designed signage materials, implemented the experiment (e.g., signage, recycling bins, reuse trays) and collected the post-treatment data; All authors analyzed the data and wrote the paper.

References

1.   Allcott, H. Social norms and energy conservation. *J. Public Econ.* **2011**, *95*, 1082–1095. [CrossRef]
2.   Glaeser, E.L. The Supply of Environmentalism: Psychological Interventions and Economics. *Rev. Environ. Econ. Policy* **2014**, *8*, 228–229. [CrossRef]
3.   Scott, J.; Marshall, G. *Oxford Dictionary of Sociology*; OUP Oxford: Oxford, UK, 2005.
4.   Ferraro, P.J.; Price, M.K. Using nonpecuniary strategies to influence behavior: Evidence from a large-scale field experiment. *Rev. Econ. Stat.* **2013**, *95*, 64–73. [CrossRef]
5.   Schultz, P.W.; Nolan, J.M.; Cialdini, R.B.; Goldstein, N.J.; Griskevicius, V. The constructive, destructive, and reconstructive power of social norms. *Psychol. Sci.* **2007**, *18*, 429–434. [CrossRef] [PubMed]
6.   Kurz, T.; Donaghue, N.; Walker, I. Utilizing a social-ecological framework to promote water and energy conservation: A field experiment. *J. Appl. Soc. Psychol.* **2005**, *35*, 1281–1300. [CrossRef]
7.   Ferraro, P.J.; Miranda, J.J.; Price, M.K. The persistence of treatment effects with norm-based policy instruments: Evidence from a randomized environmental policy experiment. *Am. Econ. Rev. Pap. Proc.* **2011**, *101*, 318–322. [CrossRef]
8.   Richetin, J.; Perugini, M.; Mondini, D.; Hurling, R. Conserving water while washing hands: The immediate and durable impacts of descriptive norms. *Environ. Behav.* **2016**, *48*, 343–364. [CrossRef]
9.   Van der Linden, S. Exploring beliefs about bottled water and intentions to reduce consumption: The dual-effect of social norm activation and persuasive information. *Environ. Behav.* **2015**, *47*, 526–550. [CrossRef]
10.   Goldstein, N.J.; Cialdini, R.B.; Griskevicius, V. A room with a viewpoint: Using social norms to motivate environmental conservation in hotels. *J. Consum. Res.* **2008**, *35*, 472–482. [CrossRef]
11.   Nolan, J.M.; Schultz, P.W.; Cialdini, R.B.; Goldstein, N.J.; Griskevicius, V. Normative social influence is underdetected. *Personal. Soc. Psychol. Bull.* **2008**, *34*, 913–923. [CrossRef] [PubMed]
12.   Kormos, C.; Gifford, R.; Brown, E. The influence of descriptive norm information on sustainable transportation behavior: A field experiment. *Environ. Behav.* **2015**, *47*, 479–501. [CrossRef]
13.   Margai, F.L. Analyzing Changes in Waste Reduction Behavior in a Low-Income Urban Community Following a Public Outreach Program. *Environ. Behav.* **1997**, *29*, 769–792. [CrossRef]
14.   Martinez, M.; Scicchitano, M. Who listens to trash talk? Education and public media effects on recycling behavior. *Soc. Sci. Q.* **1998**, *79*, 287–300.
15.   Vining, J.; Ebreo, A. Predicting recycling behavior from global and specific environmental attitudes and changes in recycling opportunities. *J. Appl. Soc. Psychol.* **1992**, *22*, 1580–1607. [CrossRef]
16.   Schwab, N.; Harton, H.C.; Cullum, J.G. The Effects of Emergent Norms and Attitudes on Recycling Behavior. *Environ. Behav.* **2014**, *46*, 403–422. [CrossRef]
17.   Abbot, A.; Nandeibam, S.; O'Shea, L. Recycling: Social norms and warm-glow revisited. *Ecol. Econ.* **2013**, *90*, 10–18. [CrossRef]
18.   Thomas, C.; Sharp, V. Understanding the normalisation of recycling behaviour and its implications for other pro-environmental behaviours: A review of social norms and recycling. *Resour. Conserv. Recycl.* **2013**, *79*, 11–20. [CrossRef]
19.   Nolan, J.M. Using Jackson's return potential model to explore the normativeness of recycling. *Environ. Behav.* **2015**, *47*, 835–855. [CrossRef]
20.   Hornik, J.; Cherian, J.; Madansky, M.; Narayana, C. Determinants of recycling behavior: A synthesis of research results. *J. Socio-Econ.* **1995**, *24*, 105–127. [CrossRef]
21.   Iyer, E.; Kashyap, R.K. Consumer recycling: Role of incentives, information, and social class. *J. Consum. Behav.* **2007**, *6*, 32–47. [CrossRef]

22. Austin, J.; Hatfield, D.B.; Grindle, A.C.; Bailey, J.S. Increasing recycling in office environments: The effects of specific, informative cues. *J. Appl. Behav. Anal.* **1993**, *26*, 247–253. [CrossRef] [PubMed]

23. Binder, K.J. The Effects of Replacing Dispersed Trash and Recycling Bins with Integrated Waste Receptacles on the Accuracy of Waste Sorting in an Academic Building. Master's Thesis, Western Michigan University, Kalamazoo, MI, USA, 2012.

24. The Economist. Available online: http://www.economist.com/blogs/graphicdetail/2012/04/daily-chart-0 (accessed on 11 January 2017).

25. Mashable. Available online: http://mashable.com/2014/04/22/earth-day-paper-infographic (accessed on 11 January 2017).

26. Paper Recycles. Available online: http://www.paperrecycles.org/statistics/recovery-of-printing-writing-papers (accessed 11 January 2017).

27. Adamowicz, W.; Louviere, J.; Williams, M. Combining Revealed and Stated Preference Methods for Valuing Environmental Amenities. *J. Environ. Econ. Manag.* **1994**, *26*, 271–292. [CrossRef]

28. Catlin, J.R.; Wang, Y. Recycling gone bad: When the option to recycle increases resource consumption. *J. Consum. Psychol.* **2013**, *23*, 122–127. [CrossRef]

29. Cecere, G.; Mancinelli, S.; Mazzanti, M. Waste prevention and social preferences: The role of intrinsic and extrinsic motivations. *Ecol. Econ.* **2014**, *107*, 163–176. [CrossRef]

30. The New York Times. Available online: https://www.nytimes.com/2015/02/17/world/americas/drought-pushes-sao-paulo-brazil-toward-water-crisis.html (accessed on 11 January 2017).

31. Reuters. Available online: http://www.reuters.com/article/2014/10/31/us-brazil-water-idUSKBN0IK1RL20141031 (accessed on 11 January 2017).

32. Waite, S.; Cox, P.; Tudor, T. Strategies for local authorities to achieve the EU 2020 50% recycling, reuse and composting target: A case study of England. *Resour. Conserv. Recycl.* **2015**, *105*, 18–28. [CrossRef]

13

Energy Payback Time of a Solar Photovoltaic Powered Waste Plastic Recyclebot System

Shan Zhong [1], Pratiksha Rakhe [1] and Joshua M. Pearce [1,2,*]

[1] Department of Materials Science & Engineering, Michigan Technological University, Houghton, MI 49931, USA; szhong@mtu.edu (S.Z.); prakhe@mtu.edu (P.R.)

[2] Department of Electrical & Computer Engineering, Michigan Technological University, Houghton, MI 49931, USA

* Correspondence: pearce@mtu.edu

Academic Editor: Michele Rosano

Abstract: The growth of both plastic consumption and prosumer 3-D printing are driving an interest in producing 3-D printer filaments from waste plastic. This study quantifies the embodied energy of a vertical DC solar photovoltaic (PV) powered recyclebot based on life cycle energy analysis and compares it to horizontal AC recyclebots, conventional recycling, and the production of a virgin 3-D printer filament. The energy payback time (EPBT) is calculated using the embodied energy of the materials making up the recyclebot itself and is found to be about five days for the extrusion of a poly lactic acid (PLA) filament or 2.5 days for the extrusion of an acrylonitrile butadiene styrene (ABS) filament. A mono-crystalline silicon solar PV system is about 2.6 years alone. However, this can be reduced by over 96% if the solar PV system powers the recyclebot to produce a PLA filament from waste plastic (EPBT is only 0.10 year or about a month). Likewise, if an ABS filament is produced from a recyclebot powered by the solar PV system, the energy saved is 90.6–99.9 MJ/kg and 26.33–29.43 kg of the ABS filament needs to be produced in about half a month for the system to pay for itself. The results clearly show that the solar PV system powered recyclebot is already an excellent way to save energy for sustainable development.

Keywords: energy payback time; distributed manufacturing; life cycle analysis; photovoltaic; recycling; solar energy; recyclebot; 3-D printing; polymer filament; EPBT

---

1. Introduction

Global plastic production is growing by 3.86% per year and is expected to increase to 850 million tons per year by 2050 [1,2]. This growth aggravates the challenges of waste plastics disposal, especially in remote areas [3]. Landfill and incineration methods induce several negative environmental issues [4–6], and this linear model of resource consumption with a "take-make-dispose" pattern has increasingly significant economic limits [7]. To mitigate the contradiction between the rapid economic growth and the shortage of virgin materials and energy, the circular economy was first proposed in 1998 to build up the circular flow of materials and the use of resources and energy through multiple phases [8,9]. Following the goals of a circular economy, recycling is becoming the mainstream method to dispose of waste plastics [10]. The conventional recycling method is to collect and transport waste plastic to a collection center and reclamation facility for separation and recycling [11]. This method usually consumes large amounts of energy for transportation [12], and needs considerable labor to separate the waste plastics [13]. In developing regions, this labor is provided by waste pickers, which collect post-consumer plastic in landfills [14].

Compared to conventional recycling methods, the distributed recycling of plastic has the potential to conserve energy. For example, plastic air-filled bottles have been used as building units to replace traditional concrete blocks and have demonstrated superior thermal insulation [15]. This conserves the energy used for the resultant building HVAC (heating, ventilation, and air conditioning), as well as the embodied energy of concrete and conventional recycling of waste plastic. Another example uses plastic containers converted into bio-gas digesters, which has demonstrated higher gas yields in black-coated plastic containers than other materials [16]. Those studies indicate that distributed plastic recycling has the potential to conserve energy for sustainable development. In this study, another distributed recycling method using a recyclebot is investigated in detail.

The recyclebot, an open source waste plastic extruder, offers a new approach to plastics recycling, which can be distributed and operated as a small business or even at home [17]. The recyclebot contains a feeding zone, heating pipe, and extrusion section. Plastic melts in the heating pipe and is extruded through a nozzle to form a filament for 3-D printing [17]. This recycling method is not difficult to operate and is supported for many thermo-plastic products, which are identified with recycling codes [18]. The system is automated, although the plastic containers must be cleaned and shredded before processing in the recyclebot. Using a recyclebot in the location that plastic waste is generated not only saves the energy required for transportation [19,20], but can also increase personal income when the filament is sold [14]. As 3-D printing technology is developing to be of wide applicability for distributed manufacturing throughout the world [21,22], an expensive commercial filament is one of the remaining impediments to the extended popularity of 3-D printing. The application of a recyclebot, which can produce a filament for about 10 cents per kg of electricity [17], can further improve the economics of 3-D printing and extend distributed manufacturing [23].

The conventional recyclebot is powered by grid-provided electricity (referred to here as an AC recyclebot for clarity). For an AC recyclebot, previous studies have shown that the embodied energy for shredding waste plastic is trivial, so the energy for producing a filament is equivalent to the electricity consumption of the recyclebot alone [24,25]. The emissions from recycling waste plastic into a 3-D printer filament are thus dependent on the greenhouse gas emissions of the electric grid that varies widely, from 0.00019 to 1.94 kg/kWh [26].

As a potential source of wide-scale renewable energy, solar electricity generation is increasing in popularity globally because of technical advances and reductions in costs [27]. Solar photovoltaic (PV) technology has been found to be particularly appropriate in the developing world [28–33]. Recent developments in the RepRap 3-D printer community [34] to make PV-powered 3-D printer designs [35,36] can be directly transferred to direct current (DC)-based recyclebot technology [37]. These solar-powered recyclebots would have a double effect on energy and emissions savings. First, they offset grid electricity to make commercial 3-D printer filaments, and then again, by reducing the energy used and emissions with distributed recycling itself.

The solar-powered recyclebot has not been quantified previously and to do this, a life cycle energy analysis is needed. Energy analysis is the process of determining the energy required directly and indirectly to allow a system to produce a specific good or service. Energy payback time (EPBT) is one metric adopted by several analysts in characterizing the energy sustainability of various technologies [38].

This paper quantifies the embodied energy of a vertical DC solar-powered recyclebot based on life cycle energy analysis and compares it to a horizontal AC recyclebot, conventional recycling, and the production of virgin 3-D printer filaments. The EPBT is calculated using the embodied energy of the materials making up the recyclebot itself and the calculations are detailed in the Methods section below. The mass of a 3-D printer filament that the recyclebot must produce to offset the embodied energy for creating the recyclebot device is calculated. In addition, after combining a recyclebot and solar PV system, the amount of filament needed to pay for the whole system is also calculated, as well as the energy payback time. These results are compared to previous studies that only investigated the energy payback time for PV alone and are discussed in the context of distributed recycling, energy conservation, and greenhouse gas (GHG) emissions mitigation.

2. Results

2.1. Embodied Energy of Recyclebot

The recyclebot can be separated into five key components by their function: barrel, frame, motor, electrical components and wiring, and feeder attachment and hopper. A detailed breakdown of the embodied energy of the five parts of the recyclebot is presented in Tables 1–5, respectively.

Table 1. Embodied energy for the barrel assembly of the recyclebot.

Part	Quantity	Material	Mass (kg)	Embodied Energy	
				Min (MJ)	Max (MJ)
Heating Tube	1	steel	0.30	9.240	10.170
Feed Tube	1	steel	0.28	8.624	9.492
Feed Screw	1	galvanized steel	0.15	5.715	6.300
Floor Flange	3	steel	0.75	23.100	25.425
Brass Nozzle	1	brass	0.15	8.610	9.495
Rod	4	galvanized steel	0.148	5.639	6.216
$1\frac{1}{2}''$ Bolt	2	galvanized steel	0.036	1.372	1.512
$2\frac{1}{4}''$ Bolt	2	galvanized steel	0.044	1.676	1.848
Nut	16	galvanized steel	0.086	3.277	3.612
Washer	64	stainless steel	0.128	4.877	5.376
Kapton Tape	1	aromatic polyimide	0.008	1.408	1.552
Nichrome Wire	17 ft	nickel chromium	0.004	0.756	0.832
Bearing	1	chrome steel	0.13	6.552	7.228
Bearing House	1	PLA	0.03	1.476	1.626
Brass Spacer	1	brass	0.12	6.888	7.596
Total				89.209	98.28

Table 2. Embodied energy for the frame assembly of the recyclebot.

Part	Quantity	Material	Mass (kg)	Embodied Energy	
				Min (MJ)	Max (MJ)
Strut Channel	2	steel	3.2	98.56	108.480
Barrel Bracket	2	steel	0.8	24.64	27.120
Motor Mount	2	steel	0.8	24.64	27.120
Strut Channel T-nut	6	galvanized steel	0.066	2.515	2.772
Socket Head Bolt	6	galvanized steel	0.132	5.029	5.544
M6 Bolt	2	galvanized steel	0.044	1.676	1.848
Split Washer	2	stainless steel	0.004	0.202	0.222
Flat Washer	2	stainless steel	0.004	0.202	0.222
Deep Well Socket	1	galvanized steel	0.1	3.810	4.200
Total				161.273	177.529

Table 3. Embodied energy for the motor assembly of the recyclebot.

Part	Quantity	Material	Mass (kg)	Embodied Energy	
				Min (MJ)	Max (MJ)
Rotor	2	galvanized steel	0.56	21.336	23.52
Ball Bearing	2	chrome steel	0.26	13.104	14.456
Shaft	1	galvanized steel	0.17	6.477	7.14
Stator	1	steel	0.15	4.62	5.085
		copper	0.08	4.488	4.952
Winding	1	copper	0.12	6.732	7.428
Cable	1	copper	0.016	0.898	0.990
		PVC	0.012	0.727	0.802
Total				58.382	64.373

Table 4. Embodied energy for the electrical components and wiring part of the recyclebot.

Part	Quantity	Material	Mass (g)	Embodied Energy	
				Min (MJ)	Max (MJ)
Speed Controller	1	silicon	0.2	0.252	0.448
		copper	29.5	1.655	1.826
		lead	6.0	0.221	0.244
		steel	150	4.62	5.085
		PVC	1.0	0.061	0.067
		epoxy	2.5	0.315	0.3475
		plastic	8.5	0.766	0.858
Temperature Controller	1	silicon	0.15	0.189	0.336
		copper	29.0	1.627	1.795
		lead	12.5	0.461	0.508
		platinum	0.005	1.37	1.51
		ferrites	2.5	0.039	0.043
		Lead-antimony	0.1	0.006	0.007
		PC	66.0	7.194	7.92
		epoxy	7.5	0.945	1.042
Solid State Relay Kit	1	silicon	0.06	0.076	0.134
		copper	17.0	0.9537	1.0523
		lead	6.0	0.221	0.244
		ferrites	2.5	0.039	0.043
		PC	12	1.308	1.44
		epoxy	2.5	0.315	0.348
Terminal Strip	1	galvanized steel	10	0.381	0.42
		copper	6.0	0.337	0.371
		PC	12.0	1.308	1.44
K-Type Thermocouple	1	copper	45.0	2.524	2.786
		nickel	40.0	6.36	7.00
Power Cord and Plug	1	copper	32.0	1.795	1.981
		PVC	36.0	2.182	2.405
Solderless Connector	2	copper	6.0	0.337	0.371
		PVC	4.0	0.242	0.267
Insulated Copper Wire	10 ft	copper	16.0	0.898	0.990
		PVC	3.0	0.182	0.200
Kapton Tape	1	aromatic polyimide	8.0	1.408	1.552
Hose Clamp	1	stainless steel	22.7	1.144	1.262
Total				41.731	46.343

Table 5. Embodied energy for the feeder attachment and hopper part of the recyclebot.

Part	Quantity	Material	Mass (kg)	Embodied Energy	
				Min (MJ)	Max (MJ)
Feeder Attachment	1	PLA	0.045	2.214	2.439
Hopper	1	HDPE	0.015	1.124	1.238
Total				3.338	3.676

2.2. Embodied Energy of Solar Photovoltaic System

The energy needed for the small PV system to power the recyclebot can be divided into two parts: the fabrication of the PV module and its balance of system. The procedure for the fabrication of a solar PV module can be described in brief as the purification and growing of crystal silicon, the cell fabrication from a silicon wafer, and module assembly. The balance of system (BOS) for a PV system includes the foundation, support structure, battery, inverter, electronic components, installation, wiring, and cables [39]. In this study, the foundation and support structure is not necessary because it is just a temporarily positioned device and PV modules can be propped against a wall or a rock in the field. Detailed information about the embodied energy of the single crystal solar PV system needed to power the recyclebot is shown in Table 6. It should be noted that the embodied energies in Table 6 all are scaled to the PV sizes used to power the recyclebot. The area of the PV modules used in this project is 0.5226 m^2, so the embodied energy of this PV system is about 2276.44 MJ or 632.34 kWh.

Table 6. Embodied energy of the single-crystal solar PV system [39].

Process & Item	Silicon Purification and Processing	Cell Fabrication	Module Assembly	Balance of System			Total
				Battery	Inverter	Electronic Components	
Embodied Energy (MJ/m^2)	2397.6	432	684	165.6	118.8	558	4356

2.3. PLA and ABS Filament

In PLA filament production, the DC recyclebot needed 17 min and 64.8 kJ for initial heating. During the extrusion process, the DC recyclebot consumed 0.01 kWh to extrude 22.6 g of PLA filament. The energy for initial heating is inconsequential compared to the energy for whole day extrusion, so the average energy used for PLA filament production is 1.59 MJ/kg. The extrusion rate is not a constant, but the average extrusion rate is 0.19 kg/h. Thus, it takes a little more than 5 h to generate a kg of PLA filament with a DC vertical recyclebot.

In ABS filament production, 8 min and 36.0 kJ were needed for initial heating by the AC vertical recyclebot. The average energy used for ABS filament production is 1.24 MJ/kg, and its average extrusion rate is 0.22 kg/h.

The process parameters and results comparing the processing of PLA pellets and ABS recycled pieces are summarized in Table 7.

Table 7. Recyclebot process parameters and results comparing the filament manufacturing of PLA pellets and ABS recycled shards.

Material/Process Parameter	PLA (Virgin Pellets)	ABS (Recycled Shards)
Auger Rotation Speed (rpm)	15	15
Recyclebot Power	DC	AC
Extrusion Set Temperature (°C)	155	158
Initial Heating Phase (min)	17	8
Extrusion Rate (kg/h)	0.19	0.22
Energy used (MJ/kg)	1.59	1.24

3. Discussion

Based on the data in Tables 1–5, the embodied energy of the recyclebot is 353.933–390.201 MJ, which is equivalent to the energy for producing two coffee makers [40]. Figure 1 is a pie chart showing the percentage of minimum embodied energy of each core component in a recyclebot. There is not much difference between the minimum embodied energy percentage and the maximum embodied energy percentage: the maximum drops from 46 to 45% for the frame, while the motor increases to 17%. In all cases, the frame part consumes nearly half of the total embodied energy. The strut channel of the frame has 98.56–108.48 MJ embodied energy, which is equivalent to the energy of a barrel part or double the energy of the electrical components and the wiring part. The frame contains many heavy components made of metal, which is the cause of the high embodied energy. Thus, to improve the sustainability of the device, this indicates that the design effort should focus on minimizing the use of metal in the frame in the future. In the electrical components and wiring parts, there are several materials which contain very high embodied energy, such as electrical grade silicon in the speed controller and platinum in the temperature controller. These materials, however, have low masses in the components so they do not contribute much to the total embodied energy. There are some small components created by a 3-D printer, such as the bearing house and the feeder attachment. When the 3-D printer is powered by the solar PV system, the embodied energy of these printed components can be reduced further [29].

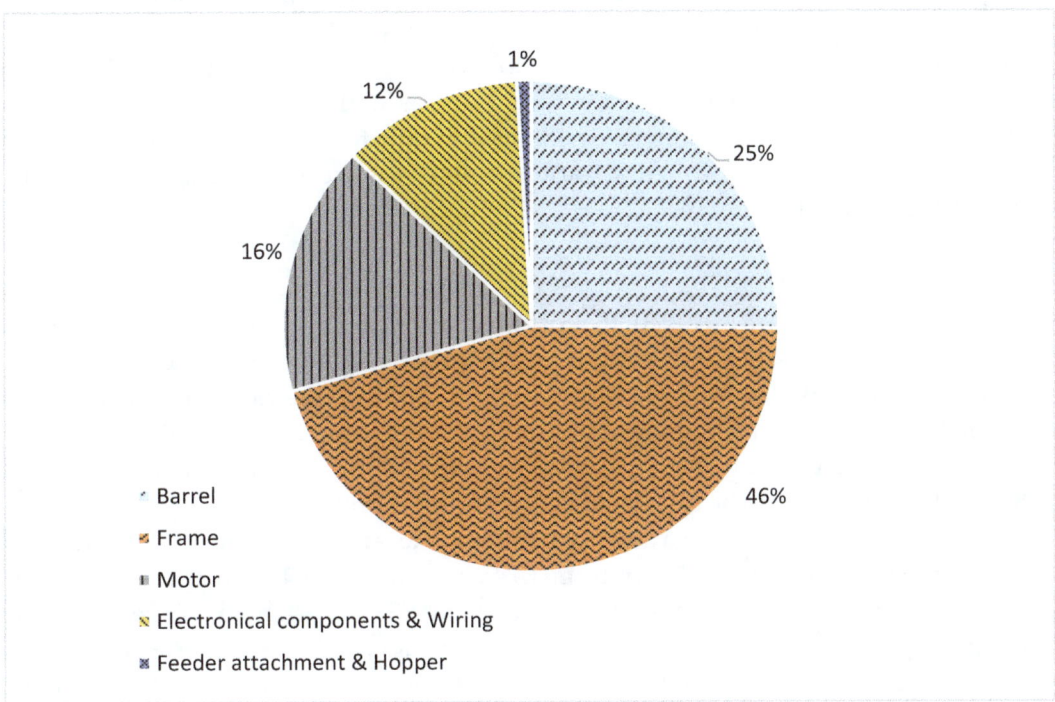

Figure 1. The percentage of the minimum embodied energy of each major component class in a recyclebot.

As can be seen by comparing Table 6 with Tables 1–5, the energy needed to make a solar PV system is 2276 MJ, which is 484–523% more than that of a recyclebot. The purification and growing crystal silicon process consumes half of the total energy. In the purification process, silicon dioxide is reduced to silicon with carbon and purified in the furnace repeatedly to metallurgical grade silicon, and then further purified to obtain solar-grade silicon, which needs a lot of energy [41]. An approach to reduce this embodied energy of the PV component in the future would be to use thin film PV [42]. In the balance of systems, the electrical components have the largest embodied energy in this case. If the solar powered recyclebot needs to be designed as a permanent device, the foundation and support structures are required. Then, the support structure will consume the largest energy in the balance of systems, which is 1800 MJ/m^2 in the open field and 720 MJ/m^2 on the roof top [39,43]. However, it should be noted that there are low-mass racking systems that may be appropriate, which will reduce these values [44].

3.1. Energy Payback Time of Recyclebot

The EPBT of a recyclebot varies with the material of the filament. In this project, it takes 1.59 MJ for a DC recyclebot to extrude 1 kg of the PLA filament, while 1 kg of commercial filament made from virgin PLA consumes 49.2–54.2 MJ. Thus, the PLA filament produced by a recyclebot from waste plastic can save 47.61–52.61 MJ/kg. Compared to the embodied energy of a recyclebot, which is 353.94–390.20 MJ, producing 6.73–8.20 kg of the PLA filament from waste plastic can pay for a recyclebot in terms of energy. Given the extrusion rate of the DC recyclebot, this could be accomplished conservatively in one week.

From this study on an AC recyclebot, the average energy used for ABS filament production from the recyclebot is 1.24 MJ/kg. Compared to a commercial filament made from virgin ABS, which needs 90.6–99.9 MJ/kg, the filament produced by the recyclebot from consumed ABS saves 89.36–98.66 MJ/kg of energy. Therefore, the recyclebot needs to produce 3.59–4.37 kg of ABS filament, and the energy saved from it will be equal to the energy for creating the recyclebot.

During the vertical DC recyclebot extrusion process, the average extrusion rate of the PLA filament was found to be 0.19 kg/h, while that of the ABS filament was 0.22 kg/h on a vertical AC-based recyclebot. It is found that the recyclebot needs to work 35.42–43.16 h with the PLA filament or 16.32–19.86 h with the ABS filament to pay for itself. Here, assuming that the recyclebot works eight hours per day, the energy payback time can be obtained in only five days based on a PLA filament or about 2.5 days based on an ABS filament. Clearly, the potential to conserve energy with distributed waste plastic recycling is substantial.

3.2. Cost of Recyclebot-Made Filament

If labor and capital costs are excluded, the cost to produce a recycled waste plastic filament can be determined by the energy use of the recyclebot. In a single recyclebot system, the energy used in the extrusion process was provided from the electricity grid. The average electricity price in the U.S. is $0.12/kWh or 3 cents/MJ. The energy consumption required to produce 1 kg of PLA and 1kg of ABS is 1.59 MJ and 1.24 MJ, respectively. With the average electricity price, the estimated cost for producing 1 kg of PLA and 1 kg of ABS can be calculated as 5 cents and 4 cents, respectively.

However, the electricity price varies from different locations. From the EIA database [45], the electricity prices in January 2017 vary from 1.74 cents/MJ to 7.04 cents/MJ in different states. The estimated costs for producing 1 kg of PLA and 1 kg of ABS can vary from 2.77 cents to 11.19 cents and 2.16 cents to 8.73 cents, respectively. Even though the device is operated in Hawaii, which has the most expensive electricity price in the U.S., the cost to produce 1 kg of PLA filament and 1 kg of ABS filament are still very inexpensive, at just 11.19 cents and 8.73 cents. The price for 1 kg of PLA filament on Amazon ranges from $14.95 to $89.99 and the price for 1 kg of ABS filament is from $13.99 and $58.51 [46–49]. Compared to the commercial filament, the recycled filament produced by the recyclebot can save significant amounts of money. Figure 2 is a comparison of the general cost ranges for producing recycled filaments and buying commercial filaments in the market, and it can be seen that the cost for the recycled filament is negligible.

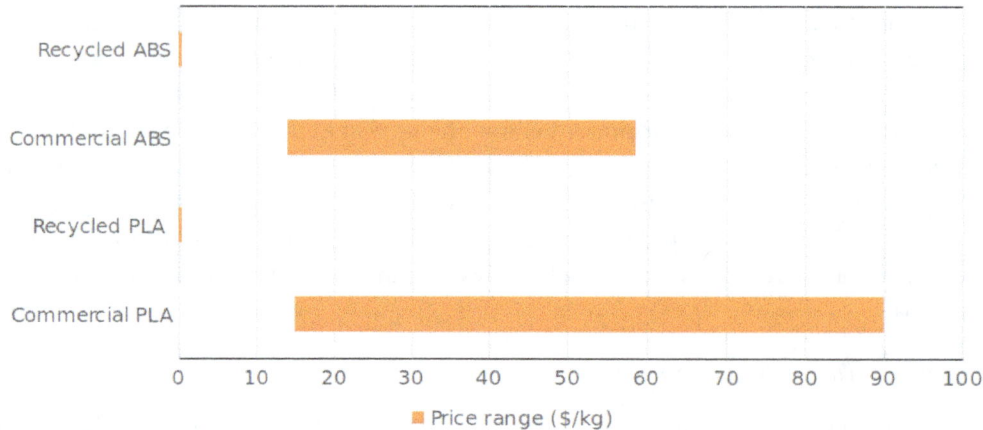

Figure 2. The general cost ranges for the recycled filament and commercial filament.

3.3. Energy Payback Time of Solar PV System

The EPBT of the solar PV system depends on the materials of the module, the balance of systems, and the geographic location [50]. Among them, the type of material determines the energy conversion efficiency and the geographical location determines the solar flux, and the energy generated by the solar PV system can be obtained by the product of solar flux and energy conversion efficiency. The EPBT can be calculated by dividing the embodied energy by the annual energy generated by the system [50]. In order to compare the effect of the solar powered recyclebot on saving energy, the general case of a monocrystalline silicon PV system is chosen for comparison, which has an energy conversion efficiency of 16.1% [51] and is located in a place which has a global average solar flux of 8 $kWh/m^2/day$. Then, the annual average insolation is 2920 $kWh/m^2/year$ and the energy generated by PV system in this project is 245.68 kWh/year. Thus, the EPBT of the solar PV system used in this project, which has an embodied energy of 632.34 kWh, is about 2.57 years. The fact that PV systems are an extremely favorable energy and emissions performer is well established. However, the results here show that these values can be improved further when PV power is used for recycling.

3.4. Energy Payback Time of Solar PV-Powered Recyclebot System

Due to two monocrystalline solar panels being used, the embodied energy of the solar PV system is 2276.44 MJ and the embodied energy of the whole system is 2630.37–2666.64 MJ (i.e., the whole system is the sum of the recyclebot and the solar PV system). When the recyclebot is powered by the solar PV system, the electricity used in producing the filament from waste material can also be saved, which means that the energy saved is equal to the embodied energy of the commercial filament. Thus, 48.53–54.2 kg of PLA filament needs to be produced from consumed plastic to pay for the whole system in terms of energy. This means that the energy payback time of the whole system is 255.42–285.26 h in terms of PLA, which is about one month, based on the assumption that the whole system works eight h per day. When compared to the EPBT of a single crystal solar PV system, whose EPBT is about 2.57 years, the EPBT of the solar PV system combined with the recyclebot can be decreased by at least 96.20% if the PLA filament is produced. Thus, the effect on saving energy of the whole system is clear. Likewise, if the ABS filament is produced from a recyclebot powered by the solar PV system, the energy saved is 90.6–99.9 MJ/kg. A total of 26.33–29.43 kg of ABS filament needs to be produced from waste plastic to pay for the whole system. As the average extrusion rate of an ABS filament is 0.22 kg/h, the EPBT of the whole system based on the ABS filament is 119.68–133.77 h, which is about 0.04 years or about half a month.

The results clearly show that the solar PV system powered recyclebot is already an excellent way to save energy for sustainable development as the EPBT is even greater than the substantial benefits from well-known sustainable technologies such as PV alone. Among PV modules, thin film modules have the lowest EPBT, which is 0.5 years, while considering the balance of system, its EPBT is 1.5 years [42]. However, the EPBT of the single-crystal silicon PV system can decrease from 2.57 years to 0.09 years when it is combined with a recyclebot, which produces the PLA filament. Thus, the single crystal silicon PV system combined with recyclebots, has about one thirtieth of the EPBT of thin film PV systems alone. However, the performance can be improved with the use of solar powered recyclebots consisting of a low-embodied energy thin film PV.

Furthermore, the EPBT of the solar PV system powered recyclebot can be further reduced with device improvements itself. In a recyclebot, the embodied energy of the strut channel is a substantial fraction of the total energy of the recyclebot. If the strut channel uses other materials which contain less embodied energy instead of steel, the total energy of the system will decrease. Properly designed platic struts could offer a good choice because of the strength and impact resistance. The size of the individual components would need to be augmented to withstand the load when materials with a greater strength are used, similar to previous work on the brackets for PV modules [24,43]. The plastic strut channel, which is a little larger in size in this project, could be strong enough to hold the recyclebot and motor. Future work is needed to optimize such a design. This would decrease the embodied energy and accelerate the EPBT, as for example, ABS has 95,300 MJ/m³ of embodied energy, while steel has 285,400.5 MJ/m³ of embodied energy. In addition to the strut channel, the barrel bracket and motor mount can also be made from ABS, which can also reduce the total embodied energy. Future recyclebots should strive to meet the RepRap model and be primarily composed from 3-D printed parts.

3.5. Economic Payback Time of Solar Powered Recyclebot

Compared to the individual recyclebot, the energy for producing filaments from waste plastics by the PV powered recyclebot can also be saved from the electricity grid, so the cost from the electricity grid can be saved by the whole system, which means that the total cost saving by the whole system is equal to the price of the commercial filament. The total cost for building the entire system is around $1000, which is the cost sum of a recyclebot and a solar PV system. The normal price of 1 kg of PLA or ABS filament is $20, so 50 kg of filament needs to be produced by the whole system to pay for itself in terms of monetary cost. With the PLA and ABS filament extrusion rates, 263.16 h and 227.27 h are needed to produce 50 kg of PLA filament and 50 kg of ABS filament, respectively. Therefore, the monetary payback time of the whole system is 263.16 h (32.9 days) with PLA filament production, or 227.27 h (28.4 days) with ABS filament production.

3.6. Applications for Developing Countries

For those developing countries whose energy access promotions have not met the requirement of sustainable development [52], the results of this study indicate that solar-powered recycling may be beneficial from an energy perspective in the future. In addition, solar-powered recycling can help developing countries to reduce carbon emissions, which is necessary because the carbon emissions from developing countries are greater than from developed countries with the same unit of value-added [53]. Finally, it is well established that access to modern energy can increase the income for families in developing countries [54,55]. This study has shown one application of modern energy using a solar-powered recyclebot that has the potential to profitably produce filaments, while avoiding the consumption of raw materials and grid electricity. This filament can be sold or higher value items can be printed to expand an entrepreneur's or community's income.

In addition, when a 3-D printer creates products, further energy is needed to melt filaments and form products. If a solar PV system not only powers the recyclebot, but also powers the 3-D printer to manufacture products, further energy and emissions are saved. Moreover, if some small components used in a recyclebot, such as a barrel bracket, motor mount, bearing house, and feeder attachment, are produced by a 3-D printer powered by solar PV system, the embodied energy of the recyclebot can be further reduced. The solar PV system powered recyclebot and 3-D printer systems are an excellent method to manufacture products from waste plastics by consumers anywhere in the world. This method can turn waste plastics into useful high-value products, so it effectively decreases the expenditure from the cost of products and transportation. In conclusion, this method not only saves energy and money, but also reduces the emission of greenhouse gas, which is in accordance with sustainable development.

4. Materials and Methods

4.1. Energy Payback Time

The goal of this study is to investigate the energy payback time of a solar PV powered vertical DC recyclebot, which consists of the PV module, a small battery system, and all of the parts for the thermo-mechanical system of the recyclebot. The EPBT can be mathematically determined by first determining the energy saved by a system,

$$E_S = E_F - E_G \tag{1}$$

where E_S (MJ) is the energy saved by the system, E_F (MJ) is the energy needed to form the system, and E_G (MJ) is the energy generated or conserved by the system. Equation (1) is the generalized form of the equation.

The input energy E_F in the system can be classified as the respective embodied energies of each material present in the whole device. The embodied energy is the amount of energy required to produce the material in its product form [39]. Thus, here, E_F is different, depending on whether the system scope is limited to only the recyclebot (E_R) or the entire system with a solar powered recyclebot (E_{whole}). In the case of a PV system alone, E_G is the energy generated by the PV, and in the case of a recyclebot, it is the difference between the energy used to make a commercial filament and the recyclebot filament, which is $E_V - E_W$, shown in the denominator of Equation (2) (note that these two terms are in terms of energy per unit mass).

The EPBT for the recyclebot alone will be calculated as:

$$EPBT_{recyclebot} = \frac{E_R}{(E_V - E_W) \times v} \text{ (Unit : h)} \tag{2}$$

where E_R (MJ) is the embodied energy of the recyclebot, E_V (MJ/kg) is the energy for producing the filament from virgin material, E_W (MJ/kg) is the energy for producing the filament by the recyclebot from waste plastic, and v (kg/h) is the extrusion rate (measured and discussed in Section 2.3). The energy for producing the recycled filament (E_R) and the filament extrusion rate (v) is obtained from the filament production experiment, and the detailed process is introduced below. Specifically, the E_R values for Equation (2) are taken from Tables 1–5 for the embodied energy of the recyclebot. The energy for producing the commercial filament (E_V) is estimated with the plastic embodied energy, which is searched from the CES EduPack, which provides a comprehensive database of materials and process information, powerful materials software tools, and a range of supporting resources [56,57]. The CES EduPack database is populated with data from the peer reviewed literature. Finally, E_W is a measured experimental value.

The EPBT of the solar PV system powered recyclebot can be calculated as,

$$EPBT_{whole} = \frac{E_{whole}}{E_V \times v} \text{ (Unit : h)} \tag{3}$$

where E_{whole} (MJ) is the embodied energy of the whole system, which is the sum of the embodied energies of the recyclebot (from Equation (2)) and the solar PV system (from Table 6).

It should be noted that the embodied energy of the recyclebot (E_R) is calculated by the sum of the embodied energies of all components, but the manual energy (e.g., human labor) is not included. The embodied energy of the solar PV system is estimated by the PV module area, and the energy consumption to produce the PV system in unit size is also determined from prior LCA studies. A previous study investigating a horizontal AC-powered recyclebot found that the energy required for producing a filament is low (8.74 MJ to produce 1 kg HDPE filament) compared to the production from virgin resin (76.7 MJ/kg) and waste HDPE processed in a conventional recycling center (48.9 MJ/kg) [25]. It should be noted that the energy required during the extrusion differs with thermoplastic materials and insulation used on a particular recyclebot machine.

The solar modules use solar insolation for the generation of electricity, which is stored in batteries. The power from the battery is then used to power the recyclebot for the extrusion of the filament. The total energy used in the whole process from solar panels to extrusion can be calculated with the consumed energy during extrusion. These calculated energies can now be used in obtaining the EPBT of the whole system.

It should be pointed out here that this study is not a full cradle to grave LCA. The recyclebot technology is new and not yet widespread, so it is difficult to account for the wide divergence in the expected environmental impacts related to the end-of-life phases of the product. However, this study does go beyond the common cradle to gate analysis as the energy related to the use of the recyclebot over the lifetime is included in order to obtain a true EPBT.

4.2. Embodied Energy

4.2.1. Solar Photovoltaic System

Solar PV is a clean, sustainable, renewable energy conversion technology that can help meet the energy demands of the world's growing population, while reducing the adverse anthropogenic impacts of fossil fuel use [58]. Solar PV growth has been rapid and by 2018, the worldwide PV capacity is predicted to double to 430 GW [59]. As the AC recyclebot extrusion of an HDPE filament requires 21.13 W power, the initial heating needs 0.06 kWh [25]. In this project, it was assumed that the DC recyclebot would use approximately the same power for initial heating, so two small monocrystalline silicon solar modules were used in the design of the system. Each module has an effective area of 0.2613 m^2 and produces 30 W of power. According to the energy requirement of initial heating, it requires a battery that has an output of 0.06 kW to finish within 1 h. An identical small battery system for the off-grid 3-D printer [36] was used. The storage battery access allows the recyclebot to work, even in the absence of solar energy (e.g., during cloudy weather). PV modules need a support structure for the setup, but this can be improvised in the field from the materials found as the system can be mobile. For example, PV modules can be propped against a wall or a rock in the field. The energy output of the solar module depends on the radiation and the proper placement of the panels, so as to receive the maximum solar radiation for maximum efficiency. Insolation varies by location and also affects the energy output of the system.

With the size of the solar PV system, its embodied energy can be estimated by the embodied energy of the solar PV system in unit size, which is found from the literature. In the solar PV system of this study, the energy consumptions for all of the production processes and accessories are assumed to be scaled to the size of the solar PV system.

4.2.2. Recyclebot (Recyclebot v4.0)

Both AC powered and DC powered vertical recyclebots v4.0 [37] are used in this study (Figure 3). The devices consist of a high power motor, feed tube, heating tube, frame, electrical components, and wiring. All of the components of the recyclebot are determined from the complete bill of the materials and their corresponding mass is obtained. The embodied energies of all the materials used in the recyclebot are tabulated from the CES database [56,57]. Then, the embodied energy of each component can be calculated by multiplying the mass by the corresponding material's embodied energy. For example, the recyclebot requires a support structure, which consists of two 60.96 cm long, 1.6 kg weight strut channels made of steel whose embodied energy is 30.8–33.9 MJ/kg. According to the product of the mass of the strut channels and embodied energy of steel, the embodied energy of the strut channels is 98.56–108.48 MJ. The embodied energies of the other components are calculated in the same way. Table 8 shows the embodied energy of the materials used in a DC recyclebot v4.0. It should be noted that these values of embodied energy are just for the materials and do not include the energy for forming products. Hence, the actual embodied energy might be about 26% larger than the value calculated in this project based on the comparison of energy between materials and forming products. The motor is a 120 W gear motor with 15 rpm combined with the heating tube and feed tube, which helps in the melting and extrusion of the filament. The embodied energy of the motor is obtained by the sum of the embodied energy of its various components within it, which is 58.382–64.373 MJ. Except for the metallic parts, the recyclebot also consists of 3-D printed parts such as a bearing house and feeder attachment, which are made of PLA. The 3-D printed parts in this system were printed from the Rep-Rap 3D printer. The requirement and the dimensions of the object are analyzed. The respective material of every component is important as it determines the strength and efficiency of the whole system.

Figure 3. Vertical recyclebot. The white hopper is on the upper left and the black spooler is on the upper right. The motor in the upper middle drives an auger that feeds plastic into the hot zone (insulated in yellow). Filament is produced and looped through a light sensor to maintain the loop length, and thus the filament diameter, automatically. In this setup, the length and diameter can be measured continually.

Table 8. The embodied energy of the materials needed in a recyclebot used in the input of the LCA (CES database [57]).

Material	Embodied Energy (MJ/kg)		CO_2 Footprint (kg/kg)
	Primary Production	Recycling	
Steel	30.8–33.9	8.1–8.98	2.26–2.49
Galvanized Steel	38.1–42	9.53–10.5	2.87–3.16
Brass	57.4–63.3	13–14.4	3.64–4.01
Stainless Steel	50.4–55.6	11.8–13	3.63–4
Chromium Steel	50.4–55.6	11.8–13	3.63–4
Copper	56.1–61.9	12.8–14.1	3.44–3.79
PVC	60.6–66.8	20.6–22.7	2.63–2.89
Silicon	1260–2240	-	94.5–168
Plastic	90.1–101	31.1–34.4	4.32–4.76
Lead	36.9–40.6	9.3–10.3	2.84–3.13
PC	109–120	36.9–40.8	7.03–7.75
HDPE	74.9–82.5	25.4–28.1	2.65–2.92
ABS	90.6–99.9	30.7–34	3.45–3.81
Ni	159–175	28.2–31.2	11.2–12.3
Pt	274,000–302,000	8140–8990	14,000–15,500
PLA	49.2–54.2	16.7–18.4	2.65–2.93
Nickle Chromium	189–208	32.2–35.6	10.4–11.4
Lead-Antimony	62–68.4	13.8–15.3	4.95–5.46
Epoxy	126–139	-	6.12–6.75
Ferrites(Fe_3O_4)	15.6–17.2	-	0.84–0.929
Aromatic Polyimide	176–194	-	9.61–10.6

4.3. Filament Production

PLA pellets and waste plastic ABS shards were used to produce a filament with a DC recyclebot and an AC recyclebot, respectively. Before the filament production, the temperature ranges of 150–180 °C and 158–190 °C with a step of 2 °C were set to find out the best temperatures for PLA and ABS filament extrusion, respectively. The minimum temperature in the ranges was determined by the limit of the mobility of plastics for extrusion, and the temperatures were increased until the plastic materials started to smoke (the maximum temperature). These test experiments were performed under three auger rotation speeds: 6 rpm, 10 rpm, and 15 rpm. It was found that the extrusion rate is largest and the melted plastic has rather higher plasticization properties when the auger rotation speed is 15 rpm. It was also found that the PLA filament extruded at 155 °C and the ABS filament extruded at 158 °C have a rather higher surface gloss and mechanical properties.

In PLA filament production, the DC recyclebot heating tube temperature was set as 155 °C, and 17 min were needed in the initial heating phase of production. In ABS filament production, the AC recyclebot heating tube temperature was set as 158 °C, and 8 min were needed in the initial heating phase. As the temperature reached the set points, the respective motors were activated to rotate the augers. The rotation speed of the augers in both recyclebots was about 15 rpm. The initial 0.5 m of filament was discarded because of poor mechanical properties as the feedback loop was established. Then, the filament was collected in an auto spooler with the help of a light sensor. The filament diameter in this study was 3.00 mm. A watt meter and timer were used to record the power and time during the process.

5. Conclusions

This study presented the embodied energy of a DC vertical recyclebot powered with a mono-crystalline solar PV-battery system, and calculated the energy payback time of the recyclebot and the whole system. The results show that using a recyclebot to create a 3-D printing filament from post-consumer plastics is an effective way to save energy, and the EPBT of a recyclebot is five days based on a PLA filament, or just 2.5 days based on an ABS filament. When the recyclebot is powered by a solar PV system to produce the filament, the energy can be further saved and is equal to the energy for producing a commercial filament from virgin materials. The EPBT of the whole system is just several weeks depending on the material used. When a solar powered recyclebot produces a PLA filament from waste plastics, the EPBT of the whole system is about one month, which decreased the EPBT of a single crystal PV system by over 96%. It is clear that solar PV powered recyclebots are an effective method to reduce energy use and protect the environment to meet the requirement of sustainable development.

Acknowledgments: The authors would like to acknowledge helpful discussions with S. Kampe.

Author Contributions: J.M.P. conceived and designed the experiments; S.Z. and P.R. performed the experiments; all authors analyzed the data; J.M.P. contributed materials/equipment/analysis tools; all authors wrote the paper.

References

1.   Shen, L.; Haufe, J.; Patel, M.K. *Product Overview and Market Projection of Emerging Bio-Based Plastics PRO-BIP 2009*; Report for European Polysaccharide Network of Excellence (EPNOE) and European Bioplastics; Utrecht University: Utrecht, Netherlands, June 2009; p. 243.
2.   Plastics Europe. Production of Plastics Worldwide from 1950 to 2014 (in Million Metric Tons). Available online: http://www.statista.comstatistics282732global-production-of-plastics-since-1950 (accessed on 16 March 2016).
3.   Barnes, D.K.A.; Galgani, F.; Thompson, R.C.; Barlaz, M. Accumulation and fragmentation of plastic debris in global environments. *Philos. Trans. Roy. Soc. London B: Biol. Sci.* **2009**, *364*, 1985–1998. [CrossRef] [PubMed]

4. Zhang, J.; Wang, X.; Gong, J.; Gu, Z. A study on the biodegradability of polyethylene terephthalate fiber and diethylene glycol terephthalate. *J. Appl. Polym. Sci.* **2004**, *93*, 1089–1096. [CrossRef]

5. Astrup, T.; Møller, J.; Fruergaard, T. Incineration and co-combustion of waste: accounting of greenhouse gases and global warming contributions. *Waste Manag. Res.* **2009**, *27*, 789–799. [CrossRef] [PubMed]

6. Tansel, B.; Yildiz, B.S. Goal-based waste management strategy to reduce persistence of contaminants in leachate at municipal solid waste landfills. *Environ. Dev. Sustain.* **2011**, *13*, 821–831. [CrossRef]

7. MacArthur, E. Towards the circular economy. *J. Ind. Ecol.* **2006**, *10*, 4–8.

8. Zhu, D.J. Sustainable development calls for circular economy. *Sci. Technol. J.* **1998**, *9*, 39–42.

9. Yuan, Z.; Bi, J.; Moriguichi, Y. The circular economy: A new development strategy in China. *J. Ind. Ecol.* **2006**, *10*, 4–8. [CrossRef]

10. Bicket, M.; Guilcher, S.; Hestin, M.; Hudson, C.; Razzini, P.; Tan, A.; Ten Brink, P.; Van Dijl, E.; Vanner, R.; Watkins, E. *Scoping Study to Identify Potential Circular Economy Actions, Priority Sectors, Material Flows and Value Chains*; European Commission: Luxembourg, 2014.

11. Alsema, E.A.; Frankl, P.; Kato, K. *Energy Pay-Back Time of Photovoltaic Energy Systems: Present Status and Prospects*; Utrecht University Repository: Utrecht, Netherlands, 2006.

12. Craighill, A.L.; Powell, J.C. Lifecycle assessment and economic evaluation of recycling: A case study. *Resour. Conserv. Recycl.* **1996**, *17*, 75–96. [CrossRef]

13. Themelis, N.J.; Castaldi, M.J.; Bhatti, J.; Arsova, L. *Energy and Economic Value of Non-Recycled Plastics (NRP) and Municipal Solid Wastes (MSW) that are Currently Landfilled in the Fifty States*; Columbia University: New York, NY, USA, 2011.

14. Feeley, S.R.; Bas, W.; Pearce, J.M. Evaluation of potential fair trade standards for an ethical 3-D printing filament. *J. Sustain. Dev.* **2014**, *7*, 1. [CrossRef]

15. Mansour, A.M.H.; Ali, S.A. Reusing waste plastic bottles as an alternative sustainable building material. *Energy Sustain. Dev.* **2015**, *24*, 79–85. [CrossRef]

16. Kumar, K.V.; Kasturi Bai, R. Plastic biodigesters—A systematic study. *Energy Sustain. Dev.* **2005**, *9*, 40–49. [CrossRef]

17. Baechler, C.; Matthew, D.; Pearce, J.M. Distributed recycling of waste polymer into RepRap feedstock. *Rapid Prototyp. J.* **2013**, *19*, 118–125. [CrossRef]

18. Hunt, E.J.; Chenlong, Z.; Nick, A.; Pearce, J.M. Polymer recycling codes for distributed manufacturing with 3-D printers. *Resour. Conserv. Recycl.* **2015**, *97*, 24–30. [CrossRef]

19. Arena, U.; Mastellone, M.L.; Perugini, F. Life cycle assessment of a plastic packaging recycling system. *Int. J. Life Cycle Assess.* **2003**, *8*, 92–98. [CrossRef]

20. Ross, S.; Evans, D. The environmental effect of reusing and recycling a plastic-based packaging system. *J. Clean. Prod.* **2003**, *11*, 561–571. [CrossRef]

21. Pearce, J.M.; Morris Blair, C.; Laciak, K.J.; Andrews, R.; Nosrat, A.; Zelenika-Zovko, I. 3-D printing of open source appropriate technologies for self-directed sustainable development. *J. Sustain. Dev.* **2010**, *3*, 17. [CrossRef]

22. Gwamuri, J.; Wittbrodt, B.T.; Anzalone, N.C.; Pearce, J.M. Reversing the Trend of Large Scale and Centralization in Manufacturing: The Case of Distributed Manufacturing of Customizable 3-D-Printable Self-Adjustable Glasses. *Chall. Sustain.* **2014**, *2*, 30–40. [CrossRef]

23. Wittbrodt, B.; Laureto, J.; Tymrak, B.; Pearce, J.M. Distributed manufacturing with 3-D printing: a case study of recreational vehicle solar photovoltaic mounting systems. *J. Frugal Innov.* **2015**, *1*, 1–7. [CrossRef]

24. Kreiger, M.A.; Mulder, M.L.; Glover, A.G.; Pearce, J.M. Life cycle analysis of distributed recycling of post-consumer high density polyethylene for 3-D printing filament. *J. Clean. Prod.* **2014**, *70*, 90–96. [CrossRef]

25. Kreiger, M.; Anzalone, G.C.; Mulder, M.L.; Glover, A.; Pearce, J.M. *Distributed Recycling of Post-Consumer Plastic Waste in Rural Areas*; In MRS Proceedings; Cambridge University Press: Cambridge, UK, 2013; Volume 1492, pp. 91–96.

26. Brander, M.; Sood, A.; Wylie, C.; Haughton, A.; Lovell, J. Electricity-specific emission factors for grid electricity. Ecometrica: Edinburgh, UK, 2011.

27. El Chaar, L.; Lamont, L.A.; El Zein, N. Review of photovoltaic technologies. *Renew. Sustain. Energy Rev.* **2011**, *15*, 2165–2175. [CrossRef]

28. Foley, G. Rural electrification in the developing world. *Energy Policy* **1992**, *20*, 145–152. [CrossRef]

29. Foley, G. *Photovoltaic Applications in Rural Areas of the Developing World*; World Bank Publications: Washington DC, USA, 1995; p. 304.

30. Acker, R.H.; Kammen, D.M. The quiet (energy) revolution: Analysing the dissemination of photovoltaic power systems in Kenya. *Energy Policy* **1996**, *24*, 81–111. [CrossRef]

31. Chambouleyron, I. Photovoltaics in the developing world. *Energy* **1996**, *21*, 385–394. [CrossRef]

32. Lorenzo, E. Photovoltaic rural electrification. *Prog. Photovolt. Res. Appl.* **1997**, *R5*, R3–R27. [CrossRef]

33. Khoury, J.; Mbayed, R.; Salloum, G.; Monmasson, E.; Guerrero, J. Review on the integration of photovoltaic renewable energy in developing countries—Special attention to the Lebanese case. *Renew. Sustain. Energy Rev.* **2016**, *57*, 562–575. [CrossRef]

34. Jones, R.; Haufe, P.; Sells, E.; Iravani, P.; Olliver, V.; Palmer, C.; Bowyer, A. RepRap—The replicating rapid prototyper. *Robotica* **2011**, *29*, 177–191. [CrossRef]

35. King, D.L.; Babasola, A.; Rozario, J.; Pearce, J.M. Mobile Open-Source Solar-Powered 3-D Printers for Distributed Manufacturing in Off-Grid Communities. *Chall. Sustain.* **2014**, *2*, 18–27. [CrossRef]

36. Gwamuri, J.; Franco, D.; Khan, K.Y.; Gauchia, L.; Pearce, J.M. High-Efficiency Solar-Powered 3-D Printers for Sustainable Development. *Machines* **2016**, *4*, 3. [CrossRef]

37. Appropedia. 2016. Available online: http://www.appropedia.org/Recyclebot (accessed on 14 September 2016).

38. Knapp, K.; Jester, T. Empirical investigation of the energy payback time for photovoltaic modules. *Solar Energy* **2001**, *71*, 165–172. [CrossRef]

39. Nawaz, I.; Tiwari, G.N. Embodied energy analysis of photovoltaic (PV) system based on macro-and micro-level. *Energy Policy* **2006**, *34*, 3144–3152. [CrossRef]

40. Lenzen, M. Primary energy and greenhouse gases embodied in Australian final consumption: An input–output analysis. *Energy Policy* **1998**, *26*, 495–506. [CrossRef]

41. Ciceri, N.D.; Gutowski, T.G.; Garetti, M. A tool to estimate materials and manufacturing energy for a product. In Proceedings of the 2010 IEEE International Symposium on Sustainable Systems and Technology (ISSST), Arlington, VA, USA, 17–19 May 2010.

42. Sherwani, A.F.; Usmani, J.A. Life cycle assessment of solar PV based electricity generation systems: A review. *Renew. Sustain. Energy Rev.* **2010**, *14*, 540–544. [CrossRef]

43. Raugei, M.; Silvia, B.; Ulgiati, S. Life cycle assessment and energy pay-back time of advanced photovoltaic modules: CdTe and CIS compared to poly-Si. *Energy* **2007**, *32*, 1310–1318. [CrossRef]

44. Wittbrodt, B.T.; Pearce, J.M. Total US cost evaluation of low-weight tension-based photovoltaic flat-roof mounted racking. *Sol. Energy* **2015**, *117*, 89–98. [CrossRef]

45. Energy Information Administration, USA. *Annual Energy Outlook 2016*; US Energy Information Administration: Washington, DC, USA, 2017; pp. 60–62.

46. Amazon. 2016a. Available online: https://www.amazon.com/Printing-Filament-Dimentional-Accuracy-0--03mm/dp/B01M2UHPPD/ (accessed on 31 March 2017).

47. Amazon. 2016b. Available online: https://www.amazon.com/HATCHBOX-3D-PLA-1KG1--75-WHT-Filament-Dimensional/dp/B01NAZJ18C/ (accessed on 31 March 2017).

48. Amazon. 2016c. Available online: https://www.amazon.com/TechOrbits-ABS-1KG1--75-Green-Filament-Dimensional-Accuracy/dp/B01FGCBV72/ (accessed on 31 March 2017).

49. Amazon. 2016d. Available online: https://www.amazon.com/IC3D-Natural-1--75mm-Printer-Filament/dp/B0192EE9OC/ (accessed on 31 March 2017).

50. Fthenakis, V. Solar cells: Energy payback times and environmental issues. In *Encyclopedia of Sustainability Science and Technology*; Meyers, R.A., Ed.; Springer New York: New York, NY, USA, 2012; Volume 1, pp. 9432–9448.

51. Bhandari, K.P.; Jennifer, M.C.; Randy, J.E.; Apul, D.S. Energy payback time (EPBT) and energy return on energy invested (EROI) of solar photovoltaic systems: A systematic review and meta-analysis. *Renew. Sustain. Energy Rev.* **2015**, *47*, 133–141. [CrossRef]

52. Louwen, A.; van Sark, W.; Schropp, R.E.I.; Turkenburg, W.C.; Faaij, A. Life-cycle greenhouse gas emissions and energy payback time of current and prospective silicon heterojunction solar cell designs. *Prog. Photovolt. Res. Appl.* **2015**, *23*, 1406–1428. [CrossRef]

53. Bhattacharyya, S.C. Energy access programmes and sustainable development: A critical review and analysis. *Energy Sustain. Dev.* **2012**, *16*, 260–271. [CrossRef]

54. Sovacool, B.K. The political economy of energy poverty: A review of key challenges. *Energy Sustain. Dev.* **2012**, *16*, 272–282. [CrossRef]
55. Jiang, X.; Liu, Y. Global value chain, trade and carbon: Case of information and communication technology manufacturing sector. *Energy Sustain. Dev.* **2015**, *25*, 1–7. [CrossRef]
56. Appropedia. 2017. Available online: http://www.appropedia.org/How_to_get_embodied_energy_from_CES_database (accessed on 4 April 2017).
57. EduPack, C.E.S. Granta Design Limited: Cambridge, UK, 2016.
58. Pearce, J.M. Photovoltaics—A path to sustainable futures. *Futures* **2002**, *34*, 663–674. [CrossRef]
59. EPIA-Publications; 2016, European Photovoltaic Industry Association: Brussels, Belgium.

14

Relationship among Vulcanization, Mechanical Properties and Morphology of Blends Containing Recycled EPDM

Aline Zanchet [1],*, **Aline L. Bandeira Dotta** [2] **and Fabiula D. Bastos de Sousa** [3]

[1] Center of Applied Engineering, Modeling and Social Sciences—CECS, Universidade Federal do ABC—UFABC, Rua Santa Adélia, 166, Santo André-SP 09210-170, Brazil

[2] Center of Exact Science and Technology, Department of Physics and Chemistry, Universidade de Caxias do Sul, Rua Francisco Getúlio Vargas, 1130, Caxias do Sul-RS 95070-560, Brazil; alinebandotta@gmail.com

[3] Technology Development Center—CDTec, Universidade Federal de Pelotas—UFPel, Rua Gomes Carneiro, 1, Pelotas-RS 96010-610, Brazil; fabiuladesousa@gmail.com

* Correspondence: alinezanchet@gmail.com

Abstract: The production of consumption goods made of elastomer generates large amounts of vulcanized residues. The final proper environmental disposal of this material is a serious problem, which involves high costs and a possible waste of a material with high added value. The recycling of elastomers is a very important alternative since it is related directly to the protection of the environment, energy conservation, and sustainability. An option for companies that produce elastomeric residues is their incorporation in the formulations by producing polymeric blends. Thus, this work aims to prepare polymeric blends composed of ethylene-propylene diene monomer rubber (EPDM) and raw EPDM/EPDM residue (EPDM-r) in different concentrations, when the residue is ground at room temperature. The morphology of the residue, vulcanization characteristics, mechanical properties, and morphology of the blends were analyzed, showing promising results that point to the feasibility of using EPDM-r in the production of polymeric blends and as a possible solution to the problem of the final disposal of solid residues.

Keywords: ethylene-propylene diene monomer rubber; elastomeric residues; polymeric blends; vulcanization characteristics; mechanical properties; morphology

1. Introduction

Creating an eco-friendlier product, environmental issues should be dealt with in the early stage of production, during the conception and design phase, by using recycled secondary raw materials, which have lengthened useful lives and also would be able to be recycled once they have been discarded [1]. The recycling of materials, especially polymeric materials, should considered due to the limited resources that human beings face nowadays. In addition, the use of recycled materials in new applications is a sustainable action, as it saves raw materials, often polymers derived from petroleum, which are finite. The use of waste in other applications helps also in the solution of the big global problem of the final disposal of solid residues. Vulcanized elastomers are materials with difficult processes of natural degradation due to their cross-linking structure and the presence of additives on their formulation, which can generate serious problems for to the environmental and public health. Additionally, recycling is considered a category of green chemistry, i.e., the use of renewable or recycled sources of raw material, and is also a source of income for many families around the world [1,2], especially those facing an economic crisis.

Research and technological initiatives are increasingly focused on the development of new methodologies of recycling, which has forced companies to seek the development of green products [3].

The disposal of elastomer wastes is one of the largest costs to the industry because the raw material used has high added value, and therefore their disposal is considered a waste. Nowadays, the situation is even more difficult since that not following the laws is considered an environmental crime, which can generate hard punishments. The ideal solution to this problem, therefore, is that the recycled material is added into the process within the industry itself, which is not so simple in the case of elastomers [4]. Elastomers like ethylene-propylene diene monomer rubber (EPDM), with unsaturations in their side groups, are less reactive and therefore more difficult to recycle [5].

A well-established way to recycle vulcanized elastomers is through the production of polymeric blends, i.e., a mixture of two or more polymers that can be miscible or not. As two or more properties of the polymers can be combined, the blends are widely studied, with the aim of improving their physical properties compared to those of neat polymers, obtaining materials that have additional properties, and the minimization of the loss of their original properties [6], in addition to being economically more viable by uniting two existing polymers to synthesize another non-existent polymer [7] through the creation of a new molecule. A plethora of polymeric blends composed of recycled elastomers can be obtained, which is considered a sustainable action.

Recovery is the use of vulcanized elastomer (as a powder) as a filler in elastomer compositions with a raw matrix or in the production of polymeric blends through its incorporation and subsequent vulcanization. It is a process that uses only mechanical processes, without changing the chemical composition of the material [8]. Although there is a loss of mechanical properties during the process, since the interaction between the vulcanized elastomer and the raw one is generally weak, its recovery may be advantageous when incorporated into new formulations, given the reduction in the cost of the final product, lower consumption of energy and raw materials, and the non-generation of hazardous residues [9–13]. In this way, some authors have studied blends containing at least one phase composed of a recycled elastomer [14–20] as a viable economic alternative and as an environmentally friendly solution to solid residues, obtaining satisfactory results. Some authors have shown that the vulcanization of blends containing a ground recycled phase is complex [14]. Thus the vulcanization stage must be carefully analyzed since the physical properties of the blends are influenced as a consequence.

However, even knowing that the recycling of elastomers is widespread, a gap concerning the feasibility of EPDM recycling by milling at room temperature and its incorporation into a raw phase, resulting in a polymeric blend, is observed in the literature. It is important to understand the influence of the milling process on mechanical properties in order to come across a sustainable production process, which results in final products with useful properties. Thus, this work proposes to analyze the feasibility of grinding ethylene-propylene diene monomer rubber residue (EPDM-r) at room temperature and preparing polymeric blends composed of raw EPDM/EPDM-r in different concentrations by analyzing the relationship between vulcanization characteristics, mechanical properties, and morphology.

2. Experimental

2.1. Materials

Vulcanized residues of ethylene-propylene diene monomer rubber (EPDM) from expanded profile trims, called EPDM-r, and compounds containing a blend of raw EPDM (DSM South América Ltd., (São Paulo, Brazil) were kindly supplied by Ciaflex Rubber Industry Ltd. (Caxias do Sul, Rio Grande do Sul, Brazil). Both the residue and raw elastomer compounds contain sulphur (Intercuf industry and trade Ltd., (São Paulo, Brazil); zinc oxide (ZnO) (Agro Zinco industry and trade Ltd., São Paulo, Brazil); stearic acid ($C_{18}H_{36}O_2$) (Proquiec chemical industry S/A, (Vargem Grande Paulista, São Paulo, Brazil); tetramethyl thiuram disulfide (TMTD) (Proquiec chemical industry S/A, (Vargem Grande Paulista, São Paulo, Brazil); and 2-Mercaptobenzothiazole (MBT) (Proquiec chemical industry S/A, (Vargem Grande Paulista ,São Paulo, Brazil). The exact formulation of EPDM-r is not known.

2.2. Milling and Characterization of EPDM-r

Initially, the scraps were heterogeneous in size and form. First, they were cut to lengths of around 10 cm using a belt saw (self-construction). After this, the materials were submitted twice to agglutination to increase their surface area. Then two types of agglutinators, with different numbers of knives (two and and) and operating speeds were used; the one with the highest speed (MH, model MH-4) was used in the end. Finally, the residue was ground in a knife mill (Marconi, model MA 580, Piracicaba, SP, Brazil). The distribution of particle size was determined by granulometric analysis, according to American Society for Testing and Materials (ASTM) D5644-01, using the sieves with 20, 25, 28, 35, 48, and 65 mesh.

Ground EPDM-r was analyzed by thermogravimetric analysis (TGA) and by Scanning Electron Microscopy (SEM). TGA was performed using a thermogravimetric analyzer from Shimadzu, model TGA-50, according to ASTM D6370-03. Approximately 10 mg of EPDM-r was heated from 25 to 450 °C under a nitrogen atmosphere to monitor the weight loss of oil and elastomer. At 450 °C, the gas flow was changed by oxygen, and the samples were heated to 800 °C to monitor the carbon black degradation. Both ramps were heated at a rate of 10 °C/min and with a gas flow of 50 mL/min. The surface characterization was carried out using a Philips XL 30 Scanning Electron Microscope. The samples were cryogenically fractured, and the surfaces to be analyzed were coated with gold by a sputter coater. The semi-quantitative analysis of the fillers was carried out by Energy Dispersive Spectroscopy (EDS).

2.3. Preparation and Characterization of the Blends

A blend of two different types of EPDM, called 4770 and 4703, containing 70.3% and 53.0% of ethylene, respectively, comprise the raw EPDM phase [21]. Thus raw EPDM is composed of the blend 4770/4703 64/36 wt %, and the formulation is shown in Table 1. The real formulation of EPDM-r is not known because it is confidential information.

Table 1. Raw ethylene-propylene diene monomer rubber (EPDM) phase composition recipe.

Compound	Amount (phr) [a]
EPDM (4770/4703)	100
ZnO	5
Stearic acid	1
TMTD	1
MBT	0.5
Sulphur	1.5
Carbon black	14

[a] parts per hundred of rubber.

The blends were prepared in an open two roll mill, model MH-600. The total mixing time for each blend was approximately 30 min at 60 °C. The blends were composed of raw EPDM/EPDM-r in the following concentrations (wt %): 100/0, 83/17, 71/29, 62/38, 50/50, 45/55, 42/58, and 38/62.

The samples were molded in a hydraulic press from Shulz, model PHS 15 T, at 160 °C with a pressure of 7.5 MPa. The vulcanization time of the sheets corresponds to the optimum cure time (t_{90}) derived from the vulcanization characteristics.

The vulcanization characteristics of the blends were studied using an oscillatory dual cone Tech Pro Rheometer Rheotech OD+, at 160 °C, according to ASTM D2084-06. From the curves of torque versus time, the vulcanization characteristics of the samples were obtained: minimum torque (ML), maximum torque (MH), torque variation ΔM ($\Delta M = MH - ML$), security time of the process or scorch time (ts_1), optimum cure time (t_{90}), and Cure Rate Average (CRA). The CRA values were calculated by using the Equation (1) [22]:

$$CRA = \frac{1}{t_{90} - ts_1} \tag{1}$$

The absolute density test was conducted by the hydrostatic method, according to ASTM D297-93. The density of the samples was calculated using Equation (2):

$$\rho = \frac{0.9971 \times m_a}{m_a - m_b} \tag{2}$$

where ρ is the density of the sample at 25 °C (g·cm^{-3}), m_a is the mass of the sample in the air (g), and m_b is the mass of the sample in the water (g).

The mechanical properties of the blends were obtained by performing hardness, tensile, and tear strength tests and by determining the compression set (CS).

Hardness tests were carried out on a Shore A durometer Teclock GS709, according to ASTM 2240-05.

Tensile tests were performed on a universal machine, EMIC DL-3000, with a rate of grip separation of 500 mm·min^{-1} and with a 20 kN load cell, according to ASTM D412-06. The following properties were obtained: tensile strength and elongation at break. The tear strength tests were carried out on a universal machine EMIC DL-3000, with a rate of grip separation of 500 mm·min^{-1} and with a 20 kN load cell, according to ASTM D624-00.

The compression set is related to the elastic recovery of the material after the prolonged action of compressive forces. Tests were performed according to ASTM D395 B-03, method A. Cylindrical samples (Ø 28.6 mm × 13 mm) were compressed by 22 h at 23 °C. The residual deformation was measured after 30 min of compression removal, and CS was determined using Equation (3):

$$CS = \left(\frac{t_0 - t_f}{t_0} \right) \times 100 \tag{3}$$

where CS is the compression set (%), t_0 is the original thickness (mm), and t_f is the final thickness (mm).

A morphological analysis of the blends was carried out using a Shimadzu SSX-550 Superscan Scanning Electron Microscope (SEM). The samples were cryogenically fractured, and the surfaces to be analyzed were coated with gold by a sputter coater.

This section is divided by subheadings. It should provide a concise and precise description of the experimental results, their interpretation, and the experimental conclusions that can be drawn.

3. Results and Discussion

3.1. Characterization of EPDM-r

3.1.1. Milling and Particle Size Distribution

The results of the granulometric analysis are shown in Figure 1.

In the recovery of elastomers, the first step of any recycling process is the milling of the material, which is necessary to increase the surface contact area and to produce a more uniform product [23–25]. According to the literature [26], ground particles in the range of 28 to 35 mesh (0.425–0.600 mm) are ideal for incorporation in compounds. This strategy aims at reducing the price of production, the replacement of mineral fillers, and the environmental impact [11]. Milling at room temperature is a low-cost process that produces rough and irregular particles [21]. The particle size and particle size distribution depend on the number of times that the powder is processed by the mill and the mill type used. In general, the first milling reduces large pieces of elastomer to sizes in the range of 10 to 40 mesh. The second one can reduce the particle size to up to 80 mesh [25]. The powder obtained is restricted to applications that require low mechanical strength due to the presence of great amounts of large particles [27].

The particle size distribution of EPDM-r showed that approximately 47% of the EPDM-r was retained on 35 and 48 mesh sieves, which demonstrated that the residue used consists mostly of 0.43 to 0.30 mm sized particles. According to the literature [26], these particle sizes are ideal for incorporation into new compounds. However, the particle size distribution ranged between 20 and 65 mesh. According to some authors [23], the mechanical grind produces particles with enough roughness and consequently increases their surface area, resulting in greater adhesion to the elastomeric matrix. In addition, as described above, the powder obtained is the appropriate size for posterior incorporation to compounds [26].

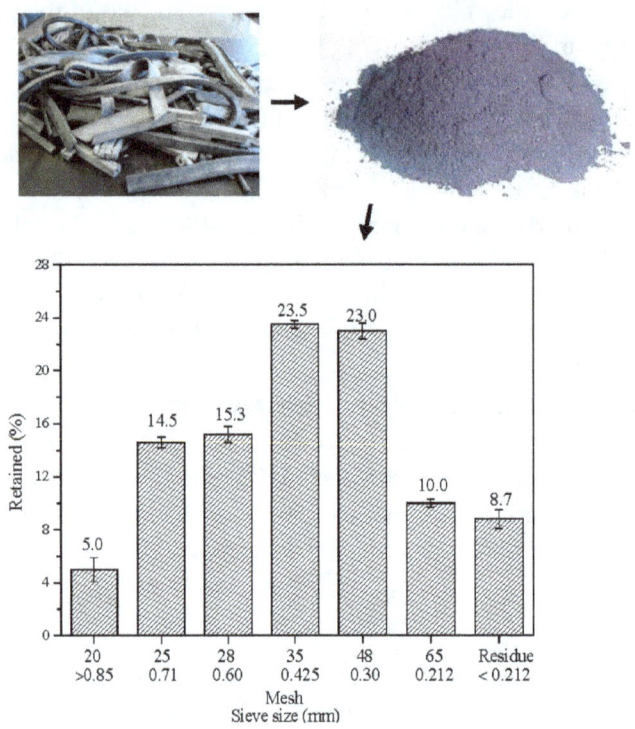

Figure 1. Particle size distribution of ethylene-propylene diene monomer rubber residue (EPDM-r).

3.1.2. Thermo-Oxidative Degradation

Figure 2 and Table 2 show the mass loss values and the temperature ranges from TGA.

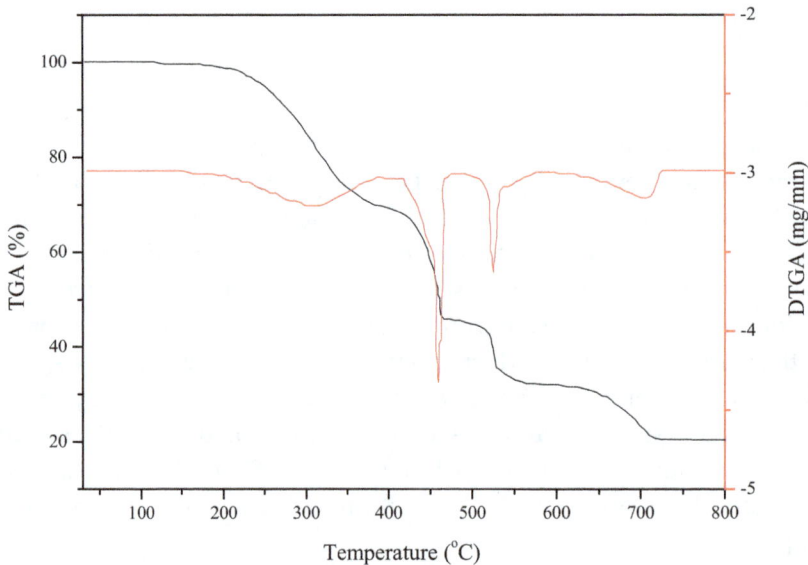

Figure 2. Thermogravimetric analysis (TGA) and derivative thermogravimetry (DTG) curves of the EPDM-r.

Table 2. Partial composition of the EPDM-r determined by TGA analysis.

Partial Composition	Weight Loss (%)	Temperature (°C)
Oil	30.10	291.6
EPDM	23.87	447.4
Carbon black	13.28	522.5
CaCO$_3$	11.42	682.5
Residue at 800 °C	21.33	—

According to Weber et al. [23], the first mass loss occurs at 291.6 °C, and it is related to the oil decomposition present in the sample (30.1%) and other organic polymeric additives [4,28,29]. At 447.4 °C, the corresponding degradation of EPDM (23.87%) occurs (under nitrogen flow) [23]. After the change of the gas flow, the mass loss at 522.5 °C is due to carbon black decomposition [23,28]. In the next stage, at 682.5 °C, CO$_2$ is released due to the decomposition of the CaCO$_3$ (11.42%) [23]. CaCO$_3$ is a filler commonly used in the rubber industry.

3.1.3. Morphology

Figure 3 shows the SEM micrograph of the EPDM-r surface particle and the semi-quantitative analysis performed by EDS.

Based on the results, it is possible to observe that, as well as irregular shape, the particles have a high surface roughness (Figure 3a). According to Gibala et al. [16], rubber ground at room temperature is convoluted, spongy, and porous in nature. It is known that powders suitable for utilization in new formulations must be smaller than 0.6 mm and present high superficial area [23,27]. For regeneration, their size should be in the range of 0.1 to 0.5 mm [23]. The semi-quantitative analysis (Figure 3b) revealed the presence of CaCO$_3$ concentrated on the analyzed surface, which was proven by the TGA results (Table 1).

However, these fillers tend not to influence the adhesion among the phases in polymeric blends. Since they are inorganic fillers, incompatible with both polymeric phases, and knowing that the mixing process provides high shear rates, they probably were easily removed from the surface of the residue phase during processing.

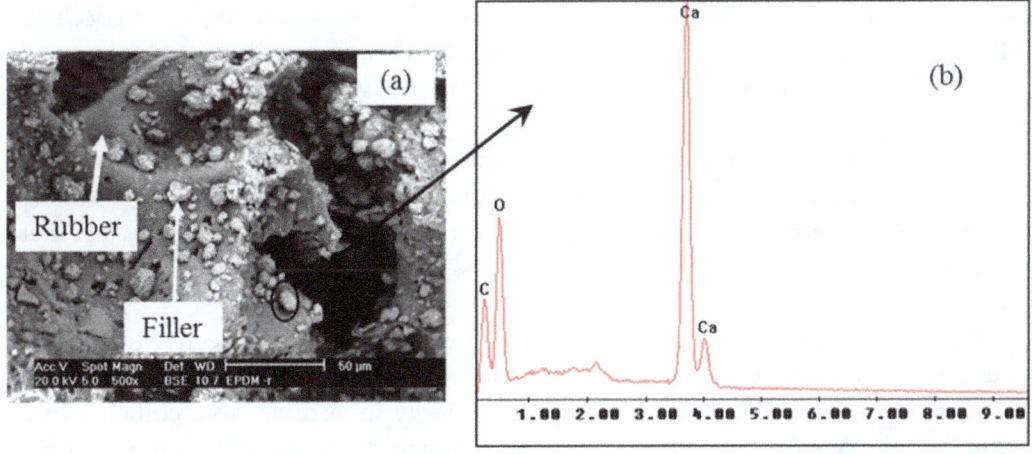

Figure 3. (**a**) SEM micrograph of EPDM-r and (**b**) Energy Dispersive Spectroscopy (EDS).

3.2. Characterization of the Blends

3.2.1. Vulcanization Characteristics

Table 3 summarizes the vulcanization characteristics of the analyzed samples.

Table 3. Vulcanization characteristics of the analyzed blends.

Sample	ML (dN·m)	MH (dN·m)	ΔM (dN·m)	ts_1 (min)	t_{90} (min)	CRA (min^{-1})
100/0	3.58	15.28	11.70	1.86	4.46	0.38
83/17	3.59	16.92	13.33	1.57	4.81	0.31
71/29	4.24	17.97	13.73	1.49	10.00	0.12
62/38	5.28	18.34	13.06	1.46	10.03	0.12
50/50	8.09	19.47	11.38	1.16	12.59	0.09
45/55	8.88	20.25	11.37	1.24	10.43	0.11
42/58	9.35	23.01	13.66	1.13	25.20	0.04
38/62	9.95	21.35	11.40	1.26	24.18	0.04

In general, the ML values increased as the concentration of EPDM-r increased in the blends. These values reflect the initial viscosity of the compounds and provide information about the processability of the compounds [24,25]. It is known that the residue is vulcanized, which increases the initial viscosity of the blend. The high viscosity may restrict the flow of the elastomeric mass, consequently making processing difficult. According to the literature [16], vulcanized particles do not deform and thus cause the rubber to continue to move at a higher velocity between particles for a given macroscopic shear rate, resulting in a higher shear stress and viscosity. Concerning the MH values (related to the stiffness of the compounds), in general, they increased as the EPDM-r concentration present in the blend increased. EPDM-r contributed to an increase the viscosity of the compounds and reduced their fluidity since it was vulcanized. In the case of the samples containing high concentrations of residue, the possible formation of agglomerates (clusters) may also have influenced the increase of the MH values observed [30,31].

ΔM values did not show a tendency as the samples 83/17, 71/29, 62/38, and 42/58 presented higher values than the control sample (100/0). The parameter correlates to the cross-linking density of the compound after the end of the vulcanization reaction [14]. Thus, as mentioned earlier, the presence of certain residue concentrations, by reducing the blend fluidity, possibly hampered in some way the vulcanization of the compounds and probably increased the number of clusters. According to Zanchet et al. [21], the more agglomerated the filler is, the higher the amount of rubber occluded into the aggregates, resulting in a greater hydrodynamic effect, which reduces the interaction filler-matrix. In the same way, the higher the agglomeration of the residue phase, the higher the occlusion of the matrix phase. As also stated by the literature [16], an elastomeric matrix can become occluded into the residue phase due to the format of the ground particles at room temperature, which present void spaces.

Even the agglomeration can cause the confinement of the vulcanization additives within the residues, leading to the decrease of the matrix vulcanization [14]. Moreover, the low cross-linking density observed in the samples containing high concentrations of residue may also be attributed to the relatively fewer reaction sites available for further cross-linking formation on the residue phase [32], since it is vulcanized.

So that a given chemical reaction happens, it is necessary to have effective shocks in order to form an intermediate structure between reagents and products, known as an activated complex. The molecules of the reagents must have sufficient energy in addition to a collision in favorable geometry [33]. Therefore, the viscosity increase and the consequent decrease in the fluidity probably hampered the occurrence of these shocks, which directly influenced the vulcanization reactions of these compounds. Still, the increase of residue clusters probably increased the amount of matrix occluded, hindering the homogenous formation of cross-linkings in the matrix.

ts_1 indicates the security time of the process before the start of the cross-linking formation. The results showed that this parameter decreased as the residue content increased in the blends. Such a reduction can be attributed to the presence of residual vulcanization additives related to the first vulcanization reaction, which is a characteristic behavior of recycled rubbers [25,34–36]. According to

Ismail et al. [36], the residual accelerator can act as a sulphur donor, which speeds up the early stages of the vulcanization process and reduces the value of ts_1. This reduction can also be related to the increased viscosity and consequent reduction in the fluidity of the compounds, as well as the increase of residue agglomerates in the matrix. As mentioned by some authors [15,19], the reduction of ts_1 values is due to the migration of accelerator fragments from the residue phase to the matrix.

Regarding t_{90} values, in general, they increased considerably with the increase of the residue concentration in the blend. This result means that, for the materials reach the optimum vulcanization time, greater periods of time were necessary, probably due to the difficulty in the generating of effective shocks, as cited previously, since the high concentrations of vulcanized residue possibly make it more difficult for these materials to flow. The higher viscosity of the samples also hinders the diffusion and dispersion of vulcanization additives, affecting the occurrence of the reaction. In other words, it seems that the rate of diffusion of the vulcanization additives decreased in the matrix phase as the residue concentration increased; consequently the t_{90} values increased [37]. As previously mentioned, the increase in the concentration of residue clusters in the samples may have hampered the vulcanization reaction due to the possible increase in the amount of occluded rubber.

In general, CRA values decreased with the increase of residue concentration in the blend. In other words, the vulcanization got more difficult due to the presence of large residue concentrations, since the CRA value equates to the speed of the reaction. As previously pointed out, the viscosity of the compounds was enlarged with the increase of the residue concentration, making the diffusion of the vulcanization additives more difficult. Some authors found similar results [14]. According to them, it is possible that the vulcanized EPDM-r phase acted as a physical barrier, slowing down the cross-linking rate of the matrix.

3.2.2. Density and Hardness

The hardnesses and densities of the blends before and after vulcanization of the blends are presented in Table 4.

Table 4. Hardness and density of the analyzed blends.

Sample	Shore A Hardness	Density (g·cm^{-3})	
		Unvulcanized	Vulcanized
100/0	41.3 ± 1	0.23 ± 0.02	0.67 ± 0.09
83/17	48.1 ± 2	0.36 ± 0.05	0.64 ± 0.06
71/29	48.3 ± 1	0.37 ± 0.10	0.55 ± 0.01
62/38	51.4 ± 1	0.39 ± 0.07	0.51 ± 0.01
50/50	53.3 ± 1	0.47 ± 0.07	0.47 ± 0.06
45/55	55.2 ± 1	0.49 ± 0.06	0.52 ± 0.05
42/58	55.4 ± 1	0.52 ± 0.05	0.58 ± 0.01
38/62	60.3 ± 1	0.54 ± 0.08	0.55 ± 0.03

According to the results, the hardness increased as the EPDM-r content increased. Therefore, the greatest concentration of hard residue (vulcanized) present in the sample and the greatest concentration of clusters possibly influenced the hardness values obtained. These values corroborated the increase of MH previously observed in Table 2.

Regarding the density, the results of the unvulcanized blends are less than those of the vulcanized the blends containing low residue concentrations. This difference is due to reticulation since the cross-linkings approach the chains and reduce the volume of voids in the samples. As in the blends containing fewer residues, the matrix phase is more concentrated and the difference between the vulcanized and unvulcanized samples is sharper.

In the case of compounds containing high residue contents, the residue influenced more strongly the density of the blend, since there was no significant variation between the density values of the vulcanized and unvulcanized samples. In short, the analysis of the results demonstrated the great influence of the reticulation in the density of blends, but this varied according to the concentration of residue present in the samples. Also, in the blends containing high concentrations of EPDM-r, less vulcanizable matrix is present, so the difference between the vulcanized and unvulcanized samples became not significant.

3.2.3. Mechanical Properties

This section brings about the discussion of the results of compression set, tensile and tear strength, and elongation at break.

The compression set results of the blends are shown in Table 5. Only the results of the samples with up to 38 wt % of EPDM-r are presented because the other compositions did not properly fill the mold. The high viscosity has restricted the flow of the elastomeric mass of the blends, damaging the filling of the mold, as previously found with the vulcanization characteristics of the samples.

Table 5. Compression set of the analyzed blends.

Sample	CS
100/0	1.9 ± 0.3
83/17	3.9 ± 1.8
71/29	5.6 ± 4.2
62/38	4.5 ± 0.9

Concerning the compression set, it may be noted that the increase of the residue content present in the blends resulted in the increase of this mechanical property. This result can be attributed to the reduced elasticity of the elastomeric matrix, which carries a plastic deformation (irreversible) and hinders the elastic recovery of the material from the imposed deformation [37].

The samples containing 17, 29, and 38 wt % of residue showed increased values of ΔM, which are related to cross-linking density. This increase consequently reduces the elasticity of these blends and, as a result, leads to the increase of the compression set values observed [38]. Still, the greatest concentration of vulcanized residue can, likewise, have influenced the obtained values.

The tensile strength and elongation at break of the blends are shown in Figure 4.

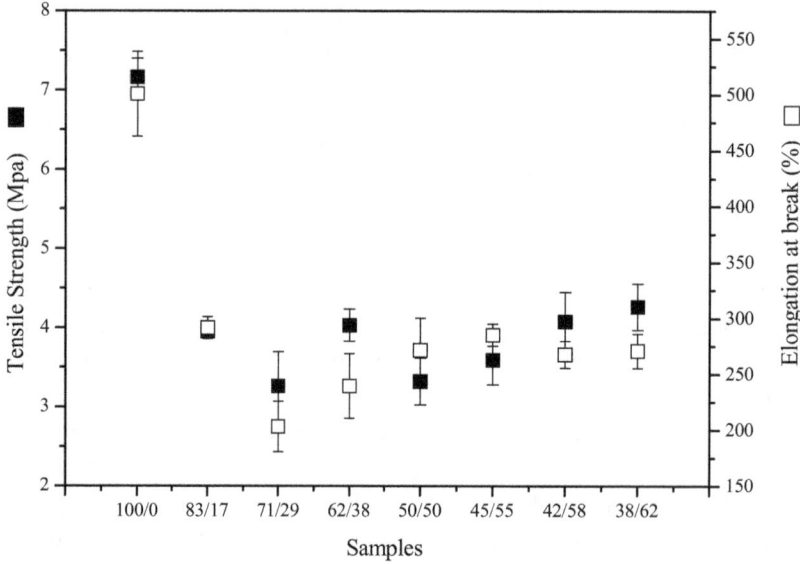

Figure 4. Tensile strength and elongation at break of the blends.

The values of the tensile properties analyzed were much smaller in the blends containing residue, no matter its concentration, since the values of the properties have not been significantly influenced by the residue concentration incorporated or due to the incompatibility among the phases [12]. This fact demonstrates the low adhesion among the phases, which reduces the values of the mechanical properties studied. According to some authors [39,40], poor mechanical properties in polymeric blends, especially elongation at break, are the result of poor adhesion and compatibility among the phases, which occurred in the present work.

The dispersion and distribution levels of the recycled phase in the matrix affect the mechanical properties as well, as will be shown. The literature has shown [41] that the lack of adhesion among phases in blends containing recycled elastomers is due to the large particles of elastomers, their superficial characteristics, and the structure of the cross-linkings, which hinder its absorption by molecules of the matrix as the use of only ground elastomer in blends makes the processing a difficult step [42], as observed in this work.

Even knowing that the residue presents an irregular structure that increases the superficial area and tends to increase the adhesion to the matrix (Figure 3), the higher amounts adopted made processing it difficult due to the increase of the viscosity, which tended to increase the agglomeration and particle-particle interaction (recycled phase), which may have caused the reduction in the observed mechanical properties. According to Gibala et al. [17], the ground residue in the tensile test is a low elongation inclusion, which undergoes multiple cracking and acts as a stress-raising flaw. Since there is weak adhesion among the phases when a mechanical request is imposed on the compound, the recycled phase is unable to transmit the tensions to the matrix.

According to Nabil et al. [12,13], a tensile strength decrease is also associated with the restriction of chain segments. The second vulcanization is considered to be responsible for the increase of the cross-linking density, taking into consideration the recycled phase. The restricted mobility of the chain segment due to the average molar mass of the rubber chain between two successive cross-linking points limits the orientation of the network chain. This would eventually reduce the number of effective network chains as a consequence of the decrease in the tensile strength [43]. In order to increase the tensile properties, surface activation of the residue phase may be performed [20].

The tear strength results can be analyzed from Figure 5.

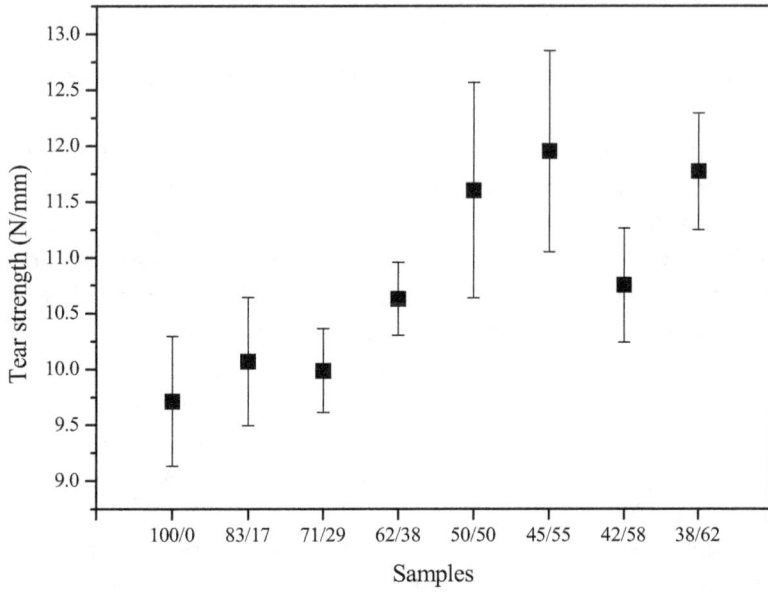

Figure 5. Tear strength of the analyzed blends.

On the whole, an increase in the tear strength of the samples can be observed as the EPMD-r content increased in the blends. The results are in agreement with the hardness results previously analyzed. Bearing in mind that the test measures the energy required to tear the sample, the harder the material, the more difficult it is to tear it.

In the same way, the influence of the residue phase on the results can be observed, given that this phase is vulcanized. The cross-linking density also influenced the results. As depicted by some authors [17], the increase in the tearing energy with the increase in the residue content may be attributed to a less cross-linked matrix, which presents a close agreement with the vulcanization characteristics presented before.

As stated earlier, for the tensile test, the discontinuity of the matrix caused by the vulcanized residue may act as a stress-raising flaw. However, in the tear test, it can increase the average tear strength. Consequently, the increase of the discontinuities (EPDM-r content) increased the tear strength.

Figure 6 shows the influence of ΔM, which is proportional to the cross-linking density, on the mechanical properties of the analyzed blends. The curves represent the trends observed during the analysis of the results.

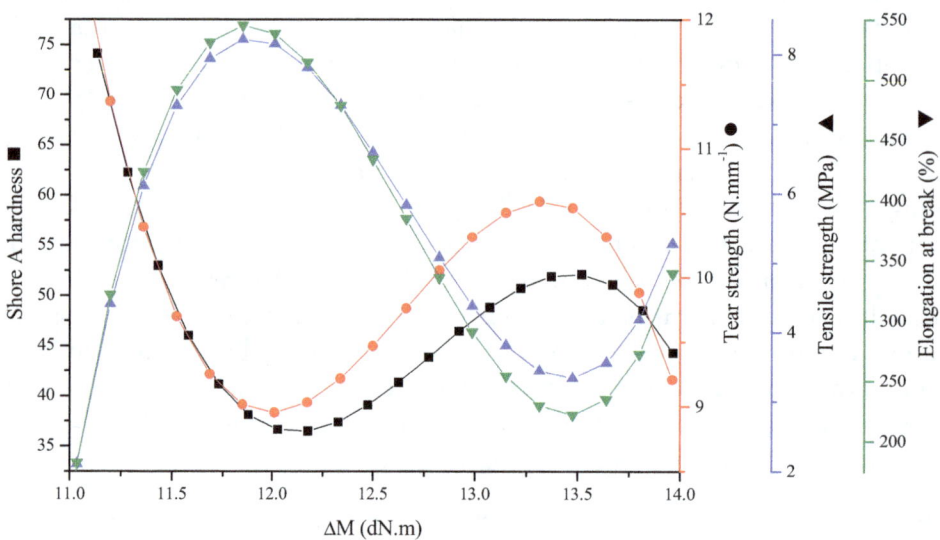

Figure 6. Influence of ΔM on the mechanical properties of the blends.

According to de Sousa [44], the addition of new factors such as the introduction of a recycled phase in another raw one is able to influence the vulcanization reaction of the raw phase, as is verified in this work. Consequently, its physical properties may also be influenced; in particular, the mechanical properties. On the other hand, the final properties of the composition also depend on the formulation used, the type and density of the cross-linkings, the type and amount of filler, and the interaction among phases (when the compound is a polymeric blend), among other factors.

In accordance with Coran [38,45], tear strength is related to the energy necessary to break the vulcanized composition. The author [38] showed that the tear strength increases with small increases in the cross-linking density to a limit, after which the property is reduced by further cross-linking formation. Thus, for the optimum tear strength result in the present work, it seems that a low cross-linking density is required.

Relating to the hardness, it is known that an increase in cross-linking density promotes the progressive increase of the hardness of the elastomeric materials [21]. However, hardness followed the same behavior as tear strength, showing that the cross-linking density increase was not significant for this property. As depicted before, the lack of adhesion among the phases and the amount of raw phase occluded into the clusters may have influenced this result.

Concerning the tensile strength and elongation at break, both showed the same trend. Elongation at break is usually reduced by the cross-linking density increase, since the cross-linkings require more energy to be broken, consequently increasing the tensile strength. However, the tensile strength was reduced as the ΔM increased, due to the lack of adhesion among the phases and the increase of residue agglomerations, as will be shown. In order to reduce this problem, some devulcanization techniques have been largely used in the recycled phase such as microwave devulcanization. It is a promising way of recycling since the results presented by the literature depicted the ability of the devulcanized elastomers to flow and to be remolded [9,23,28,29,46,47].

3.2.4. Morphology

Figure 7 shows the SEM micrographs of some samples.

According to the results, it can be observed that the increase of the amount of EPDM-r in the blends (white particles in the Figure 7b–h) increased the formation of clusters (i.e., the agglomeration size), which resulted in the difficulties faced during the vulcanization reaction. In the same way, a lack of adhesion among the phases can be observed (details in Figure 7h) since both the formation of clusters and a lack of adhesion influenced the mechanical properties previously observed.

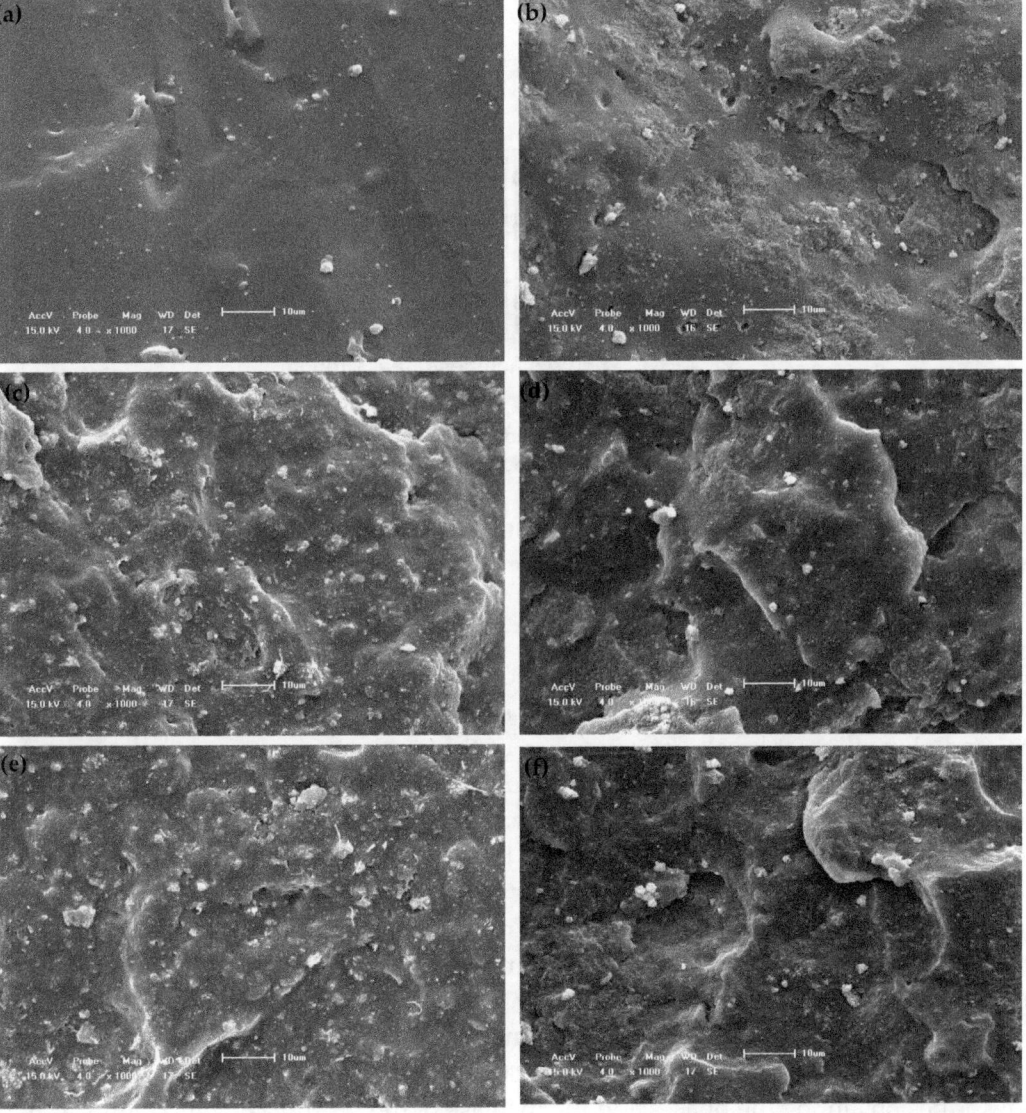

Figure 7. *Cont.*

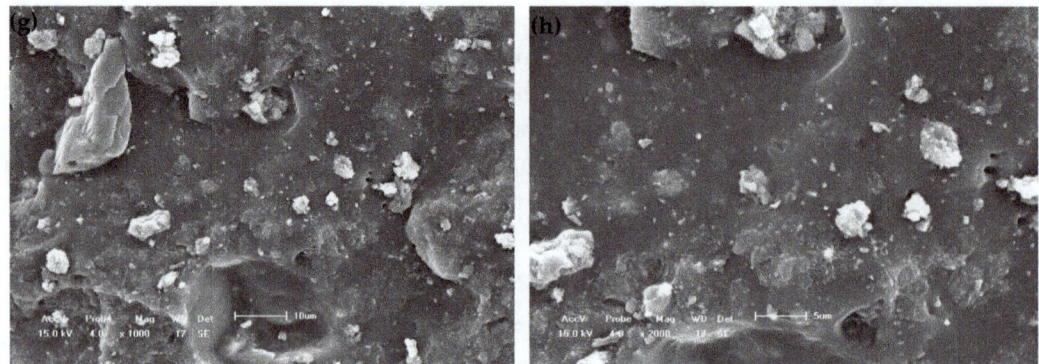

Figure 7. SEM micrographs of the samples: (**a**) 100/0; (**b**) 83/17; (**c**) 71/29; (**d**) 62/38; (**e**) 50/50; (**f**) 45/55; (**g**) 42/58; and (**h**) 42/58 (2000×).

Based on all the results presented in this work, Figure 8 proposes a reflection on the influence of the residue content on the vulcanization characteristics and mechanical properties.

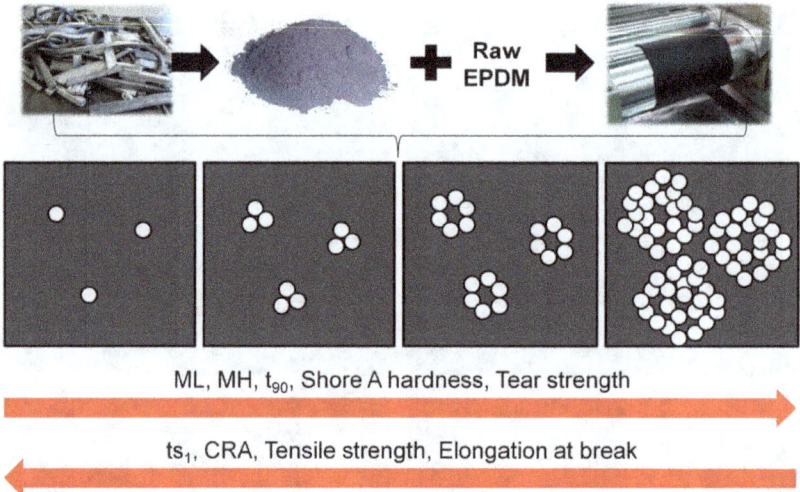

Figure 8. Schema of the influence of the residue content on the vulcanization characteristics and on the mechanical properties of the blends. The gray color represents the matrix phase (raw EPDM), and the white circles represent the ground residue phase.

According to the schema and the results previously analyzed, the increase of the EPDM-r content increased the ML, MH, shore A hardness, and tear strength, whereas it decreased the ts_1, CRA, tensile strength, and elongation at break. The increase of the residue content consequently reduced the matrix content and increased the formation of clusters by EPDM-r particle agglomerations, which increased the possibility of the unvulcanized matrix and the vulcanization additives becoming occluded into these clusters. Accordingly, the reduction in the matrix and the material occluded affected the vulcanization reaction.

The increase of the viscosity of the blends as a result of the increase of the residue content brought about the increase of the ML and MH and also decreased the diffusion rate of vulcanization additives in the matrix, increasing the t_{90}. However, vulcanization additives from the residue phase possibly influenced the reduction of the scorch time, influencing as well the vulcanization rate.

Regarding the mechanical properties, tensile strength and elongation at break were reduced due to the low adhesion among the phases and the increase of the residue clusters (Figure 7), while tear strength and hardness got the advantage of the low cross-linking density hand the high residue content.

When the real conditions and work involving elastomeric materials are considered, mechanical properties such as those analyzed in this work are necessary. The optimizing process of a large number of variables in order to obtain the best fit of a range of properties for a particular application is a normal practice and part of the development of a formulation for a particular artifact [48]. Therefore, even knowing that the use of recycled elastomers in new materials and applications presents huge challenges, the results indicate that the recovery proposed can be used to obtain artifacts with fewer mechanical requirements.

As the main conclusion, the production of polymeric blends composed of raw EPDM/EPDM-r is viable, resulting in final materials with useful properties. Nevertheless, it is important to correlate the properties of the particular application of the final materials with those of a blend with a given concentration of residue, making the production process sustainable and a possible solution to the problem of solid urban residues.

4. Conclusions

By means of milling at room temperature, an elastomeric powder with a particle size distribution in the ideal range for incorporation into compounds by compression molding processes was obtained. SEM imaging indicated the irregular shape and large surface roughness of the ground particles, which made them favorable for incorporation into the production of polymeric blends composed of raw EPDM/EPDM-r in different concentrations. The vulcanization characteristics showed that the amount of residue influenced the vulcanization reaction of the samples. By analyzing the mechanical properties, results, and morphology, especially tensile strength and elongation at break, it was concluded that there were weak interactions and adhesion among the elastomeric matrix and the residue, due to the reduction of these properties in relation to neat EPDM, no matter the residue concentration. In relation to the hardness and tear strength, it was observed that these properties increased with the increase of the residue content in the blend.

The results pointed to the potential application of the milling method used in the production of polymeric blends in the industry. However, it is important to know the mechanical properties of the final product when using the blend as a raw material in order to use the ideal residue concentration (which does not do harm when it is used), thus resulting in a sustainable production process, both for possible waste reduction and the reuse of recycled materials within the process itself.

Acknowledgments: The authors are grateful to Ciaflex Rubber Industry Ltd. for supplying all the materials used in this work.

Author Contributions: The authors Aline Zanchet, Aline L. Bandeira Dotta and Fabiula D. Bastos de worked together to perform the analyses and to write this manuscript.

References

1. Imbernon, L.; Norvez, S. From landfilling to vitrimer chemistry in rubber life cycle. *Eur. Polym. J.* **2016**, *82*, 347–376. [CrossRef]
2. De Sousa, F.D.B.; Zanchet, A.; Scuracchio, C.H. Influence of reversion in compounds containing recycled natural rubber: In search of sustainable processing. *J. Appl. Polym. Sci.* **2017**, *134*, 5. [CrossRef]
3. Adhikari, B.; De, D.; Maiti, S. Reclamation and recycling of waste rubber. *Prog. Polym. Sci.* **2000**, *25*, 909–948. [CrossRef]
4. Scuracchio, C.H.; Waki, D.A.; Da Silva, M.L.C.P. Thermal analysis of ground tire rubber devulcanized by microwaves. *J. Therm. Anal. Calorim.* **2007**, *87*, 893–897. [CrossRef]
5. Polgar, L.M.; Van Duin, M.; Broekhuis, A.A.; Picchioni, F. Use of Diels-Alder Chemistry for Thermoreversible Cross-Linking of Rubbers: The Next Step toward Recycling of Rubber Products? *Macromolecules* **2015**, *48*, 7096–7105. [CrossRef]

6.   Da Costa, H.M.; Ramos, V.D.; da Silva, W.S.; Sirqueira, A.S. Analysis and optimization of polypropylene (PP)/ethylene–propylene–diene monomer (EPDM)/scrap rubber tire (SRT) mixtures using RSM methodology. *Polym. Test.* **2010**, *29*, 572–578. [CrossRef]

7.   Bhadane, P.A.; Cheng, J.; Ellul, M.D.; Favis, B.D. Decoupling of reactions in reactive polymer blending for nanoscale morphology control. *J. Polym. Sci. Part B Polym. Phys.* **2012**, *50*, 1619–1629. [CrossRef]

8.   Papautsky, D. Recuperação e Regeneração. *Borracha Atual.* **2003**, *46*, 43–50.

9.   Zanchet, A.; Carli, L.N.; Giovanela, M.; Crespo, J.S.; Scuracchio, C.H.; Nunes, R.C. Characterization of Microwave-Devulcanized Composites of Ground SBR Scraps. *J. Elastomers Plast.* **2009**, *41*, 497–507. [CrossRef]

10. Gujel, A.A.; Bandeira, M.; Giovanela, M.; Carli, L.N.; Brandalise, R.N.; Crespo, J.S. Development of bus body rubber profiles with additives from renewable sources: Part I—Additives characterization and processing and cure properties of elastomeric compositions. *Mater. Des.* **2014**, *53*, 1112–1118. [CrossRef]

11. Fang, Y.; Zhan, M.; Wang, Y. The status of recycling of waste rubber. *Mater. Des.* **2001**, *22*, 123–128. [CrossRef]

12. Nabil, H.; Ismail, H.; Azura, A.R. Comparison of thermo-oxidative ageing and thermal analysis of carbon black-filled NR/Virgin EPDM and NR/Recycled EPDM blends. *Polym. Test.* **2013**, *32*, 631–639. [CrossRef]

13. Nabil, H.; Ismail, H.; Azura, A.R. Optimisation of accelerators and vulcanising systems on thermal stability of natural rubber/recycled ethylene-propylene-diene-monomer blends. *Mater. Des.* **2014**, *53*, 651–661. [CrossRef]

14. Carli, L.N.; Bianchi, O.; Mauler, R.S.; Crespo, J.S. Crosslinking kinetics of SBR composites containing vulcanized ground scraps as filler. *Polym. Bull.* **2011**, *67*, 1621–1631. [CrossRef]

15. Gibala, D.; Hamed, G.R. Cure and mechanical behavior of rubber compounds containing ground vulcanizates. Part I-Cure behavior. *Rubber Chem. Technol.* **1994**, *67*, 636–648. [CrossRef]

16. Gibala, D.; Laohapisitpanich, K.; Thomas, D.; Hamed, G.R. Cure and mechanical behavior of rubber compounds containing ground vulcanizates. Part II-Mooney viscosity. *Rubber Chem. Technol.* **1996**, *69*, 115–119. [CrossRef]

17. Gibala, D.; Thomas, D.; Hamed, G.R. Cure and mechanical behavior of rubber compounds containing ground vulcanizates: Part III. Tensile and tear strength. *Rubber Chem. Technol.* **1999**, *72*, 357–360. [CrossRef]

18. Ravichandran, K.; Natchimuthu, N. Vulcanization characteristics and mechanical properties of natural rubber-scrap rubber compositions filled with leather particles. *Polym. Int.* **2005**, *54*, 553–559. [CrossRef]

19. Jacob, C.; De, P.P.; Bhowmick, A.K.; De, S.K. Recycling of EPDM waste. I. Effect of ground EPDM vulcanizate on properties of EPDM rubber. *J. Appl. Polym. Sci.* **2001**, *82*, 3293–3303. [CrossRef]

20. Sutanto, P.; Picchioni, F.; Janssen, L.P.B.M.; Dijkhuis, K.A.J.; Dierkes, W.K.; Noordermeer, J.W. State of the art: Recycling of EPDM rubber vulcanizates. *Int. Polym. Process.* **2006**, *21*, 211–217. [CrossRef]

21. Zanchet, A.; Dal'Acqua, N.; Weber, T.; Crespo, J.S.; Brandalise, R.N.; Nunes, R.C. Propriedades reométricas e mecânicas e morfologia de compósitos desenvolvidos com resíduos elastoméricos vulcanizados. *Polim E Tecnol.* **2007**, *17*, 23–27. [CrossRef]

22. De Sousa, F.D.B.; Scuracchio, C.H. Vulcanization behavior of NBR with organically modified clay. *J. Elastomers Plast.* **2012**, *44*, 263–272. [CrossRef]

23. Weber, T.; Zanchet, A.; Brandalise, R.N.; Crespo, J.S.; Nunes, R.C. Grinding and Characterization of Scrap Rubbers Powders. *J. Elastomers Plast.* **2008**, *40*, 147–159. [CrossRef]

24. Weber, T.; Zanchet, A.; Crespo, J.S.; Oliveira, M.G.; Suarez, J.; Nunes, R.C. Caracterização de Artefatos Elastoméricos obtidos por Revulcanização de Resíduo Industrial de SBR (Copolímero de Butadieno e Estireno) Characterization of Elastomeric Artifacts obtained by Revulcanization of SBR Industrial Waste. *Polim E Tecnol.* **2011**, *21*, 429–435. [CrossRef]

25. Zanchet, A.; Carli, L.N.; Giovanela, M.; Brandalise, R.N.; Crespo, J.S. Use of styrene butadiene rubber industrial waste devulcanized by microwave in rubber composites for automotive application. *Mater. Des.* **2012**, *39*, 437–443. [CrossRef]

26. Gomide, R. *Operações Unitárias: Operações com Sistemas Sólidos Granulares*; Editora LTC: São Paulo, Brazil, 1983.

27. Bilgili, E.; Arastoopour, H.; Bernstein, B. Pulverization of rubber granulates using the solid state shear extrusion process Part II. Powder characterization. *Powder Technol.* **2001**, *115*, 277–289. [CrossRef]

28. Garcia, P.S.; de Sousa, F.D.B.; de Lima, J.A.; Cruz, S.A.; Scuracchio, C.H. Devulcanization of ground tire rubber: Physical and chemical changes after different microwave exposure times. *eXPRESS Polym. Lett.* **2015**, *91*, 1015–1026. [CrossRef]

29. De Sousa, F.D.B.; Scuracchio, C.H.; Hu, G.H.; Hoppe, S. Devulcanization of waste tire rubber by microwaves. *Polym. Degrad. Stab.* **2017**, *138*, 169–181. [CrossRef]

30. Sunthonpagasit, N.; Duffey, M.R. Scrap tires to crumb rubber: Feasibility analysis for processing facilities. *Resour. Conserv. Recycl.* **2004**, *40*, 281–299. [CrossRef]

31. Baranwal, K.C.; Stephens, H.L. *Basic Elastomer Technology*; Rubber Division: Akron, OH, USA, 2001.

32. Nelson, P.A.; Kutty, S.K.N. Cure Characteristics and Mechanical Properties of Maleic Anhydride Grafted Reclaimed Rubber/Styrene Butadiene Rubber Blends. *Polym. Plast. Technol. Eng.* **2004**, *43*, 245–260. [CrossRef]

33. Usberco, J.; Salvador, E. *Química*; Editora Saraiva: São Paulo, Brazil, 1999.

34. Oh, J.S.; Isayev, A.I. Continuous ultrasonic devulcanization of unfilled butadiene rubber. *J. Appl. Polym. Sci.* **2004**, *93*, 1166–1174. [CrossRef]

35. Oh, J.S.; Ghose, S.; Isayev, A.I. Effects of ultrasonic treatment on unfilled butadiene rubber. *J. Polym. Sci. Part B Polym. Phys.* **2003**, *41*, 2959–2968. [CrossRef]

36. Ismail, H.; Ishak, S.; Hamid, Z.A.A. Effect of blend ratio on cure characteristics, tensile properties, thermal and swelling properties of mica-filled (ethylene-propylene-diene monomer)/(recycled ethylene-propylene-diene monomer) (EPDM/r-EPDM) blends. *J. Vinyl Addit. Technol.* **2015**, *21*, 1–6. [CrossRef]

37. Maridass, B.; Gupta, B.R. Effect of Carbon Black on Devulcanized Ground Rubber Tire—Natural Rubber Vulcanizates: Cure Characteristics and Mechanical Properties. *J. Elastomers Plast.* **2006**, *38*, 211–229. [CrossRef]

38. Coran, A.Y. Vulcanization. In *Science and Technology of Rubber*, 3rd ed.; Mark, J.E., Erman, B., Eirich, F.R., Eds.; Elsevier: Boston, MA, USA, 2005; pp. 321–366.

39. De Sousa, F.D.B.; Gouveia, J.R.; de Camargo Filho, P.M.F.; Vidotti, S.E.; Scuracchio, C.H.; Amurin, L.G.; Valera, T.S. Blends of ground tire rubber devulcanized by microwaves/HDPE—Part A: Influence of devulcanization process. *Polím. Ciênc. Tecnol.* **2015**, *25*, 256–264. [CrossRef]

40. De Sousa, F.D.B.; Scuracchio, C.H.; Hu, G.H.; Hoppe, S. Effects of processing parameters on the properties of microwave-devulcanized ground tire rubber/polyethylene dynamically revulcanized blends. *J. Appl. Polym. Sci.* **2016**, *133*. [CrossRef]

41. Canavate, J.; Casas, P.; Colom, X.; Nogues, F. Formulations for thermoplastic vulcanizates based on high density polyethylene, ethylene-propylene-diene monomer, and ground tyre rubber. *J. Compos. Mater.* **2011**, *45*, 1189–1200. [CrossRef]

42. Magioli, M.; Sirqueira, A.S.; Soares, B.G. The effect of dynamic vulcanization on the mechanical, dynamic mechanical and fatigue properties of TPV based on polypropylene and ground tire rubber. *Polym. Test.* **2010**, *29*, 840–848. [CrossRef]

43. Kader, M.A.; Bhowmick, A.K. Thermal ageing, degradation and swelling of acrylate rubber, fluororubber and their blends containing polyfunctional acrylates. *Polym. Degrad. Stab.* **2003**, *79*, 283–295. [CrossRef]

44. De Sousa, F.D.B. Vulcanization of Natural Rubber: Past, Present and Future Perspectives. In *Natural Rubber: Properties, Behavior and Applications*; Hamilton, J.L., Ed.; Nova Science Publichers: New York, NY, USA, 2016; pp. 47–88.

45. Coran, A.Y. Vulcanization: Conventional and Dynamic. *Rubber Chem. Technol.* **1995**, *68*, 351–375. [CrossRef]

46. De Sousa, F.D.B.; Scuracchio, C.H. The role of carbon black on devulcanization of natural rubber by microwaves. *Mater. Res.* **2015**, *18*, 791–797. [CrossRef]

47. Weber, T.; Zanchet, A.; Crespo, J.S.; Oliveira, M.G.; Suarez, J.; Nunes, R.C. Caracterização de artefatos elastoméricos obtidos por revulcanização de resíduo industrial de SBR (copolímero de butadieno e estireno). *Polim E Tecnol.* **2011**, *21*, 429–435. [CrossRef]

48. Carli, L.N.; Boniatti, R.; Teixeira, C.E.; Nunes, R.C.; Crespo, J.S. Development and characterization of composites with ground elastomeric vulcanized scraps as filler. *Mater. Sci. Eng. C* **2009**, *29*, 383–386. [CrossRef]

15

Hydrocyclone Separation of Hydrogen Decrepitated NdFeB[†]

Muhammad Awais [1,2,]* (ID), **Fernando Coelho** [1], **Malik Degri** [2], **Enrique Herraiz** [2], **Allan Walton** [2] and **Neil Rowson** [1]

[1] School of Chemical Engineering, University of Birmingham, Edgbaston, Birmingham B15 2TT, UK; F.A.P.Coelho@bham.ac.uk (F.C.); N.A.Rowson@bham.ac.uk (N.R.)

[2] School of Metallurgy and Materials, University of Birmingham, Edgbaston, Birmingham B15 2TT, UK; M.J.Degri@bham.ac.uk (M.D.); eherraiz@pa.uc3m.es (E.H.); A.Walton@bham.ac.uk (A.W.)

* Correspondence: M.Awais@bham.ac.uk

[†] This paper is an extended version of our paper published in 6th International Conference Quo Vadis. Recycling, High Tatras, Slovak Republic, 6–9 June 2017.

Abstract: Hydrogen decrepitation (HD) is an effective and environmentally friendly technique for recycling of neodymium-iron-boron (NdFeB) magnets. During the HD process, the NdFeB breaks down into a matrix phase ($Nd_2Fe_{14}BH_x$) and RE-rich grain boundary phase. The grain boundary phase in the HD powder is <2 μm in size. Recycled NdFeB material has a higher oxygen content compared to the primary source material. This additional oxygen mainly occurs at the Rare Earth (RE) rich grain boundary phase (GBP), because rare earth elements oxidise rapidly when exposed to air. This higher oxygen level in the material results in a drop in density, coercivity, and remanence of sintered NdFeB magnets. The particle size of the GBP is too small to separate by sieving or conventional screening technology. In this work, an attempt has been made to separate the GBP from the matrix phase using a hydrocyclone, and to optimise the separation process. HD powder, obtained from hard disk drive (HDD) scrap NdFeB sintered magnets, was used as a starting material and passed through a hydrocyclone a total number of six times. The X-ray fluorescence (XRF) analysis and sieve analysis of overflows showed the matrix phase had been directed to the underflow while the GBP was directed to the overflow. The optimum separation was achieved with three passes. Underflow and overflow samples were further analysed using an optical microscope and MagScan and matrix phase particles were found to be magnetic.

Keywords: hydrocyclone; centrifugal separation; fine particle separation; NdFeB; recycling; rare earth elements

1. Introduction

The growing need for sustainable technologies is resulting in an increasing emphasis on the recycling of materials. This development is particularly important in the case of rare earth elements (REEs). These rare earth-based magnets—especially Neodymium-iron-boron (NdFeB) magnets—possess the highest energy product of all permanent magnets, which makes them highly efficient and suitable for lightweight mobile applications [1]. Therefore, they are widely used in computer hard drives (HDDs), loudspeakers, medical imaging, household electrical appliances, hybrid and electric vehicles (HEVs and EVs), wind turbines, and many other small consumer electronic devices. About 48% of total NdFeB magnets produced in 2012 were used in the production of motors, generators, HDD, CDs and DVDs. The amount being used varies between a few grams (e.g., loudspeakers) to tonnes of materials (e.g., wind turbines) [2].

In recent years, the increasing popularity of hybrid and electric cars and wind turbines has caused an increase in the demand for rare earth. China is currently producing more than 95% of these rare earth elements [3]. Growing demand for REEs and restrictions on the supply from the China have already triggered a rare earth crisis in the past. The reason being that the supply of these rare earths is dependent on single source, i.e., China, and this supply risk can be felt outside China. In the report on critical raw materials for the EU published in 2014, the European Commission considers rare earth elements as the most critical raw materials group, with the highest supply risk [4]. Moreover, it is anticipated that the demand for neodymium and dysprosium for production of magnets could rise by 700% and 2600%, respectively, over the next 25 years [5].

Previous studies have shown that hydrogen can efficiently be used to separate NdFeB magnets from HDD scrap [6,7]. NdFeB magnets become demagnetised when reacted with hydrogen, thus allowing the powder to be separated much more readily. During this hydrogen decrepitation process, the NdFeB magnets absorb hydrogen and break down into an interstitial matrix phase hydride ($Nd_2Fe_{14}BH_x$) and a grain boundary phase hydride. The average particles size of grain boundary phase is <2 μm [8], whereas the matrix phase breaks down in particle sizes ranging from 10 to 500 μm, depending on hydrogen cycling [9]. It should be noted that the HD powder is very friable due to the presence of microcracks in the particles [10]. Therefore, this HD powder can easily be milled down (using a jet mill) to the particle size required for sintering.

During sintering of NdFeB magnets, the RE-rich GBP melts down, resulting in liquid phase sintering. When the recycled HD powder is used to re-manufacture sintered NdFeB magnets, the GBP has a higher oxygen content and therefore it does not fully melt during the re-sintering process, which results in a lower-density magnet and reduction in magnetic properties. This liquid phase sintering process in recycled magnets can be improved by adding fresh neodymium hydride to the recycled material [8,11,12]. The aim of this work is, therefore, to separate the grain boundary phase (<2 μm) with higher oxygen content from the matrix phase using a hydrocyclone, and optimise the process to achieve optimum separation.

Hydrocyclones have been used in the chemical and mineral industries for many years. Their usage is very wide in the mineral, chemical and bio-industries due to the simple design, operational flexibility and low operation and maintenance costs. The devices use centrifugal forces to separate two products of different densities or sizes [13,14]. Despite their simplicity and low cost, they are very efficient for solid-liquid separations [15].

2. Experimental

2.1. Material

The starting material used in this study was a hydrogen decrepitated (HD) powder obtained from hard disk drive (HDD) scrap. A batch of NdFeB magnets, obtained from hard disk drive scrap, was processed at 2 bar hydrogen and room temperature for 2–4 h. The extraction of NdFeB magnets from hard disk drives and processing is explained in detail by Walton et al., 2015 [6]. Hydrogen decrepitated material was then exposed to air (for controlled oxidation) before mixing it with water. In this case, water acts as a medium for processing of the powder through the cyclone and it reduces the build-up of triboelectric charges between fine particles [16]. This, therefore, allows for better separation efficiency in the case of very fine particles.

2.2. Working Principle of Hydrocyclones

A hydrocyclone consists of two main parts, as shown in Figure 1. The first is a cylindrical part with feed inlet. This part also includes an outlet, located at the top of the cylinder, extends into the cylinder and is known as the vortex finder. The second main part is conical, and is connected to the cylindrical section at the top and to the underflow at the bottom end. The latter part is known as the spigot [17].

The feed slurry, under pressure, enters at the tangential inlet at the top of the hydrocyclone. As the feed enters the chamber, a rotation of the slurry inside of the hydrocyclone begins to accelerate the movement of the particles. This circular acceleration of the fluid directs the heavier particles towards the outer wall under the action of a centrifugal force. The particles migrate in a spiral pattern through the cylindrical section into the conical section. Depending on the particle size, the radial movement is hindered by the drag force as the particles move through the carrying fluid. At this point, the smaller particles migrate toward the centre of the hydrocyclone and spiral upward through the vortex finder. This product, which contains the finer particles and the majority of the water, is termed the overflow. As the flow descends in the hydrocyclone, the layer adjacent to the hydrocyclone wall becomes loaded with heavy particles, which exit the device through the spigot orifice of the cyclone, is termed as underflow [13,14].

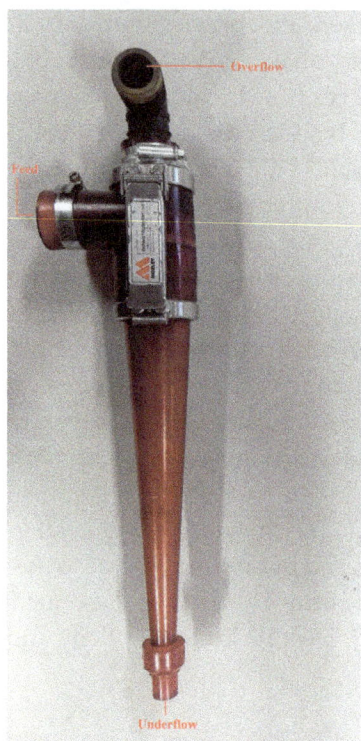

Figure 1. A typical hydrocyclone (Image taken by Malik Degri).

2.3. Methods

The experiments were carried out with a Mozley C700 rig, using a 50 mm hydrocyclone (vortex finder size was 14.3 mm and diameter of the spigot was 4.5 mm). The system was operated at a constant feed pressure of 0.8 bar. 1 kg of HD material was mixed with 10 litres of water in the tank, and circulated through the cyclone. At equilibrium, the hydrocyclone underflow and overflow were sampled for a set period, normally 2 s. Underflow and overflow sample volumes were measured. The underflow of the previous pass was mixed with 10 L water and used as the feed for the next pass. The experiment was repeated for six passes, and samples were collected after each pass.

Samples collected from each pass were then filtered and dried at 80 °C in an oven in air. These samples were then analysed using an XRF spectrometer (S8 Tiger, Haworth 503A) with ±0.50 weight % precision in measurement. The loose powder method was chosen to analyse these samples, as described in detail by Gakuto [18]. Full detection mode was selected for XRF. Oxygen analysis was performed by Less Common Metals (LCM), Cheshire, UK using a LECO instrument. 40 g of each underflow and 10 g of overflow were sieved using Retsch Vibratory Sieve Shaker AS 200. The amount on each sieve was recorded using and HR120 laboratory balance with a precision value of ±0.0002 g.

Particle size distribution of the starting material and first pass was performed using a Malvern Mastersizer 2000S and by Sympatec Qicpic 119D. The magnetic field mapping system (MagScan), supplied by Redcliffe Magtronics, was used to determine the location and magnetic strength of HD powder. The magnetic field exhibited by the sample was measured by three orthogonal hall probes situated within a probe head, allowing a resolution of 0.1×0.1 mm. The probe head was set to scan 40×40 mm, which was controlled by the computer. The MagScan produced an image showing contrasting colours, depending on the strength and direction of the field detected by the hall probes. The sensitivity of the MagScan used was 0.027 mT. A more sensitive custom-made Line Scan was also used in this work, which produced a line depending on the strength of the field above the powder. This line scan used a Mag3110 magnetometer from Freescale with a resolution of 100 nT, calibrated against a Lakeshore DSP 455 Gaussmeter and the setup is shown in Figure 2A. HD powder was filled in the acrylic block as shown in the figure as shown in Figure 2B. The recess in the acrylic block was 10 mm in diameter and 5 mm deep. The block was then placed on a horizontally moving bed controlled by the motor and moved under the magnetic field sensor. Image J software and Joel 6060 scanning electron microscope (SEM) was used to determine the particle sizes and a theoretical maximum of RE-Rich in the starting material.

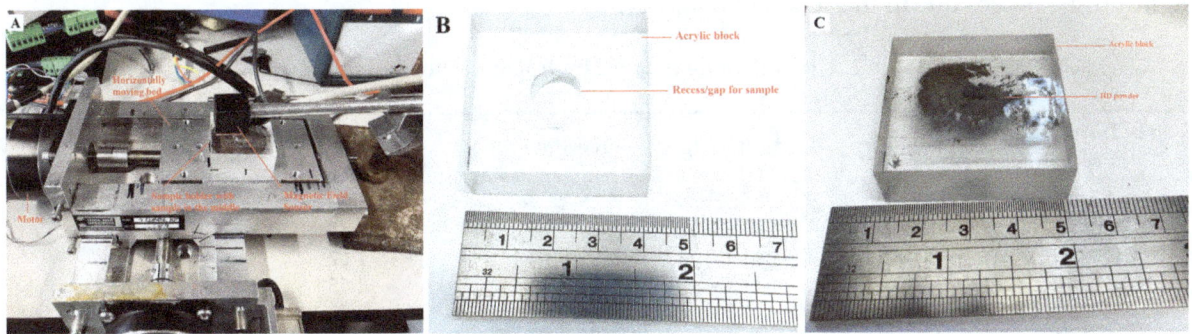

Figure 2. (A) In-house built MagScan with higher sensitivity; **(B)** acrylic block/sample holder without sample; **(C)** Prepared sample.

3. Results and Discussion

A VCM magnet was mounted in conductive bakelite, and the microstructures were analysed using SEM. The theoretical maximum amount of RE-rich (as shown in Figure 3A,B) present in the magnets was estimated by computing the total area of RE-rich in an SEM image. Image J was used for this estimation and the area fraction was then converted into mass by multiplying with density. It should be noted that the theoretical maximum of RE-rich phase was an estimation, and it was assumed that the intergranular RE-rich phase was uniformly distributed, and that the volume of RE-rich phase was the same as the area fraction. To minimise the error, and taking into account the uneven distribution of RE-rich phase, 10 different SEM images were taken at different locations. After setting the threshold in Image J, these images were then converted into a binary images, and the area fraction was calculated. The maximum and minimum value of area fraction was 9.7% and 6.6%, respectively, with a standard deviation of 1.14. The SEM and binary images used are presented in Figure 3. The average of these areas was then used for calculation. The theoretical maximum was estimated to be 8% by weight. Thus, in 1 kg of starting material, it was expected that there would be a maximum of 80 g of RE-rich phase.

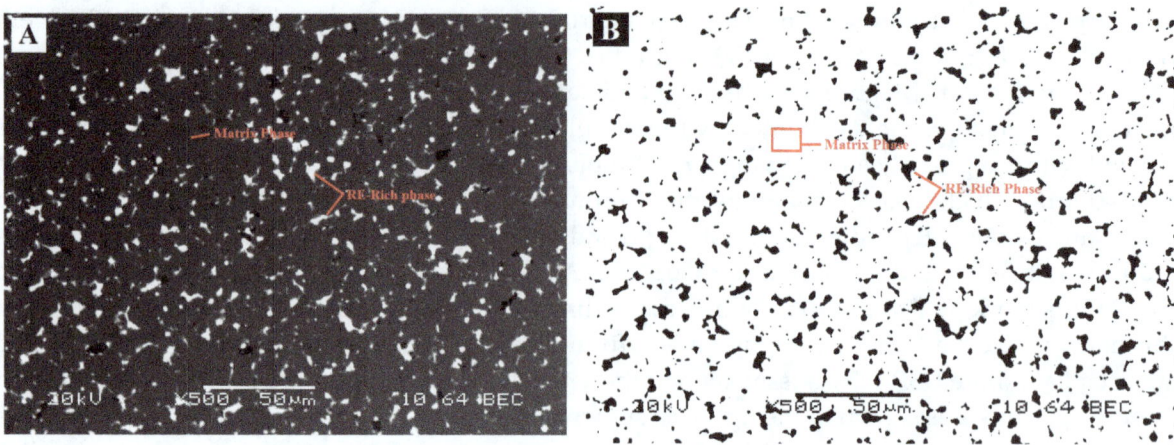

Figure 3. (**A**) Backscattered SEM image; (**B**) Binary image of SEM produced by Image J.

Results from hydrocyclone separation are presented in Table 1 below. There was a very small number of particles separated as overflow in the first three hydrocyclone passes, although this increased in last three passes due to the action of the pump.

Table 1. Hydrocyclone data for hydrogen decrepitated NdFeB.

Hydrocyclone Passes	Feed (g)	Underflow (g)	Overflow (g)	Overflow/Reject (%)	Underflow/Yield (%)
Pass 1	1000.00	987.07	12.93	1.29	98.71
Pass 2	793.67	779.93	13.74	1.73	98.27
Pass 3	653.32	642.76	10.57	1.62	98.38
Pass 4	563.27	549.45	13.82	2.45	97.55
Pass 5	486.06	474.39	11.67	2.40	97.60
Pass 6	428.64	418.30	10.34	2.41	97.59

Underflows and overflows collected from all of these passes (after drying) were then sieved again through different mesh sizes to classify the collected material based on particle sizes. Table 2 gives the percentage of underflow samples collected at each sieve. It should be noted that a sample size of 40 g was used for sieving, as it is representative of the material collected as underflow during each pass.

Table 2. Sieving results of hydrocyclone underflows in percent.

Mesh Size (μm)	Starting Material	Under-Flow 1	Under-Flow 2	Under-Flow 3	Under-Flow 4	Under-Flow 5	Under-Flow 6
125	19.53	0.60	1.89	15.51	0.16	46.78	-
90	2.86	0.38	16.32	10.58	0.10	0.38	0.61
63	3.39	0.85	69.80	39.61	67.72	9.15	24.24
45	12.12	9.54	8.02	13.53	16.33	22.93	59.49
38	7.69	56.74	2.55	17.39	14.07	13.03	14.65
−38	54.41	31.88	1.42	3.38	1.62	7.73	1.01

It is clear from Table 2 that about 20% (by weight) of the HD material used in this study was comprised of particles larger than 125 μm, and more than half of the material had particles less than 38 μm in size. As the material was passed through the hydrocyclone, the particle size distribution was changed. The HD material, being very friable, broke down inside the rotary pump and hydrocyclone, continuously generating smaller particles. These particles were then directed to the overflow with the very fine particles.

Sieving data of samples collected as overflow is presented in Table 3, below. In this case, a sample size of 10 g was used for all overflows. It can be seen that, for the first two passes, the majority of the particles were less than 45 μm, and no material with a particle size larger than 63 μm was collected. As we move on to the third pass, larger particles started to appear on higher mesh size sieves. It can be noted that particle less than 38 μm in size were removed by the hydrocyclone as overflow up to the third pass. No considerable amount of particles less than 38 μm was collected after the third pass.

Table 3. Sieving results of hydrocyclone overflow in percent.

Mesh Size (μm)	Over-Flow 1	Over-Flow 2	Over-Flow 3	Over-Flow 4	Over-Flow 5	Over-Flow 6
125	-	-	2.17	13.64	0.75	0.23
90	-	-	25.14	70.03	63.03	0.75
63	-	-	50.90	15.28	33.98	95.87
45	20.42	2.38	12.86	1.04	2.23	2.89
38	68.42	13.21	7.24	-	-	0.26
−38	11.16	84.40	1.69	-	-	-

Table 4 presents the results obtained from XRF and oxygen analysis of the feed and underflow fractions. The term $\sum$REE represents the sum of rare earth elements in the material (Nd + Dy). The hydrided matrix phase ($Nd_2Fe_{14}BH_x$), being large-sized particles, should report to the underflow; and grain boundary phase, being smaller, should report to the overflow. As both of these particles have Nd in them, the number of passes required for optimum separation is based on total rare earth elements and concentration of iron. It should be noted that XRF data is normalised to 100% and the amount of oxygen was determined using different equipment at a later stage.

Table 4. Chemical analysis of underflow streams in weight %.

Samples	Fe	Nd	Dy	$\sum$REE	Oxygen
Starting Material	60.92	35.04	1.21	36.25	1.1
Underflow 1	60.62	35.56	1.21	36.77	1.9
Underflow 2	60.21	33.60	1.18	34.78	2.1
Underflow 3	61.18	34.64	1.18	35.82	1.9
Underflow 4	61.25	34.76	1.24	36.00	2.0
Underflow 5	61.32	34.74	1.21	35.95	2.1
Underflow 6	61.94	34.40	1.16	35.56	1.9

It can be seen from the Table 4 that the concentration of iron increased with each hydrocyclone pass, whereas the amount of neodymium (Nd) decreased as expected. The slight variation in the values of Nd and Dy could be associated with the precision of the equipment.

The XRF and oxygen analysis of the overflow fractions are presented in Table 5. The overflows were found to be rich in Neodymium. The presence of Fe in the overflow indicates that some of the large particles (matrix phase) were present in overflow alongside small particles (GBP). Moreover, Dy was also present in the GBP, which means that the Dy and Nd detected in the overflow originated in both the matrix phase and the GBP. From the fourth pass, the amount of Fe suddenly started to increase in the overflow, meaning that, after the third pass, there was more matrix phase in the overflow than the grain boundary phase. Sieving data of overflows (Table 3) also agrees with this hypothesis. After the third pass, the continuous production of smaller particles affected the separation process, and no particles smaller than 38 μm were collected on the sieves, or the remaining small particles adhered to the larger particles by triboelectric charges, and could not be further separated [16].

Table 5. Chemical analysis of overflow streams in weight %.

Samples	Fe	Nd	Dy	$\sum$REE	Oxygen
Starting Material	60.92	35.04	1.21	36.25	1.1
Overflow 1	20.94	70.33	1.24	71.57	15.7
Overflow 2	23.20	68.01	1.20	69.21	15.9
Overflow 3	22.90	68.88	1.30	70.1	18.1
Overflow 4	30.91	61.06	1.34	62.40	15.6
Overflow 5	32.44	60.18	1.25	61.43	20.4
Overflow 6	36.86	56.01	1.24	57.25	17.4

The plot of $\sum$REE and Fe (Figure 4) from XRF shows that the REE-rich decreased and the Fe content increased after the 3rd pass. From this, we can assume that the overflows of first three passes were mainly RE-rich particles, and so can assume a separation efficiency close to 45% after three passes.

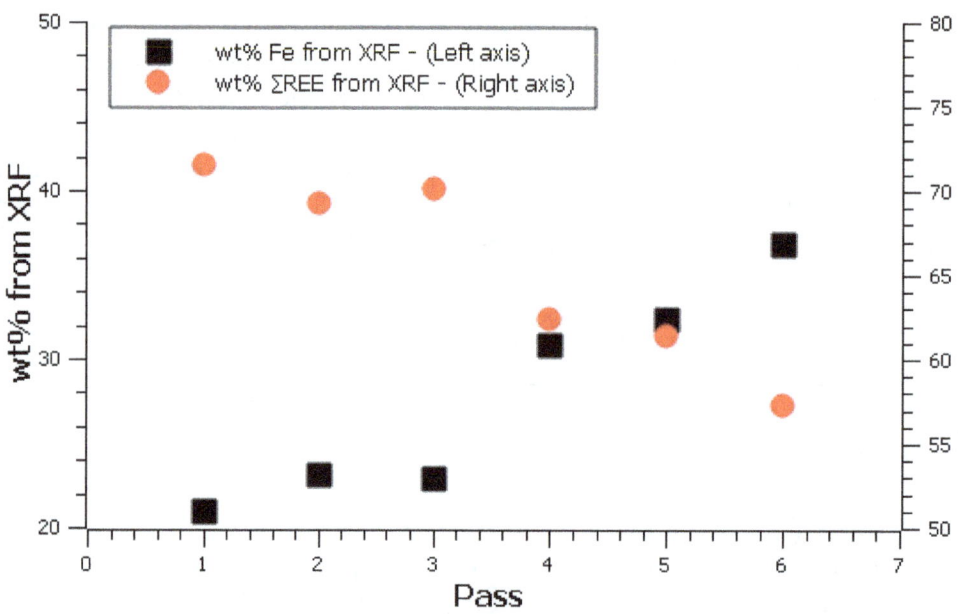

Figure 4. Change in REE and Fe concentraion in the overflows with number of passes.

The oxygen content in all underflow samples (Table 4) was slightly higher than the starting material (feed), which is likely to be due to the neodymium reacting with water to form neodymium hydroxide [8,19]. Formation of $Nd(OH)_3$ increases the amount of oxygen present in the starting material. On the other hand, there was a considerably higher amount of oxygen present in overflow streams (Table 5). This means that the smaller particles, which were mainly neodymium oxide and hydroxide, were successfully separated from the matrix phase by this separation process.

The particle size distribution results from the Mastersizer are shown in Figure 5, below. The figure shows that the particles in the starting material (feed) were distributed between 1.5 µm and 120 µm, and 10% of total particles were smaller than 14 µm.

It can be seen from the above figure that the particle size in the underflow obtained after the first pass had reduced considerably. This confirms that the larger particles present in the feed keep breaking down into smaller particles as they pass through the hydrocyclone system.

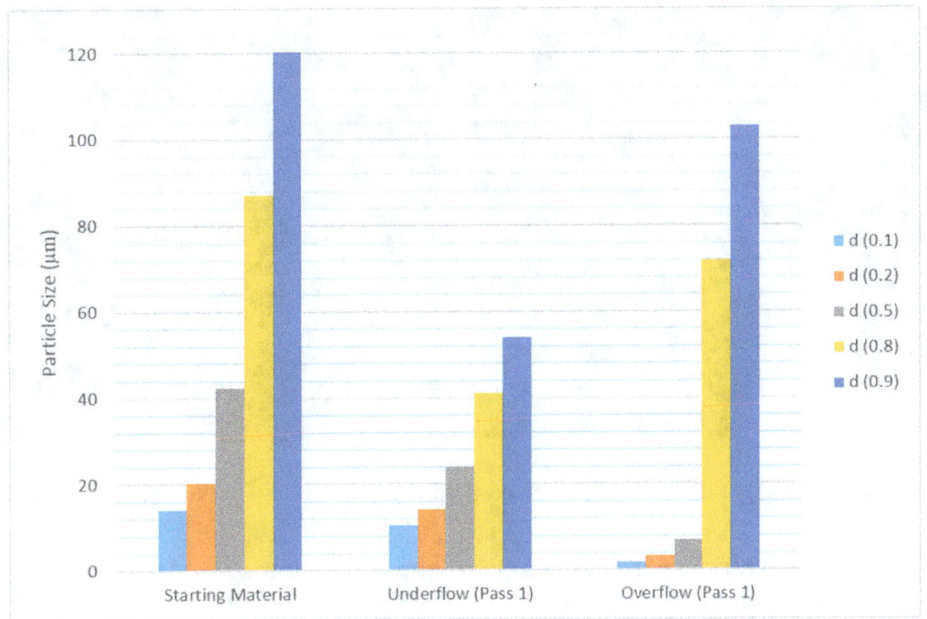

Figure 5. Particle size distribution in feed, underflow and overflow after first pass.

The overflow stream contains mainly smaller particles (GBP), along with some larger particles (matrix phase). Evidence of the presence of matrix phase in overflow was also found in the chemical analysis (Table 5) and sieving data (Table 3). Figure 5 also shows that particle size distribution (D50) had reduced from 42 μm (starting material) to 24 μm in underflow just after the first pass.

The measurement of particle size distribution was then repeated using Sympatec Qicpic 119D, and it was noticed that the none of the samples were free-flowing material. Figure 6 shows the images of some of the particles larger than 100 μm in underflow 1 taken by the Qicpic camera. From these images, it can be seen that larger particles are random in shape, and a clear void can also be noticed in some of the particles. These images suggest that these are not individual particles, but more than one particle clumped together. Thus, accurate measurement of particle size distribution in any of the samples is not possible using the optical or laser-based equipment. For all these techniques, for particle analysis to work properly, the particles must be deagglomerated.

Figure 6. Images of particles larger than 100 μm taken by built-in camera in Qicpic.

To investigate this agglomeration, underflow 1 and overflow 1 were mixed with water and observed under the Leitz optical microscope. To start with, a clear visual difference between underflow and overflow was noticed. The overflow was light coloured and the underflow was grey. Two drops of each suspension were dropped onto a mirror and spread evenly using a glass slide. The images of underflow and overflow are presented as Figure 7A,B respectively.

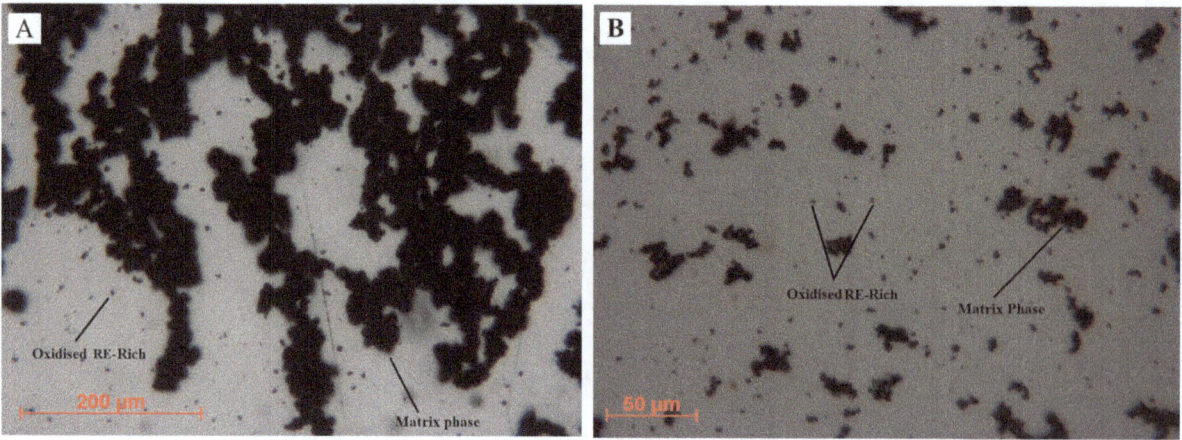

Figure 7. Underflow (**A**) and overflow (**B**) mixed with water and images taken by optical microscope.

Matrix phase and RE-rich particles are highlighted in Figure 7. Matrix phase particles are agglomerated forming a chain network. On the other hand, oxidised RE-rich particles can be seen individually on the surface of the mirror. When a magnet was brought close to these particles, the matrix phase responded to the movement of the magnet, whereas no movement was observed in oxidised RE-rich particles. Based on this information, it was suspected that the HD particles were not fully demagnetized. Rather, they attract each other with very weak magnetic force to form agglomerations.

Figure 7A shows the powder forming chain-like structures; from this, we suspected that the particles from the matrix phase still had remanent magnetization and were interacting with each other, forming chain-like agglomerations. To confirm this observation, these samples were analysed for their magnetic field using MagScan.

The contrasting colour images produced by MagScan for underflow 1 and overflow 1 are shown in Figure 8. It can be seen that the underflow has a slight magnetic field in the area where the sample was placed, as shown in Figure 2B. On the other hand, the overflow does not show any magnetic field. This confirms the observation made in Figure 6 that the matrix phase particles are still magnetic, even after exposure to water.

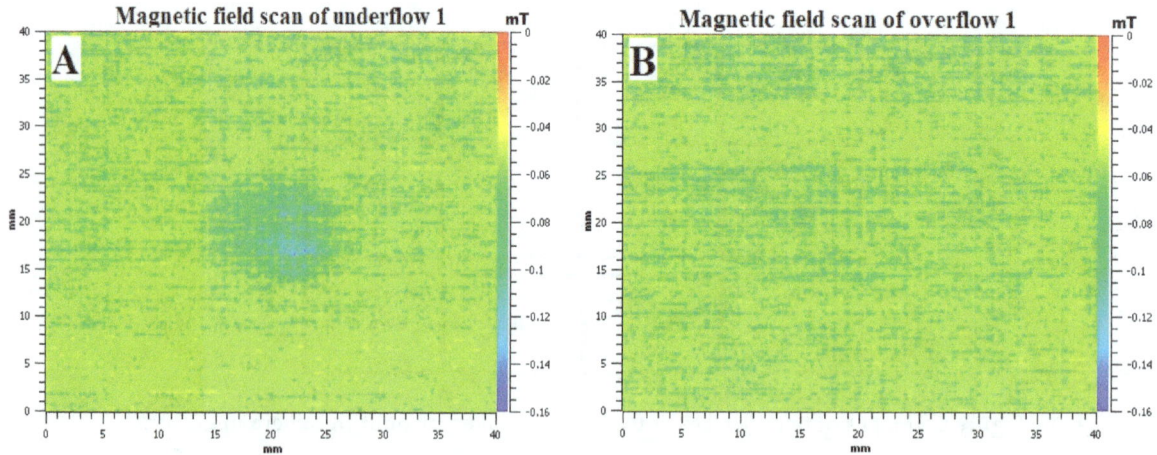

Figure 8. MagScan images of underflow (**A**) and overflow (**B**) after first hydrocyclone pass.

This was also confirmed by a more sensitive in-house-built MagScan (Figure 2A), and the Line Scan of underflow and overflow is presented in Figure 9. It can be concluded from the figure that both samples are slightly magnetic, but that the underflow has more magnetic field than the overflow. The reason for this difference in the magnetic field is the amount of matrix phase and oxidised RE-rich present in each sample. The overflow contains mostly oxidised RE-rich, with some matrix phase, as confirmed by Figure 7B.

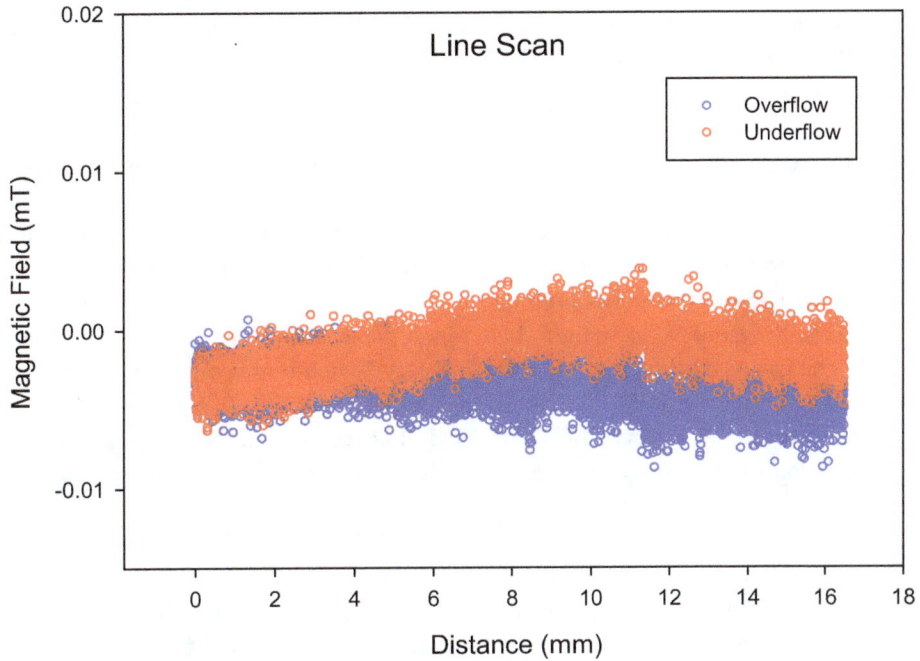

Figure 9. Line Scan of underflow and overflow after first hydrocyclone pass.

4. Conclusions

Hydrocyclone separation has been shown to be an effective method for separating the matrix $Nd_2Fe_{14}BH_x$ from the oxygen-rich grain boundary phase for HD processed NdFeB. The oxygen-rich grain boundary phase was successfully separated in the overflow along with some matrix phase particles. The chemical analysis indicated that one-third of the overflow consisted of the matrix phase. The underflows were then able to be processed into sintered NdFeB magnets with the addition of fresh neodymium hydride. This hydrocyclone separation process was also optimised, and the total number of passes required for optimum separation was found to be three. The very fine particles were removed up to the third pass. After the third pass, more matrix phase was found in the overflow, resulting in a yield loss. The HD powder was found to be slightly magnetic and friable, which limits the separation process and the use of conventional techniques used to determine particle size distribution. The difference in the magnetic field of underflow and overflow and neodymium oxide being non-magnetic open up the potential for further separation techniques, which will be published shortly.

Acknowledgments: The research leading to these results has received funding from the European Community's Horizon 2020 Programme (H2020/2014–2019) under Grant Agreement no. 674973 (MSCA-ETN DEMETER). This publication reflects only the authors' view, exempting the Community from any liability. Project website: http://etn-demeter.eu/. Enrique Herraiz received funding from the European Community's Seventh Framework Programme ([FP7/2007–2013]) under grant agreement no. 607411 (MC-ITN EREAN: European Rare Earth Magnet Recycling Network). Project website: http://www.erean.eu. The authors would also like to thank Less common metals (LCM), UK for oxygen analysis.

Author Contributions: Malik Degri and Enrique Herraiz designed the experiments. Muhammad Awais, Fernando Coelho, Malik Degri performed the experiments and analysed the data. Muhammad Awais wrote the paper and is the corresponding author. Allan Walton and Neil Rowson contributed to the design of experiments and and made a significant contribution to writing.

References

1. Jiles, D. *Introduction to Magnetism and Magnetic Materials*; CRC Press: Boca Raton, FL, USA, 2015; ISBN 9781482238877.

2. Yang, Y.; Walton, A.; Sheridan, R.; Güth, K.; Gauß, R.; Gutfleisch, O.; Buchert, M.; Steenari, B.; van Gerven, T.; Jones, P.T.; et al. REE Recovery from End-of-Life NdFeB Permanent Magnet Scrap: A Critical Review. *J. Sustain. Metall.* **2017**, *3*, 122–149. [CrossRef]

3. European Commission. *List of Critical Raw Materials for the EU*; European Commission: Brussels, Belgium, 2017.

4. European Commission. *Report on Critical Raw Materials for the EU*; European Commission: Brussels, Belgium, 2014.

5. Alonso, E.; Sherman, A.M.; Wallington, T.J.; Everson, M.P.; Field, F.R.; Roth, R.; Kirchain, R.E. Evaluating Rare Earth Element Availability: A Case with Revolutionary Demand from Clean Technologies. *Environ. Sci. Technol.* **2012**, *46*, 3406–3414. [CrossRef] [PubMed]

6. Walton, A.; Yi, H.; Rowson, N.A.; Speight, J.D.; Mann, V.S.J.; Sheridan, R.S.; Bradshaw, A.; Harris, I.R.; Williams, A.J. The use of hydrogen to separate and recycle neodymium-iron-boron-type magnets from electronic waste. *J. Clean. Prod.* **2015**, *104*, 236–241. [CrossRef]

7. Gutfleisch, O.; Güth, K.; Woodcock, T.G.; Schultz, L. Recycling Used Nd-Fe-B Sintered Magnets via a Hydrogen-Based Route to Produce Anisotropic, Resin Bonded Magnets. *Adv. Energy Mater.* **2013**, *3*, 151–155. [CrossRef]

8. Herraiz, E.D.M.; Bradshaw, A.; Sheridan, R.S.; Mann, V.S.J.; Harris, I.R.; Walton, A. Recycling of rare earth magnets by hydrocyclone separation and re-sintering. In Proceedings of the 24th International Workshop on Rare Earth Permanent Magnet and Their Applications, Darmstadt, Germany, 28 August–1 September 2016.

9. Scholz, U.D.; Kronert, W.E.; Nagel, H. A contribution to the mechanism of the hydrogenation of NdFeB alloys and the use of hydrogenated alloy for permanent magnet production. In Proceedings of the 9th International Workshop on Rare-earth Magnets and their Applications, Bad-Soden, Germany, 31 August–3 September 1987.

10. Harris, I.R.; McGuiness, P.J. Hydrogen: Its use in the processing of NdFeB-type magnets. *J. Less-Common Metals* **1991**, *174*, 1273–1284. [CrossRef]

11. Zakotnik, M.; Harris, I.R.; Williams, A.J. Multiple recycling of NdFeB-type sintered magnets. *J. Alloys Compd.* **2009**, *469*, 314–321. [CrossRef]

12. Rivoirard, S.; Noudem, J.G.; de Rango, P.; Fruchart, D.; Liesert, S.; Sobeyroux, J.L. *Proceedings of the 16th International Workshop on Rare Earth Magnets and Their Applications, Sendai, Japan, 10–14 September 2000*; Japan Institute of Metals: Sendai, Japan, 2000; p. 355.

13. Freeman, R.J.; Rowson, N.A.; Veasey, T.J.; Harris, I.R. The development of a magnetic hydrocyclone for processing finely ground magnetite. *IEEE Trans. Magn.* **1994**, *30*, 4665–4667. [CrossRef]

14. Wills, B.A. *Mineral Processing Technology*, 5th ed.; Pergamon Press: Oxford, UK, 1992; ISBN 978-0080418858.

15. Svarovsky, L. *Solid-Liquid Separation*; Butterworth Heinemann: Oxford, UK, 2000; pp. 191–243. ISBN 0750645687.

16. Dobbins, M.; Dunn, P.; Sherrell, I. Recent advances in magnetic separator designs and applications. In Proceedings of the 7th International Heavy Minerals Conference: What Next, Drakensberg, South Africa, 20–23 September 2009.

17. Habibian, M.; Pazoukib, M.; Ghanaiea, H.; Abbaspour-Sanib, K. Application of hydrocyclone for removal of yeasts from alcohol fermentations broth. *Chem. Eng. J.* **2008**, *138*, 30–34. [CrossRef]

18. Takahashi, G. Sample preparation for X-ray fluorescence analysis III. Pressed Loose Powder Methods. *Rigaku J.* **2015**, *31*, 29.

19. Fredericci, C.; de Campos, M.F.; Braga, A.P.V.; Nazarre, D.J.; Martin, R.V.; Landgraf, F.J.G.; Périgo, E.A. Nd-enriched particles prepared from NdFeB magnets: A potential separation route. *J. Alloys Compd.* **2014**, *615*, 410–414. [CrossRef]

PERMISSIONS

All chapters in this book were first published in RECYCLING, by MDPI AG; hereby published with permission under the Creative Commons Attribution License or equivalent. Every chapter published in this book has been scrutinized by our experts. Their significance has been extensively debated. The topics covered herein carry significant findings which will fuel the growth of the discipline. They may even be implemented as practical applications or may be referred to as a beginning point for another development.

The contributors of this book come from diverse backgrounds, making this book a truly international effort. This book will bring forth new frontiers with its revolutionizing research information and detailed analysis of the nascent developments around the world.

We would like to thank all the contributing authors for lending their expertise to make the book truly unique. They have played a crucial role in the development of this book. Without their invaluable contributions this book wouldn't have been possible. They have made vital efforts to compile up to date information on the varied aspects of this subject to make this book a valuable addition to the collection of many professionals and students.

This book was conceptualized with the vision of imparting up-to-date information and advanced data in this field. To ensure the same, a matchless editorial board was set up. Every individual on the board went through rigorous rounds of assessment to prove their worth. After which they invested a large part of their time researching and compiling the most relevant data for our readers.

The editorial board has been involved in producing this book since its inception. They have spent rigorous hours researching and exploring the diverse topics which have resulted in the successful publishing of this book. They have passed on their knowledge of decades through this book. To expedite this challenging task, the publisher supported the team at every step. A small team of assistant editors was also appointed to further simplify the editing procedure and attain best results for the readers.

Apart from the editorial board, the designing team has also invested a significant amount of their time in understanding the subject and creating the most relevant covers. They scrutinized every image to scout for the most suitable representation of the subject and create an appropriate cover for the book.

The publishing team has been an ardent support to the editorial, designing and production team. Their endless efforts to recruit the best for this project, has resulted in the accomplishment of this book. They are a veteran in the field of academics and their pool of knowledge is as vast as their experience in printing. Their expertise and guidance has proved useful at every step. Their uncompromising quality standards have made this book an exceptional effort. Their encouragement from time to time has been an inspiration for everyone.

The publisher and the editorial board hope that this book will prove to be a valuable piece of knowledge for researchers, students, practitioners and scholars across the globe.

LIST OF CONTRIBUTORS

Erik Stenvall and Antal Boldizar
Materials and Manufacturing Technology, Chalmers University of Technology, SE-412 96 Gothenburg, Sweden

Sandra Tostar and Mark R. St J. Foreman
Chemistry and Chemical Technology, Chalmers University of Technology, Gothenburg SE-412 96, Sweden

Erik Stenvall and Antal Boldizar
Materials and Manufacturing Technology, Chalmers University of Technology, Gothenburg SE-412 96,Sweden

Kuniko Mishima and Hidekazu Nishimura
Graduate School of System Design and Management, Keio University, Yokohama 223-8526, Japan

Michele Rosano
Sustainable Engineering Group, Curtin University, Perth, WA 6845, Australia

Nozomu Mishima
Graduate School of Engineering and Resource Science, Akita University, Akita 010-8502, Japan

Calvin Lakhan
Department of Geography, Wilfrid Laurier University, Waterloo, Ontario, ON L6S2X5, Canada

Kamran Rousta and Kim Bolton
Swedish Centre for Resource Recovery, University of Borås, Borås 501 90, Sweden

Lisa Dahlén
Waste Science and Technology, Luleå University of Technology, Luleå 971 87, Sweden

Maria Cristina Santos Ribeiro
Institute of Science and Innovation in Mechanical and Industrial Engineering (INEGI), Rua Dr. Roberto Frias, Porto 4200-465, Portugal
Faculty of Engineering of University of Porto, Rua Dr. Roberto Frias, Porto 4200-465, Portugal

António Ferreira
Faculty of Engineering of University of Porto, Rua Dr. Roberto Frias, Porto 4200-465, Portugal

António Fiúza and Maria de Lurdes Dinis
Faculty of Engineering of University of Porto, Rua Dr. Roberto Frias, Porto 4200-465, Portugal

CERENA-Polo FEUP, Center for Natural Resources and the Environnent, Porto 4100-465, Portugal

Ana Cristina Meira Castro and João Paulo Meixedo
CERENA-Polo FEUP, Center for Natural Resources and the Environnent, Porto 4100-465, Portugal
School of Engineering of Polytechnic of Porto (ISEP), Rua Dr. Bernardino de Almeida, 431, Porto 4200-072, Portugal

Mário Rui Alvim
School of Engineering of Polytechnic of Porto (ISEP), Rua Dr. Bernardino de Almeida, 431, Porto 4200-072, Portugal

Moisés Frías
Eduardo Torroja Institute for Construction Sciences (IETcc-CSIC), Madrid 28033, Spain

Rosario García and Raquel Vigil de la Villa
CSIC-UAM Associated Unit, Department of Geology and Geochemistry, Autonomous University of Madrid,Madrid 28049, Spain

Sagrario Martínez-Ramírez
Institute for the Structure of Matter (IEM-CSIC), Madrid 28006, Spain

Christian Hagelüken
Umicore, Hanau 63457, Germany

Ji Un Lee-Shin
Umicore, Brussels 1000, Belgium

Annick Carpentier and Chris Heron
Eurometaux, Brussels 1000, Belgium

Adekunle Oke
Aberdeen Business School, Robert Gordon University, Aberdeen AB10 1RT, UK

Joanneke Kruijsen
Energy and Sustainability, Robert Gordon University, Aberdeen AB10 GJ, UK

Rikka Wittstock and Alexandra Pehlken
Cascade Use, Carl von Ossietzky University of Oldenburg, Ammerlaender Heerstrasse 114-118, 26129 Oldenburg, Germany

Michael Wark
Department of Chemistry, Carl von Ossietzky University of Oldenburg, Ammerlaender Heerstrasse 114-118, 26129 Oldenburg, Germany

Pontsho Ledwaba and Ndabenhle Sosibo
Mintek, Small Scale Mining and Beneficiation, 200 Malibongwe Drive, Randburg 2125, South Africa

Bruno Wichmann, Martin Luckert, Katrina Bissonnette, Alyssa Cumberland and Claire Doll
Tanishka Gupta and Yuan Shi Department of Resource Economics & Environmental Sociology, University of Alberta, 515 General Services Building, Edmonton, AB T6G-2H1, Canada

Shan Zhong and Pratiksha Rakhe
Department of Materials Science & Engineering, Michigan Technological University, Houghton, MI 49931, USA

Joshua M. Pearce
Department of Materials Science & Engineering, Michigan Technological University, Houghton, MI 49931,USA
Department of Electrical & Computer Engineering, Michigan Technological University, Houghton,MI 49931, USA

Aline Zanchet
Center of Applied Engineering, Modeling and Social Sciences—CECS, Universidade Federal do ABC—UFABC, Rua Santa Adélia, 166, Santo André-SP 09210-170, Brazil

Aline L. Bandeira Dotta
Center of Exact Science and Technology, Department of Physics and Chemistry, Universidade de Caxias do Sul, Rua Francisco Getúlio Vargas, 1130, Caxias do Sul-RS 95070-560, Brazil

Fabiula D. Bastos de Sousa
Technology Development Center—CDTec, Universidade Federal de Pelotas—UFPel, Rua Gomes Carneiro, 1, Pelotas-RS 96010-610, Brazil

Fernando Coelho and Neil Rowson
School of Chemical Engineering, University of Birmingham, Edgbaston, Birmingham B15 2TT, UK

Muhammad Awais
School of Chemical Engineering, University of Birmingham, Edgbaston, Birmingham B15 2TT, UK
School of Metallurgy and Materials, University of Birmingham, Edgbaston, Birmingham B15 2TT, UK

Malik Degri Enrique Herraiz and Allan Walton
School of Metallurgy and Materials, University of Birmingham, Edgbaston, Birmingham B15 2TT, UK

Index